Hazards and the Built Environment

Since the built environment and urban infrastructure provide the core framework for most human activity, it is crucial to develop them with an effective measure of resilience so they can withstand, and adapt to, the threats of natural and human-induced hazards. This book sets out to explore the challenges facing the built environment and examines the strategies that must be taken if built-in resilience is to be realised in the future and built assets safeguarded.

The contributors portray a resilient built environment as providing the essential groundwork upon which the technical, organisational, social and economic frameworks so necessary for societal resilience can be founded. The range of issues covered within this book not only demonstrate the trans-disciplinary nature of the subject but also illustrate that non-structural as well as structural adaptations need to be considered to reduce the threat, and impact, of disasters and that lessons can be learnt from a range of disciplines and socio-cultural contexts. Broad conclusions are drawn and seven guiding principles are provided in relation to the ways in which construction and developmental practitioners might adapt their *modus operandi* to better address a range of hazards.

This book is essential reading for a wide range of undergraduate and postgraduate students, managers and practitioners involved with the way buildings and infrastructure are planned, designed, built, managed and operated.

Lee Bosher is a research fellow in the Department of Civil and Building Engineering at Loughborough University, UK. He is a fellow of the Royal Geographical Society, and a member of the Institute of Civil Defence and Disaster Studies.

Also available from Taylor & Francis

Geological Hazards
Their assessment, avoidance and mitigation
Fred Bell
Hb: 9780415318518

Natural Disasters
David C. Alexander
Hb: 9781857280937
Pb: 9781857280944

Environmental Hazards, 4th edn
Keith Smith
Hb: 9780415318037
Pb: 9780415318044

Hazards and the Built Environment

Attaining built-in resilience

Edited by Lee Bosher

Routledge
Taylor & Francis Group

LONDON AND NEW YORK

First published 2008
by Taylor & Francis
2 Park Square, Milton Park, Abingdon, Oxon OX14 4RN

Simultaneously published in the USA and Canada
by Routledge
270 Madison Ave, New York, NY 10016

Routledge is an imprint of the Taylor & Francis Group, an informa
business

Typeset in Sabon by
HWA Text and Data Management, Tunbridge Wells
Printed and bound in Great Britain by
Antony Rowe Ltd, Chippenham, Wiltshire

British Library Cataloguing in Publication Data
A catalogue record for this book is available from the British
Library

Library of Congress Cataloging-in-Publication Data
Bosher, Lee.
  Hazards and the built environment : attaining built-in resilience /
Lee Bosher.
     p. cm.
  Includes bibliographical references and index.
  1. Building failures–Prevention. 2. Building failures–Risk
assessment. 3. Hazard mitigation. 4. Building–Standards–
Government policy. 5. City planning. 6. Reliability
(Engineering) I. Title.
TH441.B67 2008
690'.22–dc22   2007044004

ISBN13: 978–0–415–42729–6 (hbk)
ISBN13: 978–0–415–42730–2 (pbk)
ISBN13: 978–0–203–93872–0 (ebk)

ISBN10: 0–415–42729–0 (hbk)
ISBN10: 0–415–42730–4 (pbk)
ISBN10: 0–203–93872–0 (ebk)

Contents

Figures

Tables

About the editor

Lee Bosher is a research fellow who has been based within the Department of Civil and Building Engineering at Loughborough University since June 2005 working on the Safe, Secure and Sustainable Built Environment project (S3BE). The S3BE project has investigated how a more resilient built environment can be achieved in the United Kingdom by reducing the frequency and impact of disasters that result in damage to critical infrastructure, built assets and the loss of human life. Lee is currently working on the PRE-EMPT Project that aims to ensure that a more resilient built environment is attained via the structured integration of hazard mitigation/adaptation strategies into the construction sector's decision-making processes. Lee is also involved with the RE-Design project that seeks to ensure that best practice in the design of effective and acceptable resilient public places can be more widely achieved through the structured and considered integration of counter-terrorism measures into the decision-making processes of key stakeholders involved with the planning, design, construction, operation and management of public places and transport systems in the UK.

Lee attained his PhD at the Flood Hazard Research Centre, Middlesex University. The study investigated the influence of social and institutional influences on access to the key socio-economic resources that determine levels of vulnerability to large-scale disasters and smaller scale crises in southern India. He offers specific experience on disaster risk management processes and the multi-disciplinary integration of these processes within the construction sector (amongst others). He has experience of undertaking qualitative and exploratory research that embraces a range of research methods. Lee is a Fellow of the Royal Geographical Society and a Fellow of the Institute of Civil Defence and Disaster Studies; he recently authored a book entitled *Social and Institutional Elements of Disaster Vulnerability* (2007).

Contributors

David Alexander is a professor in the University of Florence's Interdepartmental Centre for Civil Protection and Risk Studies, Italy. His books include *Natural Disasters* (1993), *Confronting Catastrophe* (2000) and *Principles of Emergency Planning and Management* (2002). He is co-editor of the international journal *Disasters*.

Guillaume Chantry is an engineer and the programme coordinator for DWF in Vietnam, and has worked on development issues for 30 years, the past 20 years of which have been in Asia.

Jon Coaffee is a senior lecturer in the School of Environment and Development at the University of Manchester, England. Since the mid-1990s he has been researching issues related to countering terrorism in UK cities. He published *Terrorism, Risk and the City* (Ashgate, 2003) and is currently undertaking a number of research projects funded by the UK research councils on urban resilience.

Andrew Dainty is Professor of Construction Sociology at Loughborough University, England. His research focuses on human social action within construction and other project-based sectors. He is co-author of *HRM in Construction Projects* (2003), *Communication in Construction* (2006) and is co-editor of *People and Culture in Construction* (2007).

Amod Dixit is the Executive Director of the National Society for Earthquake Technology in Kathmandu, Nepal and is a member of the Risk RED team.

Andrew Fox is a chartered civil engineer with more than a decade of experience in Disaster Management. His career spans the public, private and charity sectors, working both in the UK and internationally. He was engaged for six years on the Coventry University Disaster Management programme before moving in 2007 to the University of Plymouth.

Jacqueline Glass is Lecturer in Architectural Engineering at Loughborough University, England and specialises in research on environmental, sustainability and innovation issues within building design. She has written numerous industry reports, academic papers and a book, the *Encyclopaedia of Architectural Technology* (Wiley-Academy, 2002).

Rebekah Green is a research associate at the Institute for Global and Community Resilience, Western Washington University, USA. She has combined the fields of engineering and anthropology to research informal settlements and hazard vulnerability in different cities throughout the world. Her most recent work focused on disaster recovery among low-income residents of New Orleans and informal settlements in Istanbul, Turkey.

Rohit Jigyasu is a conservation architect and risk management consultant from India. After obtaining his post-graduate degree in Architectural Conservation, Rohit has undertaken his doctoral research on 'Reducing Disaster Vulnerability through Local Knowledge and Capacity – the Case of Earthquake-prone Rural Communities in India and Nepal'.

Ilan Kelman is a researcher at the Center for Capacity Building, National Center for Atmospheric Research, Colorado, USA. He is also a member of the Risk RED team.

Jessica Lamond is a doctoral student at the University of Wolverhampton, UK.

Jason Le Masurier is a senior lecturer in the Department of Civil Engineering, University of Canterbury, New Zealand.

James Lewis RIBA has worked on design, management and inspection of construction in Algeria, Bangladesh, the Caribbean, Hong Kong, the South Pacific, the United Kingdom and the USA. Formerly co-founder of the Disaster Research Unit, University of Bradford, he is a member of Development Workshop France and Visiting Fellow in Development Studies at the University of Bath, England.

Simon McCarthy is a research fellow with a background in sociology and expertise in recovery from floods and works at the Flood Hazard Research Centre at Middlesex University, England.

John Norton is President of the Development Workshop France (DWF), a French NGO, and co-founder of 'Development Workshop', which since 1973, has specialised in settlement development problems focused on disaster mitigation, environmental degradation and promoting local resources and skills in Asia and Africa.

George Ofori PhD DSc, is a Ghanaian. He is a professor at the National University of Singapore. His research area is construction industry development, focusing on developing countries. He is a consultant to international agencies and governments and has authored more than 250 international journal and conference papers, books and reports.

Stefano Pampanin is Senior Lecturer in Structural (Earthquake) Engineering at the University of Canterbury, New Zealand. His main research interests, based on experimental and analytical/numerical approaches, are related to the design (performance-based) of earthquake-resistant structures and to the assessment and retrofit of existing buildings.

Edmund Penning-Rowsell is Professor of Geography and Senior Research Fellow, interested in hazards, institutions and decision making. He works in the Flood Hazard Research Centre at Middlesex University, England.

Marla Petal is Co-Director of Risk RED a not-for-profit non-governmental organisation that aims to increase the effectiveness and impact of disaster risk reduction education and is a member of the Risk RED team.

David Proverbs is Head of the Construction and Infrastructure Department at the University of Wolverhampton, UK.

James Rotimi is a lecturer at UNITEC Institute of Technology and is currently completing his PhD in disaster construction.

Victor Samwinga is a senior lecturer and Programme Leader (Construction Management) at Northumbria University, UK.

Rajib Shaw is Associate Professor at the Graduate School of Global Environmental Studies, Kyoto University, Japan. He is also a member of the Risk RED team.

Robby Soetanto is a senior lecturer in the Department of the Built Environment at Coventry University, UK.

Sylvia Tunstall is Senior Research Fellow, interested in hazards, institutions and decision making, and works in the Flood Hazard Research Centre at Middlesex University, England.

Christine Wamsler, an architect and urban planner with a Master's degree in International Humanitarian Assistance, is currently working as a researcher in the Department of Housing Development and Management (HDM), Lund University, Sweden.

Suzanne Wilkinson is an associate professor in the Department of Civil and Environmental Engineering at the University of Auckland, New Zealand.

Kelvin Zuo is a doctoral student at the University of Auckland and is currently completing his PhD in disaster reconstruction.

Preface

Since the built environment and urban infrastructure provide the core framework for most human activity, it is crucial to develop them with an effective measure of resilience so they can withstand, and adapt to, the threats of natural and human-induced hazards. This book sets out to explore the challenges facing the built environment and examines the strategies that must be taken if built-in resilience is to be realised in the future and built assets safeguarded.

During the last few decades the prevalence of disastrous events has stimulated a growth in theoretical developments in relation to the way in which disasters are avoided and managed. A paradigmatic shift has led to a focus on disaster preparedness, hazard mitigation and vulnerability reduction rather than disaster management and relief. The discourse of resilience now resonates throughout the disciplines involved with the mitigation of disasters. This is reflected within the contents of this book, where a persuasive case for embracing disaster risk management provides the common thread through a deliberately eclectic set of contributions to the resilience debate. This book portrays a resilient built environment as providing the essential groundwork upon which the technical, organisational, social and economic frameworks so necessary for societal resilience can be founded.

The range of issues covered within this book not only demonstrate the trans-disciplinary nature of the subject but also illustrate that non-structural as well as structural adaptations need to be considered to reduce the threat, and impact, of disasters and that lessons can be learnt from a range of disciplines and socio-cultural contexts. The challenge for the reader is to assimilate these different world-views and to relate them to their personal context. Hopefully the reader will also be sufficiently engaged with the subject matter of this book to break free from their disciplinary shackles and read some of the chapters that they did not originally feel would garner their attention; after all, the underlying causes of disasters cannot be resolved by disciplines working in isolation from each other.

The afterword of this book draws some broad conclusions regarding the ways in which construction and developmental practitioners might adapt their *modus operandi* to better address the wide range of hazards that are discussed. In doing this, the final chapter provides seven guiding principles which can be used as a point of departure for the development of context-sensitive resilience frameworks in the future.

Lee Bosher
Loughborough, England

There are risks and costs to a program of action. But they are far less than the long-range risks and costs of comfortable inaction.

John F. Kennedy (1917–1963)

Part I
Introduction

Chapter 1

Introduction

The need for built-in resilience

Lee Bosher

Introduction

Recent natural and human-induced events have highlighted the fragility and vulnerability of the built environment to disasters. These physical systems have traditionally been designed, built and maintained by the myriad professions involved with the construction industry. Therefore, designing and constructing a built environment that can cope with the impacts of disasters demands an in-depth understanding of the expertise and knowledge on avoiding and mitigating the effects of hazards in order to secure a more sustainable future (Hamelin and Hauke 2005; Bosher *et al.* 2007).

This introductory chapter provides an insight into the prevalence and impact of disasters that occur globally and questions whether 'natural disasters' are really natural events. The observed shift in the way disasters are being managed is also discussed, explaining the move away from the reactive attributes of Disaster Management towards the more proactive Disaster Risk Management (DRM) approach that should be 'mainstreamed' into developmental initiatives. The function of a multitude of disciplines responsible for how the built environment is delivered is therefore a critical aspect required for the mainstreaming of DRM into long-term development. This chapter will discuss why the attainment of built-in resilience is important, highlighting the structural and non-structural aspects of hazard mitigation and justifying the need for the multi-disciplinary approaches advocated by this book.

The prevalence and impact of disasters

A 'disaster' is defined by the United Nations International Strategy for Disaster Reduction (UN/ISDR 2004: Appendix 1) as:

> A serious disruption of the functioning of a community or a society causing widespread human, material, economic or environmental losses which exceed the ability of the affected community or society to cope using its own resources. A disaster is a function of the risk process. It

results from the combination of hazards, conditions of vulnerability and insufficient capacity or measures to reduce the potential negative consequences of risk.

Threats to society and the built environment are diverse and include extreme natural hazards (such as earthquakes, floods and storms) and human induced hazards (such as terrorist attacks, explosions at industrial facilities and mass transportation accidents). Natural and human induced hazards may not only threaten the lives of those unfortunate to become affected by them but can result in disasters that can also threaten economies, businesses and in some cases political regimes. The impacts of disasters also drain millions of dollars every year in relief, rehabilitation, reconstruction and insurance costs for many nations.

Figure 1.1 illustrates the huge figures related to the estimated damages caused by disasters by continent over the period between 1991 and 2005, with the total reported damages totalling nearly US$ 1,193 billion and the worst hit continent being Asia (EM-DAT 2007).[1] The costs of these damages averaged US$ 79.5 billion per year (EM-DAT 2007) and had a particularly devastating effect upon the economies of underdeveloped nations and their development strategies.

However, the economic costs of disasters are only one part of the problem, with the impacts on human life being the most fundamental and distressing elements of the effects. Figures obtained from EM-DAT (2007) indicate that between the years 1991 and 2005, on average 231 million people per year (approximately 3.5 per cent of the world's population)[2] were affected by various types of disasters. Disasters wreak havoc on nations irrespective of a

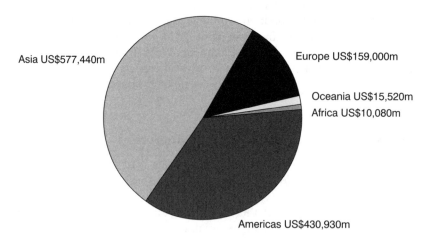

Figure 1.1   Total numbers of reported disaster related damages by continent (1991–2005), figures in US$ millions

country's wealth or resources but invariably it is the least developed nations that suffer the most.

Palakudiyil and Todd (2003) make the observation that during the decade between 1992 and 2001, the least-developed countries[3] lost on average 1,052 lives per disaster; while the average deaths per disaster for countries with medium or high development were 145 and 23 respectively. In addition, of the total number of people reported killed by disasters during the same period 75 per cent of the deaths were in Asia and 67 per cent were in the least-developed countries while only 4 per cent of the deaths were in nations with high levels of development (Palakudiyil and Todd 2003). The statistics indicate that people who live in countries with a low level of development are (on a global scale) the most likely to be affected and killed by disasters.

Natural hazards

Natural hazards are typically split into two categories, namely; (1) geo-hazards, and (2) hydro-meteorological hazards. Figure 1.2 provides a list of the key geo-hazards and hydro-meteorological hazards that occur globally. The magnitude of natural hazards tends to be determined by key factors such as meteorology (which is influenced by the changing seasons), topography, hydrology, geology, biodiversity (of flora and fauna) and tidal variations (caused by lunar and meteorological influences, coastal topography and influenced by the type and locality of coastal developments). These processes are typically benign and provide the basis for people to exist in harmony with their natural environment. However, infrequently (and some would

	Natural hazards
Geo-hazards	Earthquakes
	Volcanic eruptions
	Tsunamis
	Landslides (incl. all kinds of mass movements, cavity collapses and ground failure)
Hydro-meteorologal hazards	Floods
	Coastal erosion and flooding
	Wind storms (incl. cyclones, hurricanes and typhoons)
	Extreme temperatures
	Storm surges
	Droughts
	Wild fires

Figure 1.2 Types of natural hazards that are experienced globally

suggest more frequently) natural hazards impact upon the built environment, causing damage, deaths, disruption and financial losses. There is now a broad scientific consensus (including from the Intergovernmental Panel on Climate Change) that the global climate is changing in ways that are likely to have a profound impact on human society and the natural environment over the coming decades (Solomon *et al.* 2007). Commentators have posited that the impact of global climate change (which is arguably both natural and anthropogenic in nature) has increased the frequency of disasters, and will further increase the frequency of such events in the future (Munich Re 2003). The impacts of these natural events can be psychological, sociological and political but are typically reported in economic terms[4] and as such, economic losses due to the impacts of natural hazards have increased ten-fold in the last 40 years (Munich Re 2003).

Table 1.1 shows the average yearly occurrence, impact and death rate associated with disasters that have occurred between 2000 and 2005. The figures suggest that disasters are not rare events but that they actually occur relatively frequently (on average more than one disaster per day) and affect a significant proportion of the world's population. Floods and wind storms are the most prevalent events, droughts and floods affect the most people on average, while earthquakes and tsunamis/storm surges on average cause the most fatalities. These figures are important because, to paraphrase a popular saying, 'natural hazards don't kill people, unsafe buildings do'.[5] It is therefore clear that future construction practice needs to be more sensitive to mitigate the impacts of a wide range of hazards. This needs to be achieved through proactive measures. These proactive measures are likely to have a

Table 1.1 Average annual occurrence and impact of disasters globally, 2000–5

Type of hazard	Average occurrence	Average number of people affected	Average number of people killed
Drought	32	145,601,529	217
Earthquake	31	3,646,234	21,656
Extreme temperatures	23	721,095	9,202
Flood	162	106,190,567	5,205
Landslides	20	333,807	762
Volcano	5	147,454	34
Wave/storm surge/tsunami	3	408,465	37,739
Wild fires	17	47,843	32
Wind storm	106	38,925,602	2,821
Total	399	296,022,596	77,668

Source: EM-DAT (2007).

bearing on the professional training (formal and informal) and day-to-day activities of a vast range of construction professionals.

Are 'natural disasters' really that natural?

The human influences upon the causes of disasters are too often overlooked because sometimes these influences can be discrete and driven by very different socio-economic factors. For example, in many high-income countries, people like to live near rivers (and are prepared to pay for the benefit in many cases), for the aesthetic and recreational benefits that rivers can offer. For example, upstream, along the non-tidal stretch of the River Thames in England, some 12,000 houses are within 500 metres of the riverbank, and their riverside location adds £580m (approximately £48,000 per house) to the value of these properties (McGlade 2002). Therefore, a flood event that occurs in the non-tidal stretch of the River Thames, inundating people's homes, businesses and lifelines will typically be referred to as a 'natural disaster' but the flood event manifests itself as a disaster because society (and this is predominantly the case in high-income societies) has chosen to build homes, infrastructure and businesses in an area vulnerable to floods.

Socio-economic factors that affect people's exposure to hazards can manifest themselves differently in low-income nations, with key factors being related to poverty (low access to assets), marginalisation (poor access to public facilities) and powerlessness (low access to political and social networks) (Bosher 2007). These factors will have an influence on the choices that people have regarding where they can live (for instance the landless squatters that live on the flood plain of the Buriganga River in Dhaka, Bangladesh and the informal slums (*favelas*) situated on the steep landslide-prone hills of Rio de Janeiro in Brazil). These factors will also influence the levels to which people can provide themselves with adequate shelter to protect themselves from local conditions; therefore geographic proximity and exposure to hazards will affect levels of individual and social resilience (Wisner *et al.* 2004). Consequently, unlike the case of higher-income nations where many people choose to live in areas that are exposed to hazards, in low-income countries it is more the case of a 'lack of choice' that forces people to live in areas that are exposed to hazards.

Another human aspect that influences the prevalence of disasters is related to how the built environment is planned, designed, built, maintained and operated. Urban areas that are constructed (whether formally or informally) without due regard for natural and human induced hazards are unlikely to provide the levels of resilience that are required to ensure settlements are safe and sustainable. This concern is related not merely to residential properties but also associated infrastructure and institutional/governmental and commercial built assets and is a problem that will be discussed in more depth in the forthcoming chapters.

It is therefore pertinent, if not somewhat contentious and provocative, to suggest that 'natural disasters' are actually quite rare occurrences[6] and that *most* of the events that are labelled as 'natural disasters' are actually a combination of natural and human induced events.[7] To help clarify and illustrate the point it would be worthwhile considering the following examples:

- Carlisle, in northern England, was flooded in 2005 because it is located on a flood plain and historical attempts to restrict development on the flood plain had achieved limited success. The fact that many critical support services (such as the city council offices including the emergency planning department, plus the police station, the fire station and the electricity substation) were located on the flood plain and therefore severely affected, exacerbated the impact of the floods by impeding the ability of the city to cope with, and recover from, the event.
- New Orleans, USA, was devastated by Hurricane Katrina in August 2005 because it is located in a hurricane zone, situated on land that is largely below sea level and was protected by substandard and under-maintained flood defences. The US Army Corps of Engineers admitted that faulty design specifications, incomplete sections and substandard construction of levee segments, not the hurricane, was the primary cause of the flooding in the New Orleans area (Whittell 2005).
- Pakistan was severely affected by an earthquake in October 2005 because many modern buildings (including, as some figures have suggested, 17,000 government-built schools that collapsed killing an estimated 19,000 children)[8] were not built with sufficient aseismic measures so that they could withstand the affects of earthquakes (the threat of which had been known about for many years).

Whether or not the reader agrees with the assertion that 'natural disasters' are not actually that natural, what is important to acknowledge is that when disasters are labelled as 'natural disasters' (because of the perception that 'it was nature's fault') it becomes too easy for society, governmental institutions, and the construction sector to absolve themselves of blame. It is important to appreciate that 'natural disasters' are not as natural as they may first appear and that there are significant human-induced aspects to such events; therefore society, governmental institutions and the construction sector should be taking more responsibility for how the built environment is planned, designed, built, maintained and operated.

The shift from disaster management to disaster risk management

During the last few decades documented increases of disastrous events have combined with theoretical developments to necessitate a fresh approach to the way in which disasters are handled. Emphasis has moved away from relief and disaster preparedness, towards a more sustainable approach incorporating hazard mitigation and vulnerability reduction elements. DRM can be summarised into four phases (Figure 1.3): (1) hazard identification, (2) mitigative adaptations, (3) preparedness planning; and (4) recovery (short-term) and reconstruction (longer-term) planning. These mutually interconnected phases (see Figure 1.3) can be defined as (UN/ISDR 2004):

- **Hazard identification**
  Identification of potentially damaging physical events, phenomenon or human activity that may cause the loss of life or injury, property damage, social and economic disruption or environmental degradation.
- **Mitigative adaptations** (otherwise referred to as 'hazard mitigation')
  Structural and non-structural measures undertaken to limit the adverse impact of hazards.
- **Preparedness planning**
  Activities and measures taken in advance to ensure effective response to the impact of hazards, including the issuance of timely and effective early warnings and the temporary evacuation of people and property from threatened locations.
- **Recovery and rehabilitation**
  Decisions and actions taken after a disaster with a view to restoring or improving the pre-disaster living conditions of the stricken community, while encouraging and facilitating necessary adjustments to reduce disaster risk. Recovery (rehabilitation and reconstruction) affords an opportunity to develop and apply disaster risk reduction measures.

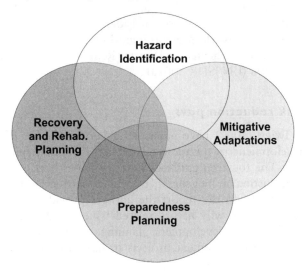

Figure 1.3 The interconnected phases of the DRM framework

DRM should be concerned with people's capacity to manage their natural, social and built environments; and take advantage of it in a manner that safeguards their future and that of forthcoming generations. DRM needs to be holistic and new initiatives need to be found in order to ensure that associated strategies are viewed as a shared responsibility that includes issues such as hazard mitigation (Pelling 2003; Trim 2004) and land-use planning (Burby 1998; Burby *et al.* 2000; Wamsler 2004).

This movement has been seen as part of a wider paradigm shift in which top-down, 'expert-led' (and often capital intensive) proposed solutions are being rejected in favour of 'bottom-up', community-based and sustainable long-term developmental initiatives. This change in focus is particularly significant, as it is an approach advocated by the United Nations along with other organisations involved in vulnerability reduction and developmental programmes and led towards the development of the *Hyogo Framework for Action 2005–2015: Building the resilience of nations and communities to disasters* (UN/ISDR 2005).

The *Hyogo Framework for Action* (HFA) was approved by 168 countries in January 2005 and is an ambitious programme of action to significantly reduce disaster risk globally. Since 2005, many countries and organisations have already realigned their policies and strategies to directly respond to the expectations and directions of the HFA. For example, HFA focal points have been established by 106 countries and 5 territories; national platforms for disaster reduction have been initiated in 38 countries; ministerial-level regional agreements and strategies have been agreed, or are being developed in several regions and sub-regions (Africa, Asia, and the Pacific Islands), and specific risk reduction strategies or initiatives have been developed by a number of international agencies, including the United Nations Development Programme, World Bank, International Federation of Red Cross and Red Crescent Societies and World Meteorological Organization, exemplified by the launch of the Global Facility for Disaster Reduction and Recovery by the World Bank in 2006 (UN/ISDR 2007a).

Disaster risk reduction pays

Benson and Twigg (2007) have published a very useful set of 'Guidance Notes' for development organisations related to the mainstreaming of disaster risk reduction. The fourteen guidance notes provide short, practical briefs supplementing existing, more general, guidelines on programming, appraisal and evaluation tools. In the report the authors provide some interesting examples that illustrate why hazard-related issues need to be considered in national and sectoral development planning, country programming and in the design of all development projects in hazard-prone countries. The examples include (Benson and Twigg 2007: 6):

- A Vietnam Red Cross mangrove planting programme implemented in eight provinces in Vietnam to provide protection to coastal inhabitants from typhoons and storms cost an average US$ 0.13 million a year over the period 1994 to 2001, but reduced the annual cost of dyke maintenance by US$ 7.1 million. The programme also helped save lives, protect livelihoods and generate livelihood opportunities.
- Spending 1 per cent of a structure's value on vulnerability reduction measures can reduce probable maximum loss from hurricanes by around a third in the Caribbean, according to regional civil engineering experts.
- One dollar spent by the Federal Emergency Management Agency (FEMA) in the USA on hazard mitigation generates an estimated US$ 4 on average in future benefits according to a study of FEMA grants (including for retrofitting, structural mitigation projects, public awareness and education and building codes).
- Only two schools were left standing in Grenada after the passage of Hurricane Ivan (September 2004). Both had been subject to retrofit through a World Bank initiative. One of the schools was used to house displaced persons after the event.

The role of the built environment

It has been acknowledged that the way the built environment is delivered can itself lead to disasters, particularly in less developed nations where building codes and planning regulations may not be as well policed as they are in other nations (Ofori 2002). It would also appear that with socio-economic progress, the built environment becomes more vulnerable as settlements become more reliant on their increasingly extended supply lines (Menoni 2001), and ever-expanding and vital distribution networks of water, power, gas and telecommunication systems. Moreover, with globalisation, major urban settlements are also inter-connected and a disaster in one of them can precipitate widespread disruption in many others. Furthermore, constructed items represent most of every nation's savings. For instance, studies show that Gross Domestic Fixed Capital Formation in construction is 45–60 per cent of the total capital formation (Hillebrandt 2000; Han and Ofori 2001); therefore the built environment is an investment that is worthy of protection.

Designing and constructing a resilient built environment demands an in-depth understanding of the expertise and knowledge of avoiding and mitigating the effects of threats and hazards (Lorch 2005; Hamelin and Hauke 2005; Bosher *et al.* 2007). Organisations such as the United Nations and the British Government's Department for International Development (DFID 2006) have highlighted the importance of 'mainstreaming' DRM as part of an initiative to build collaboration between stakeholders in order to reduce the impact of disasters by integrating disaster risk reduction into development policies. Amongst other requirements, the *Hyogo Framework for Action 2005–2015* (UN/ISDR 2005) calls on governments to mainstream disaster risk considerations into planning procedures for major infrastructure projects. However, to date little research has been done globally on how disaster risk reduction can be effectively mainstreamed into construction projects (Wamsler 2006). Although there is increasing, and wide-ranging, agreement that resilience should be systematically built in to the whole design, construction and operation process, and not simply added on as an afterthought, it is apparent that this is not being achieved sufficiently (Bosher *et al.* 2007).

A case for built-in resilience

According to the Oxford English Dictionary, resilience means 'an act of rebounding or springing back' (Simpson and Weiner 1989: 714). Holling (1996) noted that resilience of a system has typically been defined in two very different ways – the differences in definition reflecting which of two different aspects of resilience are emphasised. For instance, 'engineering resilience' is defined in Gunderson *et al.* (2002: 530), as 'the time of return to a global equilibrium following a disturbance'; this definition therefore implies only one stable state (and global equilibrium). In contrast 'ecological resilience' is based on the demonstrated property of alternative stable states in ecological systems and is defined by Gunderson *et al.* (2002: 530), as 'the amount of disturbance that a system can absorb before it changes state'. The consequences of these different aspects for ecological systems were first emphasised by Holling (1973) in drawing attention to the paradoxes between efficiency on the one hand and persistence on the other, or between constancy and change, or between predictability and unpredictability. One definition focuses on efficiency, control, constancy and predictability – all attributes at the core of desires for fail-safe design and optimal performance. The other focuses on persistence, adaptiveness, variability and unpredictability – all attributes embraced and celebrated by those with an evolutionary or developmental perspective. More recently, the term has been applied to human social systems (Agar 2000), economic recovery (Rose 2004), and urban planning and recovery after calamitous events in cities (Vale and Campanella 2005). Recent metaphors of resilience have been used to describe how cities and nations attempt to 'bounce back' from disaster, and the embedding of security and contingency features into planning systems (Coaffee 2005).

There are three themes or dimensions around which the concept of resilience can be articulated in the context of DRM (also see Chapter 15). These dimensions are:

1   A shift towards greater mitigation and preparedness rather than reactive disaster management;
2   A broadening of the DRM agenda, not just to cover the management of natural hazards but to increasingly focus on technological and human induced hazards (such as those associated with global terrorism); and
3   The concept of resilience within DRM has been broadened to encompass not only the resilience of public services and governmental mechanisms (as was the focus in the past) but also to include the resilience of private business (through Business Continuity Planning) and 'community resilience'.

The overall shift to a more holistic concept of resilience is particularly pertinent for this book because it integrates the physical (both built and natural) and socio-political aspects of resilience and goes some way to illustrating the complex nature of resilience. The socio-political aspects are arguably as important to the attainment of resilience as the physical aspects; resilient engineering also demands a more resilient infrastructural context with regards to the professions and the structures and processes which govern construction activity. Therefore, a resilient built environment should be designed, located, built, operated and maintained in a way that maximises the ability of built assets, associated support systems (physical and institutional) and the people that reside or work within the built assets, to withstand, recover from, and mitigate for the impacts of extreme natural and human-induced hazards.

However, it is worth acknowledging that the concept of resilience is also evolving to recognise that in some cases it is not sufficient for a system or built asset to merely 'bounce back' to its original state. It is essential to acknowledge that there is a need to ensure the system or built asset is a more robust version of what was there originally (this is particularly the case for assets that have been historically at risk from hazards such as flooding, windstorms, earthquakes and landslides). If built assets are affected repeatedly by particular hazards, there is a pressing need to learn lessons from this and replace or retrofit the original structure(s) with an improved version that is more resilient (in social, physical and economic terms).

Structural and non-structural approaches

Schneider (2002) stated that DRM has often been viewed as a reactive profession because activities such as risk reduction and hazard mitigation were rarely seen as urgent. DRM strives to be more holistic than previous approaches to dealing with disasters and new initiatives are required in order to ensure that associated strategies are viewed as a shared responsibility that includes issues such as hazard mitigation (Pelling 2003; Trim 2004) and land-use planning (Burby 1998; Burby et al. 2000; Wamsler 2004). The concept of hazard mitigation begins with the realisation that many disasters are not unexpected (Mileti 1999), and the impacts of many natural and human-induced hazards can therefore be reduced. With particular reference to natural hazards, it is common to discuss two types of hazard mitigation, as summarised below.

Structural mitigation such as the strengthening of buildings and infrastructure exposed to hazards (via building codes, engineering design and construction practices, etc.).

Non-structural mitigation includes directing new development away from known hazard locations through land use plans and regulations, relocating existing developments to safer areas and maintaining protective features of the natural environment (such as sand dunes, forests and vegetated areas that can absorb and reduce hazard impacts). These non-structural mitigation initiatives can have significant value in risk and cost reduction (Godschalk *et al.* 1999; Godschalk 2003). For example, in the United States the 1988 Stafford Act (amended in June 2006) was critical in advocating the proactive rather than reactive approach to disaster management (a shift from a disaster-driven system to a policy- and threat-driven system).

The construction sector can play an important role in the structural elements of mitigation, while developers and planners (Wamsler 2006) should be able to positively influence the non-structural elements of mitigation.[9] DRM should therefore be concerned with people's capacity to manage their natural, social and built environments; and take advantage of them in a manner that safeguards their future and that of forthcoming generations. Part of this shared responsibility could be achieved by integrating the work of construction professionals, who possess the knowledge and experience of how to design, build, retrofit and operate what are typically bespoke built assets, into the DRM framework (Bosher *et al.* 2007).

Brief introduction to the forthcoming chapters

Since the built environment and urban infrastructure provide the core framework for most human activity, it is crucial to develop them with an effective measure of resilience so they can withstand, and adapt to, the threats of natural and human-induced hazards. Examining multi-hazard adaptation issues from a construction sector perspective, this book also considers the non-structural elements of multi-hazard adaptation; encompassing international contexts, natural and human-induced hazards, and embracing a series of multi-disciplinary initiatives. It demonstrates how multi-hazard adaptation can be 'mainstreamed' into the construction decision-making process.

The book is split into four sections: (a) Introduction, (b) Structural adaptations, (c) Non-structural adaptations and finally, (d) Conclusions and recommendations. These sections have been selected to help guide the reader through the book but it is also important to acknowledge that there are many interlinkages between the structural and non-structural aspects covered in these sections; the structural aspects are not only structural, likewise the non-structural aspects are not only non-structural. The contents provide theoretical and empirical discussions of how built-in resilience should be attained. The forthcoming chapters present a wealth of useful case studies and research findings from the United Kingdom, Italy, Turkey, India, Nepal, Central Asia, Vietnam, Indonesia, Kenya, Peru and New Zealand. These chapters consider an array of hazards (natural and human induced) including earthquakes, floods, terrorist bombs, typhoons/hurricanes, landslides, tsunamis, low-quality construction, ineffective urban planning, high-density informal urbanisation, the ineffective enforcement of building codes and corruption in construction.

This book provides the reader with what appears to be an eclectic array of chapters. Yet, this wide range of perspectives (refer to Table 1.2) are central to relaying a holistic message about how built-in resilience can and should be attained. The perspectives that are encompassed highlight the varying levels of input that are required; embracing the formally and informally trained construction professionals and artisans as well as the 'client', the governmental policy maker and the non-governmental organisation; in fact a significant range of decision makers involved in the delivery of the built environment. The range of issues covered not only demonstrates the trans-disciplinary nature of the subject but also illustrates that non-structural as well as structural adaptations need to be considered to reduce the threat, and impact, of disasters and that lessons can be learnt from a range of disciplines and socio-cultural contexts. The challenge for the reader is to assimilate these different world-views and to relate them to their personal context.

The final chapter will summarise some of the cross-cutting themes which emerge from the individual contributions in order to synthesise the issues and challenges that the built environment faces. In addition, it seeks to join up the contributions to this book by drawing some broad conclusions with respect to the ways in which construction practitioners might adapt their *modus operandi* to better address the wide range of hazards that are discussed. In doing this, the final chapter provides seven guiding principles which can be used as a point of departure for the development of context-sensitive resilience frameworks in the future.

Notes

1 EM-DAT is the OFDA/WHO Collaborating Centre for Research on the Epidemiology of Disasters (CRED) International Disaster Database at the Université Catholique de Louvain Brussels – Belgium: www.em-dat.net.
2 Figure is based on an estimated world population of 6,602 million people (United States Census Bureau 2007).
3 A list of the world's 50 'least developed nations' can be found at http://www.un.org/special-rep/ohrlls/ldc/list.htm (accessed 25 July 2007).

Table 1.2 Summary matrix displaying the disciplinary perspectives covered in the forthcoming chapters

Chapter	Architecture	Engineering	Construction management	Urban planning	Disaster risk management	Policy and legislation	Education	Development/ social studies	Insurance/ economics
1			◆		◆			◆	
2					◆	◆			◆
3		◆	◆		◆	◆	◆	◆	◆
4		◆	◆		◆		◆	◆	◆
5	◆	◆	◆				◆	◆	
6	◆	◆							
7		◆			◆				◆
8					◆			◆	
9	◆				◆				
10		◆		◆	◆		◆		
11				◆	◆	◆	◆	◆	◆
12					◆	◆		◆	
13			◆			◆			
14		◆			◆	◆			
15				◆	◆	◆			
16		◆		◆	◆	◆			
17	◆	◆	◆	◆	◆	◆	◆	◆	◆

Disciplinary perspectives

4 Possibly because it is easier to quantify the economic costs of extreme events than the political, sociological and psychological impacts.
5 The popular saying is 'Earthquakes don't kill people: buildings do' as stated by Susan Hough and Lucile Jones of the US Geological Survey, Pasadena in the San Francisco Chronicle Op-Ed Commentary, 12/4/2002.
6 For example it could be argued that the 2004 Indian Ocean earthquake and tsunami was a 'natural disaster'. However, it is also pertinent to state that the failure to utilise an informal warning system (due to the inactivity of regional governments in sharing information and disseminating information/warnings to neighbouring countries) was a factor that contributed to the large death toll in nations such as Sri Lanka, India and further afield.
7 A recent report *Guidelines: National Platforms for Disaster Risk Reduction* by the United Nations has also identified the need to drop the adjective 'natural' in front of the word 'disaster' due to the recognition that disasters are a consequence of the combination of natural hazards and social and human vulnerability (UN/ISDR 2007b).
8 Figures obtained from Pattan Development Organisation (2005) 'Pakistan: Earthquake disaster response – Position paper No.1', Islamabad, Pattan Development Organisation, and EERI (2006) 'The Kashmir Earthquake of October 8, 2005: Impacts in Pakistan', EERI Special Earthquake Report – February 2006, Oakland, USA: Earthquake Engineering Research Institute.
9 A wide range of disciplines, such as ecologists, environmental scientists, educationalists and social scientists, can make positive contributions to non-structural mitigation but this book will focus on those disciplines most usually associated with construction decision-making processes.

References

Agar, W.N. (2000) 'Social and ecological resilience: are they related?', *Progress in Human Geography*, 24(3): 347–64.

Benson, C. and Twigg, J. (2007) *Tools for Mainstreaming Disaster Risk Reduction: Guidance Notes for Development Organisations*, Geneva: International Federation of Red Cross and Red Crescent Societies/the ProVention Consortium.

Bosher, L.S. (2007) *Social and Institutional Elements of Disaster Vulnerability: The Case of South India*, Bethesda, USA: Academica Press.

Bosher, L.S., Dainty, A.R.J., Carrillo, P.M., Glass, J. and Price, A.D.F. (2007) 'Integrating disaster risk management into construction: a UK perspective', *Building Research & Information*, 35(2): 163–77.

Burby, R. (ed.) (1998) *Policies for Sustainable Land Use: Cooperating with Nature*, Washington, DC: Joseph Henry Press.

Burby, R., Deyle, R.E., Godschalk, D.R. and Olshansky, R.B. (2000) 'Creating hazards resilient communities through land-use planning', *Natural Hazards Review*, 1(2): 99–106.

Coaffee, J. (2005) 'Urban renaissance in the age of terrorism: revanchism, social control or the end of reflection?' *International Journal of Urban and Regional Research*, 29(2): 447–54.

DFID (2006) *Reducing the Risk of Disasters*, East Kilbride: Department for International Development.

EERI (2006) 'The Kashmir earthquake of October 8, 2005: impacts in Pakistan', EERI Special Earthquake Report, February 2006, Oakland, USA: Earthquake Engineering Research Institute.

EM-DAT (2007) OFDA/WHO Collaborating Centre for Research on the Epidemiology of Disasters (CRED) International Disaster Database at the Université Catholique de Louvain Brussels, Belgium. Online. Available: HTTP: www.em-dat.net (accessed 1 June 2007).

Godschalk, D.R. (2003) 'Urban hazard mitigation: creating resilient cities,' *Natural Hazards Review*, 4(3): 136–43.

Godschalk, D.R., Beatley, T., Berke, P., Brower, D.S. and Kaiser, E.J. (1999) *Natural Hazard Mitigation: Recasting Disaster Policy and Planning*, Washington, DC: Island Press.

Gunderson, L., Holling, C.S., Pritchard, L. and Peterson, G.D. (2002) 'Resilience', in H.A. Mooney and J.G. Canadell (eds) *The Earth System: Biological and Ecological Dimensions of Global Environmental Change*, Volume 2, Chichester: John Wiley & Sons.

Hamelin, J-P., and Hauke, B. (2005) *Focus Areas: Quality of Life: Towards a Sustainable Built Environment*, Paris: European Construction Technology Platform.

Han, S.S. and Ofori, G. (2001) 'Construction industry in China's regional economy', *Construction Management and Economics*, 19: 189–205.

Hillebrandt, P.M. (2000) *Economic Theory and the Construction Industry*, third edition, Basingstoke: Macmillan.

Holling, C.S. (1973) 'Resilience and stability of ecological systems', *Annual Review of Ecol. and Syst*, 4: 2–23.

Holling, C.S. (1996) 'Engineering resilience vs. ecological resilience', in P.C. Schulze (ed.) *Engineering Within Ecological Constraints*, Washington, DC: National Academy Press.

Lorch, R. (2005) 'What lessons must be learned from the tsunami?', *Building Research and Information*, 33(3): 209–11.

McGlade, J. (2002) 'Thames navigation and its role in the development of London', *Proceedings of the 2002 London's Environment and Future (LEAF) Conference*, London: University College London.

Menoni, S. (2001) 'Chains of damages and failures in a metropolitan environment: some observations on the Kobe earthquake in 1995', *Journal of Hazardous Materials*, 86(1–3): 101–19.

Mileti, D.M. (1999) *Disasters by Design: A Reassessment of Natural Hazards in the United States*, Washington, DC: Joseph Henry Press.

Munich Re (2003) *Topics: Annual Review: Natural Catastrophes*, Munich: Munich Re Group.

Ofori, G. (1999) *The Construction Industry: Aspects of its Management and Economics*, Singapore: Singapore University Press.

Ofori, G. (2002) 'Construction industry development for disaster prevention and response', *Proceedings of the International Conference on Post-Disaster Reconstruction: Planning for Reconstruction,* May 23–25, Montreal, Canada.

Palakudiyil, T. and Todd, M. (2003) *Facing up to the Storm: How Local Communities Can Cope With Disasters: Lessons from Orissa and Gujarat*, London: Christian Aid.

Pattan Development Organisation (2005) 'Pakistan: Earthquake disaster response – Position paper No.1', Islamabad: Pattan Development Organisation.

Pelling, M. (2003) *The Vulnerability of Cities: Natural Disaster and Social Resilience*, London: Earthscan.

Rose, A. (2004) 'Defining and measuring economic resilience to disasters,' *Disaster Prevention and Management*, 13(4): 307–14.

Schneider, R.O. (2002) 'Hazard mitigation and sustainable community development', *Disaster Prevention and Management*, 11(2): 141–7.

Simpson, J.A. and Weiner, E.S.C. (1989) *The Oxford English Dictionary*, second edition, Volume XIII, Oxford: Clarendon Press.

Solomon, S., Qin, D., Manning, M., Alley, R.B., Berntsen, T., Bindoff, N.L., Chen, Z., Chidthaisong, A., Gregory, J.M., Hegerl, G.C., Heimann, M., Hewitson, B., Hoskins, B.J., Joos, F., Jouzel, J., Kattsov, V., Lohmann, U., Matsuno, T., Molina, M., Nicholls, N., Overpeck, J., Raga, G., Ramaswamy, V., Ren, J., Rusticucci, M., Somerville, R., Stocker, T.F., Whetton, P., Wood, R.A. and Wratt, D. (2007) 'Technical summary', in S. Solomon, D. Qin, M. Manning, Z. Chen, M. Marquis, K.B. Averyt, M. Tignor and H.L. Miller (eds) (2007) *Climate Change 2007: The Physical Science Basis*. Contribution of Working Group I to the Fourth Assessment Report of the Intergovernmental Panel on Climate Change, Cambridge: Cambridge University Press.

Trim P. (2004) 'An integrated approach to disaster management and planning', *Disaster Prevention and Management*, 13(3): 218–25.

UN/ISDR (2004) *Living with Risk: A Global Review of Disaster Reduction Initiatives*, Geneva: United Nations International Strategy for Disaster Reduction.

UN/ISDR (2005) *Hyogo Framework for Action 2005–2015: Building the Resilience of Nations and Communities to Disasters*, Geneva: United Nations International Strategy for Disaster Reduction.

UN/ISDR (2007a) *Disaster Risk Reduction: 2007 Global Review*, Consultation Edition, Geneva: United Nations International Strategy for Disaster Reduction.

UN/ISDR (2007b) *Guidelines: National Platforms for Disaster Risk Reduction*, Geneva: United Nations International Strategy for Disaster Reduction.

Vale, L.J. and Campanella, T.J. (eds) (2005) *The Resilient City: How Modern Cities Recover from Disaster*, Oxford: Oxford University Press.

Wamsler, C. (2004) 'Managing urban risk: perceptions of housing and planning as a tool for reducing disaster risk', *Global Built Environmental Review (GBER)*, 4(2): 11–28.

Wamsler, C. (2006) 'Mainstreaming risk reduction in urban planning and housing: a challenge for international aid organisations', *Disasters*, 30(2): 151–77.

Whittell, G. (2005) 'Warnings were loud and clear – but still city drowned', *The Times*, 8 September 2005, London.

Wisner, B., Blaikie, P., Cannon, T. and Davis, I. (2004) *At Risk: Natural Hazards, People's Vulnerability, and Disasters*, second edition, London: Routledge.

Chapter 2

Mainstreaming disaster risk management

David Alexander

Introduction

Disaster risk management (DRM) is the process of mitigating exceptional threats to society and communities by reducing and where possible avoiding the impact of catastrophe. Through DRM society is encouraged to develop coping strategies in order to achieve resilience, which is defined as the ability to deflect, absorb or resist the adverse impacts of disaster and to resume normal life as soon as possible. In the event that 'normal life' is insufficiently safe and productive, resilience must also be a direct function of security and economic development.

At the Kobe World Conference in January 2005 the United Nations' International Strategy for Disaster Reduction (UNISDR) identified five goals of DRM. These are described in the Hyogo Framework for Action (ISDR 2005) as follows:

- ensure that disaster risk reduction is a national and a local priority with a strong institutional basis for implementation;
- identify, assess and monitor disaster risks and enhance early warning;
- use knowledge, innovation and education to build a culture of safety and resilience at all levels;
- reduce the underlying risk factors; and
- strengthen disaster preparedness for effective response at all levels.

This chapter will examine the prospects for implementing these goals in the modern world and thus for mainstreaming disaster risk management – i.e. making it an integral part of governance, public administration and the general maintenance of security. It will consider the problems involved and examine some of the associated issues. These include the promotion of sustainability in disaster risk management and the democratisation of civil protection. First, a theoretical basis for examining the issues will be developed.

Towards a theoretical basis

In the last decade more than 6,000 disasters have affected more than 2.5 billion people around the world (IFRCRCS 2006). These simple statistics are a stark reminder of the need to manage risks and reduce the impact of catastrophe. However, because it remains a hypothetical phenomenon until materialising as impact, risk is a difficult phenomenon to characterise. If one subscribes to Heisenberg's principle that all events involve some final and residual uncertainty, then every risk must be considered probabilistic to a greater or lesser extent. It thus contains an element of chance and cannot fully be characterised determinately – not least because it is multifaceted and influenced by perception, communication and decision-making.

Risk is fundamentally a product of hazard and vulnerability, but there has been much debate over the relative importance of these two factors. After the publication of the 'radical critique' of Hewitt and his colleagues (Hewitt 1983) there was a tendency to emphasise the latter. Hence, the term 'vulnerability science' was coined, and this sub-field was based on the assumption that people's susceptibility to adverse events is the main key to understanding the impact of disasters, though authors such as Bankoff (2001) have noted that vulnerability can be an artefact of Western political and economic hegemony over the rest of the world – i.e. that the rich countries contribute little to its reduction in the poor nations because that suits the exploitative form that the world economy has taken.

In essence, the concept of vulnerability cannot be separated from the social and cultural conditions under which it exists. For any person or entity, vulnerability to a given threat is a holistic phenomenon that cannot be divided up into components. However, it can be considered according to seven different contexts:

(1) In *total vulnerability* life is generally precarious. For example, the populations of the Horn of Africa and Afghanistan live amid much vulnerability to disaster that, for reasons of climatic, political and military instability, and economic deprivation, has not been reduced (Comenetz and Caviedes 2002).

(2) *Economic vulnerability* means that people lack adequate employment or financial support. Unemployment due to economic stagnation following disaster, and inability to fund reconstruction may be implicated here (Ewing *et al.* 2005).

(3) *Technological (technocratic) vulnerability*, signifies that technology is the main source of risk. For instance, the vulnerability of the urban environment of Kobe, devastated by earthquake in 1995, had much to do with its technological complexity (Menoni 2001).

(4) *Social vulnerability* is intertwined with the economic and technological forms of vulnerability. Caste and class, marginalisation as a result of

excessive political hegemony and tribal conflict are some of the root causes (Bosher 2007).

(5) *Newly generated vulnerability* is caused by unmitigated changes in controlling circumstances. Climate change, as represented by rising sea levels and intensifying meteorological hazards, is probably going to be the main culprit in the future (Van Aalst 2006).

(6) In *delinquent vulnerability* corruption, negligence and criminal activity are to blame. For instance, the earthquakes of 1999 in Turkey graphically revealed the effect of ignoring building codes when putting up new buildings, many of which collapsed with heavy loss of life (Sengezer and Koç 2005).

(7) *Residual vulnerability* is caused by lack of modernisation and investment in risk reduction and is characteristic of poor countries, regions or cities (see Bankoff 2003).

Given the complexity of modern life, there are many instances in which more than one of these categories will apply. For instance, in Tokyo vulnerability to earthquakes has largely been considered a technological or technocratic problem, but, referring to the classification given above, there are sectors of society, particularly the homeless, that suffer badly from economic vulnerability and in the worst cases total vulnerability (Wisner 1998).

As noted, risk is primarily a function of hazard and vulnerability. Characterising and reducing it therefore depends on how these two phenomena are identified, defined and tackled – moreover on their importance relative to one another. The distribution of hazards is now well known at many scales, including the global one (Berz *et al.* 2001). Although all natural hazards, and many anthropogenic ones involve fluctuating impacts, many have stationary trends (Hewitt 1970). According to the work of Pielke *et al.* (1999), despite the likelihood of more intense meteorological disasters, normalised series show that so far the trend in increasing impacts is very largely accounted for by the growing vulnerability of human populations, not by changes in physical events. Hence in promoting disaster risk reduction it is probably more important to address the outstanding problems of vulnerability than to concentrate on hazard reduction.

The problem of disaster risk reduction is one of ensuring that life, health, employment and shelter are protected among populations at risk of catastrophe. This is largely a matter of applying existing knowledge on construction, organisation, management and health care, rather than generating new knowledge. It is a problem of investment and adaptation, of reducing the 'implementation shortfall'. Perhaps a priority for new research should be how to apply existing knowledge efficiently and effectively.

Some current problems of disaster risk management

Despite the obvious need to adopt disaster risk management in areas of high hazard, various factors restrict its implementation. The most obvious are shortage of funds and political will. Many authors (e.g. Özerdem and Jacoby 2006) have pointed to the importance of good governance in creating the conditions for the implementation of disaster reduction plans. However, there are various other hurdles, such as the following.

Failure to implement existing knowledge. While there is much talk of 'lessons learned' (e.g. Banipal 2006), in reality it is often more a case of 'lessons identified': whether or not they are then taken on board is another matter. Moreover, lessons can be 'unlearned' if institutional memory is short (Kletz 1993). In Turkey, for example, building codes were tightened successively after various major earthquakes, but many buildings that collapsed in the 1999 Kocaeli and Duzce earthquakes were not built to code (Sengezer and Koç 2005). It is unfortunately common for code enforcement to lapse between large events (Burby and May 1999).

The net outcome of 'lessons unlearned' is either the perpetuation of ignorance or 'reinvention of the wheel' as laborious attempts are made to design and construct systems that already exist in some form elsewhere. The solution rests with improvements in the quality and intensity of training and education. Moreover, greater efforts are needed to share knowledge and transfer technology and expertise to where it is needed. Compendia of knowledge, such as those published by the United Nations (e.g. ISDR Secretariat 2005) and International Federation of the Red Cross (e.g. IFRCRCS 2006), have been particularly valuable for disseminating appropriate knowledge and best practice. On the other hand, there is now a confusing plethora of aids and resources: web sites, 'toolboxes', organisations, acronyms, associations, networks, and so on. The process of disseminating knowledge requires that order be imposed and complexity be simplified.

Failure to agree on standards. In emergency preparedness, there are two conflicting imperatives. The first is to create interoperability and ensure quality by imposing uniform criteria upon services and products, and the second is to avoid restricting initiative and flexibility. In an ideal world there would probably be universally agreed standards, or at least benchmarks and compendia of best practice. These would be applied to emergency planning, disaster management, emergency education and training, and the acquisition, storage, retrieval, manipulation and transmission of information. However, it is difficult and sometimes impossible to design standards that fit areas with different cultures, administrative systems, traditions and resource bases. In the European Union, for example, there will probably never be a single disaster management system, as one could never be designed that would fit a collection of countries with such diverse administrative systems. Development of European crisis response will probably not go much beyond the current 'clearing house' approach, strengthened by increasingly sophisticated co-operation mechanisms (Jörgensen 1997).

Elsewhere (Alexander 2003, 2005) I have argued that standards in emergency preparedness offer both advantages and disadvantages. In synthesis, it is a highly attractive prospect to have interoperable systems whose efficiency is both guaranteed and widely recognised, but not at the expense of riding roughshod over local cultures and organisations.

Professionalisation of the field. There needs to be a greater realisation that disasters should not be managed by amateurs and untrained people. Despite this, not all of the bodies responsible for emergency preparedness are aware of the need to put the work in the hands of trained operatives, especially with regard to emergency planning and co-ordination. The field is becoming more professional but at an unacceptably slow pace (Crews 2001). In part this is because of a 'vicious circle' in which the links between training, certification, official recognition of qualifications, and appropriate employment are not yet fully established (see Figure 2.1). The field has a history of intellectual development that stretches back nearly 90 years (see Prince 1920; Barrows 1923), and for the latter half of that period disasters have been studied intensively. However, scholarly endeavour has been fragmented and at present it remains so (see Alexander 2000, ch. 3). Education and training need to be broad and holistic – either interdisciplinary

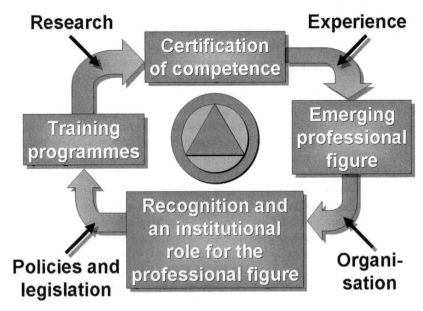

Figure 2.1   Training in disaster risk management and its role in the professionalisation of the field

or better still *non*-disciplinary, in which the practical problems to be solved determine what is taught and studied.

Insufficient transfer of knowledge and expertise. One could be forgiven for thinking that the world disaster relief system has become set in its ways to the extent of being comfortable with inefficiency. For example, when a disaster such as a major earthquake occurs, as many as 2,000 foreign rescuers may make their way to the affected area, which will also see an influx of field hospitals and all the other impedimenta of international disaster relief. However, in many cases it all arrives too late to be of much use in the primary work of saving victims' lives. For example, 1,600 foreign search-and rescue personnel arrived in Kerman Province, Iran, in the first three days after the Bam earthquake of December 2003, but they saved only 30 victims from death by entrapment, which represents a cost of perhaps US$1 million per successful rescue (Movahedi 2005).

One cannot reasonably expect to solve the problem of how to extricate people from sudden entrapment and injury by sending rescuers half way around the world. Instead, more needs to be done to encourage local preparedness. This should involve transferring knowledge and expertise to where it is needed most. The world's disaster areas are too well known to use the fig-leaf of 'unexpected events' as an excuse (Munich Re Group 1999). Moreover, the key ingredients are information and organisation: the price of the former is falling, while the latter need not be expensive. As the Chinese have shown (Lu Jingshen *et al.* 1992), where technology proves to be too expensive and unreliable, human resources can compensate for its absence at least to some extent, providing that participants are adequately organised, trained and informed. Hence there is no intrinsic reason why relatively poor countries, if they are politically stable and free from conflict, need not have effective civil protection systems. Cuba has shown the way by developing cheap but sophisticated resilience to hurricanes (Aguirre 2005).

Inconsistent investment and legislation to reduce the risks of disaster. The period after a major disaster represents the so-called 'window of opportunity' (Solecki and Michaels 1994), and one expects that the changes which occur as the window swings open will favour increased security. As political attention and social pressure tend to reach their maxima during the aftermath of a large disaster, this is the period in which the greatest investments are made. It is also the time in which something like three-quarters of all legislation relevant to disasters is passed. For example, in the central Apennines of Southern Italy more than 150 people were killed by mudflows on 5 May 1998. Five months later, a law was passed to ensure that all of the country's 20 regions quickly and thoroughly evaluated landslide risk (Repubblica Italiana 1998). Investment in landslide risk investigation therefore had to follow the legislation. This is one example of a more or less universal process in which change is achieved largely after the event, rather than before it. More foresight and prudence are needed, so that the problem of disaster risk reduction is tackled during times of quiescence, not merely in the heat of the event.

Love of technology and fear of social issues. The relentless pace of technological change tends to induce a reliance on technological solutions to problems. One modern example could well be the Indian Ocean tsunami warning system, whose implementation follows the failure of warning during the seismic sea wave disaster of 26 December 2004. All hazard warning systems consist of technical, administrative and social components (Figure 2.2). As the seismic zone that generated the 2004 tsunami is located close to the densely populated coast of Indonesia, near-field warning is required there. This means that messages must be transmitted in a very few minutes after seismic activity that may affect the ocean is detected. Everything depends on how well the process of relaying warnings to people in coastal communities is articulated, and whether it reaches the different constituencies that are at risk (Kelman 2006). If the social, perceptual and cultural aspects of the warning process are not firmly tackled, a mighty pan-national web of technology and communications could potentially be rendered ineffective at the level of local beneficiaries.

The negative political image of disaster reduction. Politicians tend not to like the negative connotations of disaster. These are only partially compensated for by the positive aspects of enhanced security, prudence and preparedness. By and large, the public does not fully appreciate the need to be ready for exceptional events. Popular culture exalts the victim, but does not disseminate a reasonable, rational, prudent model of disaster response. It prefers to reinforce myths and stereotypes (Arnold 2006).

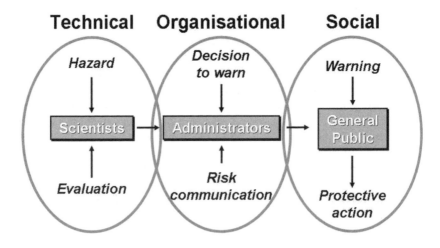

Figure 2.2 The technical, organisational and social components of the warning process

The great challenge is to achieve the transition from a dominant emphasis on reacting to disasters when they occur to a prevailing ethos of preparedness coupled with prevention. In this context, cost-benefit analyses have seldom convinced decision makers, and moreover they can be misleading if the choice and characterisation of costs and benefits are not rigorous, comprehensive and objective (Hanley 1998).

Is poverty reduction the key to resilience?

Poverty and vulnerability are often equated in the literature (Wisner *et al*. 1976; GTZ 2005) but in reality they are not exact synonyms, as poor communities are sometimes capable of extraordinary feats of mitigation, despite their lack of economic resources. For example, peasant farmers on the banks of the Brahmaputra River in Bangladesh have evolved dozens of strategies for coping with floods, droughts and bank erosion (Paul 1995). However, poverty and marginalisation are frequently major components of lack of resilience in developing countries. Marginalisation may be a tool of political repression, part of a deliberate strategy of subjugation of particular social groups, or it may be a consequence of extreme poverty and instability. In either case, it does not produce fertile ground for risk mitigation and the avoidance of disaster, especially when there is little to distinguish between occasional calamities and the 'disaster that is daily life' (Bankoff *et al*. 2004; Bosher 2007).

Concomitantly, personal resilience has fallen in the rich countries, perhaps also in some of the poor ones, though for different reasons. In the former, the self-reliance of 60 years ago, stimulated by the adverse conditions of a world conflict, has been replaced by a network of dependencies (Horlick-Jones 1995). There is a tendency to demand services rather than put up, shut up and improvise. While people have every right to expect public authorities to prepare for disaster, this does not mean that resilience should be considered exclusively an institutional phenomenon rather than also a family or a personal one.

Yet personal resilience does demand some collective effort. It cannot easily be rendered atomistic. For example, US hurricane evacuation procedures have shown some of the advantages and disadvantages of dependency on personal transport. Thus in Florida positive results were achieved with the timely and efficient evacuation of more than two million coastal residents (Wolshon 2001), but in Hurricane Katrina on the Gulf Coast, where a smaller proportion of the population was self-sufficient in transport, the outcome was much less positive (Childs 2005; Tierney 2006).

One final issue in disaster risk reduction is worth discussing here, and it may provide the solution to some of the other problems.

Sustainable emergency management

Sustainable development is now a well-recognised term in ecology and international development studies. It can be defined as 'development that respects the quality of life of future generations and that is accomplished through support for the viability of the Earth's resources and ecosystems' (Saunier 1999: 587). Its main focuses are equitable growth, poverty alleviation and food security (and to a lesser extent environmental protection, which is obviously vital to the other goals). According to the emerging paradigm, sustainability is achieved by conservation, development planning and reliance on formal mechanisms to reconcile conflicts over development actions. The last of these provisions requires the participation of a broad range of stakeholders from throughout society.

Hazards and disasters pose a threat to sustainable development, especially where they can affect the security of food production and livelihoods. It therefore follows that disaster reduction is necessarily an integral part of sustainability (Mileti 2002). From this it is only a short step to discussing *sustainable emergency preparedness*. However, this begs the question of why disaster management should not be sustainable in any case? Besides the problems discussed in the previous section, the answer lies in the vicissitudes of funding and level of political attention, and also in the processes of marginalisation and deprivation that make people and communities vulnerable to disaster.

I would argue that there are six canons of sustainable emergency management, as follows:

- *The local level:* it must be centred on the local level but the local systems must be harmonised by action taken at higher levels of public administration, principally the regional and national levels.
- *The disaster cycle:* it should be present during all the phases of the 'disaster cycle': mitigation, preparedness, emergency response, recovery and reconstruction.
- *A fundamental service:* it should be a fundamental, everyday service for the whole population and be taken seriously by political and community leaders. The service must be provided explicitly for the population, and not be focused merely upon the state or any of its agencies.
- *Emergency planning:* this should be achieved with the aid of emergency response plans that are professionally drawn up, fully disseminated, regularly tested and frequently revised. The plans need to be functionally integrated between levels of government, institutions and sectors of the economy. The plans must be generic ('all hazards') and based wherever possible on detailed, realistic scenarios. They should be designed to reduce the vulnerability and tackle the fundamental needs of the general population of the geographical area in which the plans apply. They need to be compatible with local urban and regional plans.

- *Popular support:* sustainable emergency management need to have the support and involvement of the general population, who are its main beneficiaries and who must become involved in a participatory manner. In this respect it should concentrate on the security needs as they are manifested and expressed by ordinary people. It must keep the public well informed of any risks and contingencies that may require people to take action.
- *Non-military:* sustainable emergency management should be based on principles of civil administration and be fully demilitarised.

Progress towards these goals is slow, uneven and fragmentary. However, there is a growing realisation that the alternative, involving some degree of ad hoc improvisation, is as inefficient as it is unsustainable and therefore cannot be justified.

Finally, the impact of structural collapse in terms of loss of life and livelihood ensure that the built environment will always be a key element of disaster risk reduction. In this context, sustainability must involve a broader approach than merely creating disaster-resistant structures and ensuring that building codes are observed, important though these aspects are. Yet the potential for a more holistic approach remains largely undeveloped. In the UK, for example, a survey of professionals involved with the planning, construction and operation of the built environment indicates that knowledge and awareness of integrated approaches to disaster risk management is poor, and the construction sector is not being used sufficiently as a key stakeholder and potential resource (Bosher *et al.* 2007). Moreover, resilience in the face of natural and human-induced hazards needs to be an integral part of the planning, design and construction of the built environment, not simply added on as an afterthought.

Urban planning is an essential element of resilience, as it can enable disaster risks to be avoided by land-use control (Wamsler 2006). Furthermore, Carreño *et al.* (2007) noted that not only should the process of surveying the built environment for risk assessment encompass all relevant hazards, it should also tackle social and environmental fragility. In a comparative study of seismic risk in Barcelona and Bogota, these authors combined indices of physical susceptibility to damage with social indices relating to delinquency, deprivation, population density, and so on. This enabled a picture of disaster risk to be created that was more accurate than one based purely on the propensity of buildings to collapse.

Concluding discussion

In conclusion, disaster risk management must begin and end at the local level (i.e. the upper part of Figure 2.3). This will doubtless be the theatre of operations when disaster strikes. Moreover, during the first two or three

Figure 2.3 The ingredients of disaster risk management: community resources are at the top and the public administration components are in the lower half of the diagram

days after impact, about 90 per cent of initial aid is likely to come from local sources, and if these are not forthcoming the aid will probably not arrive in time to meet critical needs such as rescue, medical care and basic shelter (Haddow and Bullock 2006). The local level of emergency management has close democratic ties with the population, for example when the municipal chief executive is also the nominal head of civil protection. However, local authorities can be subject to the pressure and counter pressures of devolution and centrism. In fact, during emergency situations, disaster relief can become a barter market, in which the local level begs for more resources, the national and international levels ration resources, and the intermediate level of government does both (Figure 2.4). In terms of both general organisation and direct response to disasters, the correct balance between centrism and devolution of powers and responsibilities needs to be worked out (Scanlon 1995). It will depend on the national political–administrative system.

Let us now reconsider the UN's Hyogo Declaration goals in the light of the foregoing discussion of disaster risk management's problems and opportunities.

Goal 1. Is disaster risk reduction a national and a local priority with a strong institutional basis for implementation? Improvements are occurring but the process is slow and uneven. Political stability, economic growth and

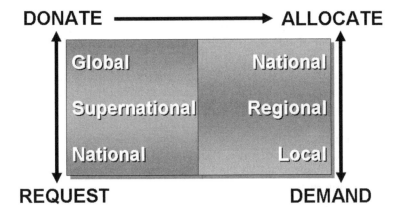

Figure 2.4   Disaster relief as a barter market for resources

devolution are necessary preconditions for successful DRM. So is awareness of the risks that society runs. The extent to which countries and their internal political units regard DRM as a priority has much to do with whether disaster impacts have occurred there in the recent past, and how severe they have been. Moreover, incumbent politicians often manifest attitudes to disaster reduction strategies that are strongly conditioned by their own political needs. Disasters thus become political events that can change the balance of power (Jalali 2002).

International and pan-national initiatives can help stimulate the formulation and application of national policies. In this age of global terrorism threats, there is a move to think in terms of the overall security of populations, and thus to tackle risk reduction comprehensively (Brown 2003). However, it is important that natural risks not be overshadowed by terrorism threats which, in many cases, are largely hypothetical, even if they are serious. The inexorable progress of climate change and rise of populations that inhabit the major hazard zones will inevitably lead to intensified natural hazard impacts. Counter-terrorism should not be treated ideologically in such a way as to relax preparedness for natural events (Beresford 2004). Moreover, with its emphasis on control, command and secrecy, counter-terrorism activities tend to be highly centralised and thus removed from local control. In contrast, the canons of sustainable emergency management listed above indicate the importance of a strong local response. For effective natural hazard management it cannot be over-emphasised.

Goal 2. Are disaster risks identified, assessed and monitored and is early warning a priority? The UN's International Strategy for Disaster Reduction has worked hard to promote this approach, and so have the institutes of the UN University (Birkmann 2006). Again, the results at this juncture are patchy, although at least the concept of early warning is now well known (Kaushik 2005). Since the 1980s the indicators of famine have been studied and developed in terms of early warning (Buchanan-Smith and Davies 1995). Early warning of major events such as tsunamis is now taken seriously, at least on a technological level (Kelman 2006).

Goal 3. Are knowledge, innovation and education used to build a culture of safety and resilience at all levels? There are signs of significant progress towards this goal, though once again there is scope for much more, especially in terms of the international sharing and transfer of expertise. One encouraging development is the rise of community-based disaster preparedness (Marsh and Buckle 2001). The Hyogo Declaration and all its support mechanisms under the UN's International Strategy for Disaster Reduction are further examples of positive efforts to share expertise. However, one could argue that representative democracy has often been less effective than participatory democracy in promoting resilience, especially at the local level (cf. Twigg 2005). Moreover, many important initiatives are still at an early stage, such as, for example, the Coalition for Global School Safety (Wisner *et al.* 2007).[1] Good governance is a vital aspect of the democratisation of resilience (Ahrens and Rudolph 2006) and it must adequately fulfil the stewardship, regulatory and management roles of government. In synthesis, the problem of disaster risk management is too great for scientists, experts and administrators to resolve on their own: it requires the public to assume some of the responsibility as knowledgeable stakeholders in the outcome (Kumar *et al.* 2005).

Goal 4. Are the underlying risk factors being reduced? In a previous discussion (Alexander 2000, Chaps 2 and 3) I examined the dialectic between factors that increase the risk of disaster and those that decrease it. Economic globalisation has tended to weaken the resistance of communities by marginalising them (Pelling 2003). A pessimistic assessment would stress that this process is not fully counterbalanced by the globalisation, or the local establishment, of disaster risk reduction measures. Earthquake scenarios for mega-cities such as Istanbul and Tehran (Nateghi 2001), which predict very high death tolls and other extremely serious consequences, suggest that there are many areas in which risk reduction has been utterly insufficient, largely through the continued perpetuation of vulnerability. The know-how is not lacking but a complex of political, economic and strategic interests has prevented the necessary levels of safety from being reached (Chan *et al.* 1998).

Goal 5. Is disaster preparedness being strengthened for effective response at all levels? Again, progress is slow. There is still a lack of consensus on what emergency planning and management should involve (Alexander 2002). Moreover, there is still an excessively unbalanced emphasis on providing aid after disaster has struck rather than preparing for it beforehand, including in the allocation of funds, which tends to respond more easily to impacts more than to threats. Moreover, in the modern 'complex emergency' (about 25 of which exist in the contemporary world), the moral and ethical difficulties of operating in highly polarised situations have severely troubled the aid agencies (Slim 1997). Yet despite the anguishing slowness of progress, there have been notable initiatives in favour of greater safety, especially in the wake of events that have had a profound impact on society. For example, the death of 26 children and three teachers in the collapse of a school in Molise Region, southern central Italy, in 2002 (Langenbach and Dusi 2004) had such a profound effect on national psyche that it led to an immediate re-evaluation of seismic safety in schools, much has had occurred in California after the 1933 Long Beach earthquake (which though it severely damaged schools fortunately happened 'out of hours' when they were empty).

Although it is difficult to generalise for almost 200 countries and many kinds of risk, it is clear that progress is being made in ensuring that disaster risk management is kept on the agenda. The occurrence of at least 700 disasters a year (IFRCRCS 2006) is one of the factors that ensures that this is so. Yet the prospects for a disaster of unprecedented proportions in one of the world's great megalopolises (e.g. refer to RMS 1995) suggest that the picture could change drastically at some time in the future.

Note

1 See http://www.interragate.info/cogss.

References

Aguirre, B.E. (2005) 'Cuba's disaster management model: should it be emulated?', *International Journal of Mass Emergencies and Disasters* 23(3): 55–72.

Ahrens, J. and Rudolph, P.M. (2006) 'The importance of governance in risk reduction and disaster management', *Journal of Contingencies and Crisis Management* 14(4): 207–20.

Alexander, D.E. (2000) *Confronting Catastrophe: New Perspectives on Natural Disasters*. Harpenden: Terra Publishing and New York: Oxford University Press.

Alexander, D.E. (2002) *Principles of Emergency Planning and Management*. Harpenden: Terra Publishing and New York: Oxford University Press.

Alexander, D.E. (2003) 'Towards the development of standards in emergency management training and education', *Disaster Prevention and Management* 12(2): 113–23.

Alexander, D.E. (2005) 'Towards the development of a standard in emergency planning', *Disaster Prevention and Management* 14(2): 158–75.

Arnold, J.L. (2006) 'Disaster myths and Hurricane Katrina 2005: can public officials and the media learn to provide responsible crisis communications during disasters?', *Prehospital and Disaster Medicine* 21(1): 1–3.

Banipal, K. (2006) 'Strategic approach to disaster management: lessons learned from Hurricane Katrina', *Disaster Prevention and Management* 15(3): 484–94.

Bankoff, G. (2001) 'Rendering the world unsafe: "vulnerability" as Western discourse', *Disasters* 25(1): 19–35.

Bankoff, G. (2003) 'Vulnerability as a measure of change in society', *International Journal of Mass Emergencies and Disasters* 21(2): 5–30.

Bankoff, G., Frerks, G. and Hilhorst, D. (eds) (2004) *Mapping Vulnerability: Disasters, Development and People*. London: Earthscan.

Barrows, H.H. (1923) 'Geography as human ecology', *Annals of the Association of American Geographers* 13: 1–14.

Beresford, A.D. (2004) 'Homeland security as an American ideology: implications for U.S. policy and action', *Journal of Homeland Security and Emergency Management* 1(3): Article 301. Online. Available: HTTP: www.bepress.com/jhsem/vol1/iss3/301 (Accessed 2 Jan 2007).

Berz, G., W. Kron, W., Loster, T., Rauch, E., Schimetschek, J., Schmieder, J., Siebert, A., Smolka, A. and Wirtz, A. (2001) 'World map of natural hazards: a global view of the distribution and intensity of significant exposures', *Natural Hazards* 23(2–3): 443–65.

Birkmann, J. (ed.) (2006) *Measuring Vulnerability to Natural Hazards: Towards Disaster Resilient Societies*. Tokyo: United Nations University Press.

Bosher, L.S. (2007) *Social and Institutional Elements of Disaster Vulnerability: The Case of South India*. Bethesda, MD: Academica Press.

Bosher, L.S., Dainty, A.R.J., Carrillo, P.M., Glass, J. and Price, A.D.F. (2007) 'Integrating disaster risk management into construction: a UK perspective', *Building Research and Information* 35(2): 163–77.

Brown, M.E. (ed.) (2003) *Grave New World: Security Challenges in the 21st Century*. Washington, DC: Georgetown University Press.

Buchanan-Smith, M. and Davies, S. (1995) *Famine Early Warning and Response: the Missing Link*. London: Intermediate Technology Publications.

Burby, R.J. and May, P.J. (1999) 'Making building codes an effective tool for earthquake hazard mitigation', *Environmental Hazards* 1(1): 27–37.

Carreño, M-L., Cardona, O.D. and Barbat, A.H. (2007) 'Urban seismic risk evaluation: a holistic approach', *Natural Hazards* 40(1): 137–72.

Chan, L.S., Chen, Y., Chen, Q., Chen, L., Liu, J., Dong, W. and Shah, H. (1998) 'Assessment of global seismic loss based on macroeconomic indicators', *Natural Hazards* 17(3): 269–83.

Childs, J.B. (ed.) (2005) *Hurricane Katrina: Response and Responsibilities*. Santa Cruz, CA: New Pacific Press.

Comenetz, J. and Caviedes, C. (2002) 'Climate variability, political crises, and historical population displacements in Ethiopia', *Environmental Hazards* 4(4): 113–27.

Crews, D.T. (2001) 'The case for emergency management as a profession', *Australian Journal of Emergency Management* 16(2): 2–3.

Ewing, B.T., Kruse, J.B. and Thompson, M.A. (2005) 'Empirical examination of the Corpus Christi unemployment rate and Hurricane Bret', *Natural Hazards Review* 6(4): 191–6.

GTZ (2005) *Linking Poverty Reduction and Disaster Risk Management*. Eschborn, Germany: Deutsche Gesellschaft für Zusammenarbeit.

Haddow, G.D. and Bullock, J.A. (2006) *Introduction to Emergency Management* (2nd edition), London: Butterworth-Heinemann.

Hanley, N. (1998) 'Resilience in social and economic systems: a concept that fails the cost benefit test?', *Environment and Development Economics* 3(2): 244–9.

Hewitt, K. (1970) 'Probabilistic approaches to discrete natural events: a review and theoretical discussion', *Economic Geography Supplement* 46(2): 332–49.

Hewitt, K. (ed.) (1983) *Interpretations of Calamity: From the Viewpoint of Human Ecology*. London: George Allen and Unwin.

Horlick-Jones, T. (1995) 'Modern disasters as outrage and betrayal', *International Journal of Mass Emergencies and Disasters* 13(3): 305–15.

IFRCRCS (2006) *World Disasters Report 2006*. Geneva: International Federation of Red Cross and Red Crescent Societies.

ISDR (2005) *Hyogo Framework for Action, 2005–2015: Building the Resilience of Nations and Communities to Disasters*. Geneva: United Nations' International Strategy for Disaster Reduction.

ISDR Secretariat (2005) *Know Risk*. Geneva: United Nations' International Strategy for Disaster Reduction and London: Tudor Rose.

Jalali, R. (2002) 'Civil society and the state: Turkey after the earthquake', *Disasters* 26(2): 120–39.

Jörgensen, K.E. (ed.) (1997) *European Approaches to Crisis Management*. Boston, MA: Kluwer Law International.

Kaushik, R. (2005) 'Early warning!' *Know Risk*. Geneva: ISDR Secretariat and London: Tudor Rose.

Kelman, I. (2006) 'Warning for the 26 December 2004 tsunamis', *Disaster Prevention and Management* 15(1): 178–89.

Kletz, T. (1993) *Lessons From Disaster: How Organizations Have No Memory and Accidents Recur*. Houston, TX: Gulf Publishing.

Kumar, A., Ragunathan, M. and Nandanwar, Y. (2005) 'Involving stakeholders', *Know Risk*. Geneva: ISDR Secretariat and London: Tudor Rose.

Langenbach, R. and Dusi, A. (2004) 'On the cross of Sant' Andrea: the response to the tragedy of San Giuliano di Puglia following the 2002 Molise, Italy, earthquake', *Earthquake Spectra* 20(S1): S341–58.

Lu Jingshen, Du Gangjian and Song Gang (1992) 'The experience, lesson and reform of China's disaster management', *International Journal of Mass Emergencies and Disasters* 10: 315–27.

Marsh, G. and Buckle, P. (2001) 'The concept of community in emergency management', *Australian Journal of Emergency Management*, 16(1): 5–7.

Menoni, S. (2001) 'Chains of damages and failures in a metropolitan environment: some observations on the Kobe earthquake in 1995', *Journal of Hazardous Materials* 86(1–3): 101–19.

Mileti, D.S. (2002) 'Sustainability and hazards', *International Journal of Mass Emergencies and Disasters* 20(2): 135–8.

Movahedi, H. (2005) 'Search, rescue, and care of the injured following the 2003 Bam, Iran, earthquake', Special issue on the 2003 Bam, Iran, earthquake. *Earthquake Spectra* 21(S1): S475–85.

Munich Re Group (1999) *Topics 2000: Natural Catastrophes, the Current Position*. Munich: Munich Re Group.

Nateghi, F.A. (2001) 'Earthquake scenario for the mega-city of Tehran', *Disaster Prevention and Management* 10(2): 95–100.

Özerdem, A. and Jacoby, T. (2006) *Disaster Management and Civil Society: Earthquake Relief in Japan, Turkey and India*, International Library of Postwar Reconstruction and Development no. 1. London: I.B. Tauris.

Paul, B.K. (1995) 'Farmers' responses to the flood action plan (FAP) of Bangladesh: an empirical study', *World Development* 23(2): 299–309.

Pelling, M. (ed.) (2003) *Natural Disasters and Development in a Globalizing World.* London: Routledge.

Pielke, R.A. Jr, Landsea, C.W., Downton, M. and Muslin, R. (1999) 'Evaluation of catastrophe models using a normalized historical record: why it is needed and how to do it', *Journal of Insurance Regulation* 18: 177–94.

Prince, S.H. (1920) 'Catastrophe and social change: based upon a sociological study of the Halifax disaster', *Studies in History, Economics, and Public Law* 94: 1–152.

Repubblica Italiana (1998) 'Decreto del Presidente del Consiglio dei Ministri, 29-9-98 'Atto di indirizzo e coordinamento...', *Gazzetta Ufficiale della Repubblica Italiana* 140(3): 8–38 (5 January 1999).

RMS (1995) *What if the 1923 Earthquake Strikes Again? A Five-Prefecture Tokyo Region Scenario.* Menlo Park, CA: Risk Management Solutions.

Saunier, R.E. (1999) 'Sustainable development', in D.E. Alexander and R.W. Fairbridge (eds) *Encyclopedia of Environmental Sciences.* Dordrecht: Springer.

Scanlon, T.J. (1995) 'Federalism and Canadian emergency response: control, co-operation and conflict', *Australian Journal of Emergency Management* 10: 18–24.

Sengezer,B. and Koç, E. (2005) 'A critical analysis of earthquakes and urban planning in Turkey', *Disasters* 29(2): 171–94.

Slim, H. (1997) 'Doing the right thing: relief agencies, moral dilemmas and moral responsibility in political emergencies and war', *Disasters* 21(3): 244–57.

Solecki, W.D. and Michaels, S. (1994) 'Looking through the post-disaster policy window', *Environmental Management* 18(4): 587–95.

Tierney, K. (2006) 'Social inequality, hazards and disasters', in R.J. Daniels, D.F. Kettl and H. Kunreuther (eds) *On Risk and Disaster: Lessons From Hurricane Katrina.* Philadelphia, PA: University of Pennsylvania Press: 109–28.

Twigg, J. (2005) 'Community participation: time for a reality check?', *Know Risk.* ISDR Secretariat, Geneva, and Tudor Rose, London: 64–5.

Van Aalst, M.K. (2006) 'The impacts of climate change on the risk of natural disasters', *Disasters* 30(1): 5–18.

Wamsler, C. (2006) 'Mainstreaming risk reduction in urban planning and housing: a challenge for international aid organisations', *Disasters* 30(2): 171–7.

Wisner, B. (1998) 'The geography of vulnerability: why the Tokyo homeless don't "count" in earthquake preparations', *Applied Geography* 18(1): 25–34.

Wisner, B., O'Keefe, P. and Westgate, K. (1976) 'Poverty and disaster', *New Society* 37: 546–8.

Wisner, B., Kelman, I., Monk, T., Bothara, J.K., Alexander, D., Dixit, A.M., Benouar, D., Cardona, O.D., Kandel, R.C. and Petal, M. (2007) 'School seismic safety: falling between the cracks?', in C. Rodrigue and E. Rovai (eds) *Earthquakes*, Routledge Hazards and Disasters Series. London: Routledge.

Wolshon, B. (2001) '"One-way-out": contraflow freeway operation for hurricane evacuation', *Natural Hazards Review* 2(3): 105–12.

Part II
Structural adaptation

Chapter 3

Construction in developing nations

Towards increased resilience to disasters

George Ofori[1]

Disasters and their effects

Disasters and developing countries

Over the years, disasters have had an increasingly heavy economic and social impact in many countries. The Independent Evaluation Group (IEG) of the World Bank (2006) reported that, in constant dollars, disaster costs during the decade 1990–9 were 17 times higher (US$ 652 billion in material losses) than they were in 1950–9 (US$ 38 billion, at 1998 values). The IEG also noted that human losses have risen: between 1984 and 2003, more than 4.1 billion people were affected by disasters. The number of affected persons grew from 1.6 billion in the first half of the period in 1984–93, to 2.6 billion in the second half, in 1994–2003. The International Federation of Red Cross and Red Crescent Societies (2006) noted that in the past two decades, direct economic losses from disasters multiplied five-fold to US$ 629 billion. Annual direct losses from weather-related events alone increased from about $ 3.9 billion in the 1950s to $ 63 billion in the 1990s. Moreover, in the 1990s, an average of 80,000 people died each year due to disasters. In 2003, there were about 700 disasters which killed about 75,000 people and caused about US$ 65 billion damage.

The international reinsurance firm, Munich Re (2006) estimated that 2005 was the most expensive year on record with overall economic losses of US$ 210 billion and total disaster-related insurance claims of over US$ 94 billion. The United Nations Human Settlements Programme (UNHSP) (2007a) reported that in 2006, some 300 disasters around the world (including droughts in China and Africa and flooding in Africa and Asia) affected 117 million people and caused some US$ 15 billion in damages. Du Plessis (2001) presents data which show that of the 100 most expensive disasters of the twentieth century, 65 occurred in the 1990s, 25 in the 1980s and 10 in the 1970s, and much fewer in the previous decades.

Many authors predict that the trend will be more severe owing to increasing urbanisation, environmental degradation, and increasing vulnerability due to climate change and variability. For example, the UNHSP (2007a) reported that in the first 6 months of 2006, there were 113 flood disasters representing a record 65 per cent of all disasters (the average for that period during the previous 10 years was 58 floods and 36.5 per cent of disasters). Some 85 per cent of the deaths from disasters were in South-East Asia. The International Federation of Red Cross and Red Crescent Societies (2006) warns that the frequency and cost of disasters is likely to increase because of: environmental degradation; climate change; population growth, especially in cities; and globalisation.

From the statistics, and as outlined in Chapter 1, the frequency and impact of disasters is greater in the developing countries than in industrialised nations. The Operations Evaluation Department (OED) of the World Bank (OED 2005) notes that injury and death rates from the devastations caused by storms, floods, earthquakes and other disasters can be up to 100 times higher in the poorer developing countries than in industrialised nations. More than 95 per cent of all deaths caused by disasters occur in developing countries; and losses due to natural hazards (as a percentage of GDP) are 20 times greater in developing countries than in industrialised nations. The OED (2005: 1) notes that 'Lack of mitigation is itself an indicator of underdevelopment'. Similarly, the IEG (2006: xix) observes that 'lack of development itself contributes to disaster impacts'. The UNHSP (2007b) notes that disasters are a development issue because they undo much of the work of development: they prevent capitalisation of the gains achieved by individuals, families, communities and nations. For this reason, disasters also tend to exacerbate poverty and inequality (Pelling 2006). Moreover, the UNHSP (2007b) notes that poverty and lack of resources increase vulnerability, weaken coping strategies and delays the recovery process.

Disasters, buildings and infrastructure

As the often dramatic descriptions, photographs and videos of the aftermath of disasters show, buildings and infrastructure bear the brunt of the damage from disasters of all kinds. A press release by the World Bank (1970) in November 1970 described the cyclone in the then East Pakistan as follows: 'The cyclone ravaged the area leaving behind destroyed houses, roads and bridges, huge holes in the streets and enormous erosion along the streams.' The World Bank describes itself as the largest funder of disaster recovery and reconstruction in the world, lending more than $ 40 billion for over 550 disaster-related projects. It cites the following examples: (i) in 1961, the Bank financed the Highway Maintenance Project in Chile to rehabilitate the roads devastated by earthquakes in May 1960; and (ii) in 1979 the Bank approved the loan for Emergency Road Reconstruction in the Dominican Republic after two major hurricanes in September 1979.

Increased urbanisation and high population growth in developing countries raise the potential for loss from disasters because they concentrate infrastructure and other physical assets in one place, and these places tend to be vulnerable to disaster (International Federation of Red Cross and Red Crescent Societies 2005). The UNHSP (2007a) reported that, in 1976, one-third of the world's population lived in cities. In 2006, the figure was half; it was expected to rise to two-thirds (or 6 billion people) by 2050. It notes, further, that, ironically, most of the largest cities in the world are in areas where earthquakes, floods, landslides and other disasters are likely to happen. The UNHSP (2007b) notes that urban authorities in developing countries are usually ill-equipped to provide sufficient infrastructure and services. As a result, most of the world's poor live in densely populated squatter settlements, on the periphery of cities, which lack the basics of life, making them increasingly vulnerable. Demand for commercial and residential land in cities has led to the use of unsuitable terrain prone to natural hazards. Many informal settlements, where the poorest people in developing countries live, are located in dangerous or unsuitable areas, such as floodplains, unstable slopes or reclaimed land. Moreover, these cities are often unable to manage rapid population growth; poorly planned urbanisation with increasing numbers of inadequately constructed and badly maintained buildings, further increases the level of vulnerabilities. Pelling (2006) suggests that continuing rapid urbanisation, the weight of accumulated failures of urban development and ineffective urban governance have placed large and growing numbers of people in cities around the world at risk.

As Moor (2002) notes, with socio-economic progress, settlements become more vulnerable as they tend to be more reliant on their increasingly extended supply lines and vital distribution networks of critical resources such as food and water, as well as water, power, gas and telecommunication systems. They also become dependent on community networks and government agencies. Social and security infrastructure such as health facilities, civil defence and the police also become crucial. In such situations, small events can trigger large-scale disasters (Pelling 2006).

Importance of constructed items and the impact of disasters

Importance of constructed items

Constructed items represent most of every nation's savings. Studies show that Gross Domestic Fixed Capital Formation in construction is 45–60 per cent of the total capital formation in all countries (Hillebrandt 2000; Han and Ofori 2001). Constructed items are vital to the pursuit of economic activity as they provide the space needed for the production of all goods and services. The physical infrastructure facilitates productive activity. The World Bank (2007a) observes that if Africa had attained infrastructure growth rates comparable to those in East Asia in the 1980s to 1990s, it could have achieved annual economic growth rates about 1.3 per cent higher. Similarly, it is estimated that the lack of investment in infrastructure in the 1990s reduced long-term economic growth in the Latin America and Caribbean region by 1–3 percentage points (World Bank 2007b).

Built items also offer people the opportunity to enhance their quality of life. Badiane (2001) estimates that investments in housing alone constitute between 2 and 8 per cent of GNP; between 10 and 30 per cent of gross capital formation; between 20 and 50 per cent of accumulated wealth; and between 10 and 40 per cent of household expenditure. Indeed, there are benefits beyond economics: Gueli (2007) suggests that improving national infrastructure networks and enhancing service delivery to ordinary citizens may be an effective strategy to help prevent, reduce or manage violent conflict.

Thus, the World Bank (2004a) aptly notes that disasters are closely linked to poverty as they can wipe out decades of development in a matter of hours. The Bank's data showed that from 1990 to 2000, disasters resulted in damage constituting between 2 per cent (in Argentina and China) and 15 per cent (in Nicaragua) of an affected country's annual GDP. It found that GDP losses for individual events could be even more devastating: in Honduras, Hurricane Mitch caused losses equal to 41 per cent of GDP, and some 292 per cent of the government's annual tax revenue.

Impact of disasters on built items in developing countries

For citizens of developing countries, losses resulting from disasters can be more severe in magnitude than for persons in industrialised nations. This is best illustrated by using residential units. First, many houses in these countries are also used by families for income-earning activities, and indeed, some houses are specifically designed for such purposes, such as the 'shophouses' in South-East Asia and South Asia. Thus, loss or damage of houses has both economic and psychological aspects. Second, a house represents several times each person's annual income, and it might be impossible for the owner in a developing country to repair one which is severely damaged, or replace one that is destroyed. Third, whereas in the industrialised countries, disaster prevention measures reduce the risk of damage, thus making insurance rates affordable, in developing countries, there is usually no suitable insurance scheme to ameliorate the financial losses resulting from the destruction of a house in a disaster; where these schemes are available, they may be beyond the reach of, or unfamiliar to, the ordinary citizens (the following chapter provides further discussion on this issue). Finally, the governments of these countries typically face budgetary constraints, and thus, are unable to offer adequate compensation for the losses suffered by the citizens in a disaster in a timely manner.

As has been discussed above, in developing countries, the built environment is already unsatisfactory. While most developing countries have formulated comprehensive policies on urban and infrastructure development, implementation levels remain unsatisfactory. The UNHSP (2007b) found that, in many cities, more than 50 per cent of the people live in slums and 'have little or no access to shelter, water, and sanitation, education and health services' (p. 4). It estimated that by 2006, more than one billion people around the world lived in slums. Tibaijuka (2006) observes that in sub-Saharan Africa, slum dwellers form 70 per cent of urban populations, and that the figure in other parts of the world is 50 per cent. For example, it is estimated that 51 per cent of the total population of Colombo, the capital of Sri Lanka, live in slums and shanties, while only 42 per cent of the national housing stock is permanent (Samaraweera 2001).

The construction process itself can lead to disasters. UNHSP (2006) observed that flood hazards are natural phenomena, but damage and losses from floods are the consequence of human action, adding that urbanisation aggravated flooding by restricting where flood waters can go, covering large parts of the ground with roofs, roads and pavements, obstructing sections of natural channels and building drains that ensure that water moves to rivers faster than it did under natural conditions. In many cities, there have been calls for big advertising billboards to be banned. As Iglesias (2006: 7) notes, in the Philippines, which are typhoon-prone, 'Too often they go flying and kill people and crush cars and buildings when they slam back into the ground'.

Construction has some environmental impacts (see Ofori 2000; du Plessis 2001). All construction requires land. However, the preparation of the land for construction can destabilise land formations, leading to landslides, mudslides or rockslides. Some construction projects such as dams for hydroelectric power generation, also require large numbers of people to be displaced, with an adverse impact on their economic and social well-being, and on some occasions, loss of livelihood. There can also be loss of heritage. The extraction, beneficiation, processing and transportation of materials for construction can also have major environmental impacts in terms of the scarring of landscapes; land, air and water pollution; and intensive energy usage (with its own pollutive consequences and climate change potential). Finally, construction and demolition waste accounts for a significant proportion of landfilled wastes in many countries.

Construction in disaster resilience and recovery

Accumulated knowledge and current motivation

Several international organisations and conferences have called for disaster management to be given a high priority in planning for development, starting with the United Nations' Yokohama Strategy and Plan of Action for a Safer World launched in 1984, which led to the 1990s being declared the International Decade for Natural Disaster Reduction (Montoya 2004). The current International Strategy for Disaster Reduction puts emphasis on disaster risk management and policy making. The UNHSP (2007b: 19) suggests: 'Greater variations in precipitation due to climate change, together with an increase in the vulnerability of populations, highlights the need to shift our emphasis from disaster response to risk management.' Rollnick (2006: 4) quotes figures from the United Nations which indicate that 'one dollar invested in disaster reduction today can save up to seven dollars tomorrow in relief and rehabilitation costs'.

The World Bank and the US Geological Survey (World Bank 2004a) estimate that economic losses worldwide from disasters in the 1990s could have been reduced by $ 280 billion if $ 40 billion had been invested in preventative measures. It is also estimated that the $ 3.15 billion spent on flood control in China over the last four decades of the twentieth century averted losses of about $ 12 billion. A study on Jamaica and Dominica in the Caribbean, calculated that the potential avoided losses compared with the costs of mitigation when building infrastructure such as ports and schools would have been two to four times. Thus, the World Bank (2004a) suggests that disaster prevention should be considered as an integral component of development, and incorporated into national development plans. However, the OED (2005: 1) observes that 'Disaster mitigation, because it is a periodic need rather than a constant one, tends to lose out to other priorities'.

Table 3.1 shows the features of constructed items which make them vulnerable to disasters. As an example, owing to the physical characteristics of constructed items, they cannot be moved elsewhere even if an imminent disaster could be accurately predicted. As many disasters cannot be prevented, the most effective action is in providing the constructed items with features which can limit the damage from the occurrence of disasters. Thus, there is a need for adequate planning, design and construction capacity and capability to attain this aim.

The concern with effective disaster risk management and recovery is not new. Rollnick (2006) notes that the wealth of experience in disaster risk management and the mitigation of hazards can be traced to the California earthquake of 1906 or even to the measurements taken of the biggest volcano explosion in recorded history on the Indonesian island of Krakatoa in 1883. Disaster risk management is at the fore in the priorities of governments and international organisations where the built environment is concerned. For example, the priority areas of the Countries of the Commonwealth Consultative Group on Human Settlements, launched in 1999, include (Member Countries of Commonwealth 2001): (i) promotion of employment opportunities in conjunction with shelter provision, particularly for the low and no-income groups, and other under-privileged sectors; (ii) training

Table 3.1 Features of constructed items and their contribution to the vulnerability of these items

Features of constructed items	Resulting vulnerability
Location specificity and immobility	The constructed items are exposed to disasters which occur where they are located; they cannot be moved as a precaution even if a disaster is predicted.
High expense	It is impossible to test the completed item by exposing it to the full force of a possible disaster. Thus, simulations and limited tests are applied which may not fully reflect the real situation.
Length and complexity of development process	Planning, design and construction of the items involve a high number of operations undertaken by several companies and organisations. The effective co-ordination of these contributions poses a challenge. There is also need for continuous enhancement of the expertise of these persons in many areas including aspects of disasters, and their impact, and appropriate actions during various stages of the process.
	There is also dispersed responsibilities and control among many government agencies.
Need to comply with regulations and codes	To ensure the safety and health of occupants and neighbours of the constructed items, nations apply regulations and codes to guide planning, design and construction. The enforcement of these regulations and codes poses a challenge.
Durability	Durability is both a requirement and a feature. The items are exposed to the elements of the constructed items and wear and tear which may weaken it. There is need for continuous maintenance.
	It is difficult and expensive to alter the items to make them disaster resistant if errors or omissions are detected after their completion, or new technologies become available.
	The bulkiness of the components of the items increases casualties and fatalities if the items are damaged during disasters. It also makes repair and rehabilitation difficult.
Form of usage	Items are occupied and utilised for various purposes (work, rest and play). Thus, disasters affecting them can lead to loss of lives.

and capacity building to support member countries' human settlement programmes; (iii) development of mutually supportive partnerships in order to fuse government, civil society and the private sector into a cohesive and efficient support mechanism; and (iv) *improving disaster preparedness and mitigating the consequences of conflict.*

The Habitat Agenda, which was passed at the City Summit in Istanbul in 1996, urged local, national and international action to enhance capabilities in disaster prevention, mitigation and preparedness (United Nations 1996). The *Draft Declaration on Cities and Other Human Settlements in the New Millennium* signed at the Istanbul+5 Summit in 2001 (UNHSP 2001) commits the international community to improving prevention, preparedness, mitigation, and response capabilities with the cooperation of national and international networks in order to reduce the vulnerability of human settlements to natural and human-made disasters, and to implement effective post-disaster programmes.

There is still inadequate knowledge, and effective action seems to take place only after disasters have occurred. A study by the ProVention Consortium, an international network of public, private, non-governmental and academic organisations (formed in 2000) dedicated to reducing the impacts of disasters in developing countries, analysed responses to five disasters: floods in Bangladesh in 1998; Hurricane Mitch in Honduras in 1998; an earthquake in Turkey in 1999; an earthquake in Gujarat, India in 2001; and floods in Mozambique in 2000 and 2001 (World Bank 2004b). The resources allocated to disaster recovery interventions came from external agencies, governments, affected people's savings and remittances from expatriates overseas. Most of the funding was spent on infrastructure and housing recovery efforts. The report notes that the recovery process after natural disasters was 'little understood' and the lack of a systematic analysis of recovery programmes meant that mistakes were being repeated. However, there is inadequate attention to enabling the affected communities to participate in disaster risk management activities; and limited cross-country learning about disaster risk management policy. The study identified some positive trends. In all the cases studied, policies developed in the wake of the disasters paid more attention to risk reduction and preparedness, with disasters being treated as part of the development process. The Turkish government was enforcing building codes in housing construction after the 1999 earthquake. In India, after the Gujarat earthquake, the government allowed the 'self-build' of housing to speed up the recovery process. In Mozambique, opportunities were taken in recovery programmes to construct infrastructure that could better withstand future flooding.

Construction process and disasters

Planning and building regulations and codes

The statutes which relate to buildings are formulated and enforced to safeguard the users and the community. Land-use planning regulations determine the location of items (zoning); the intensity of development (density); the heights of constructed items (massing); and the distances of the items from one another (setbacks). Building control statutes seek to ensure the safety and health of the users and the neighbours. In each of these statutory provisions, measures can be taken to avoid serious impacts in disasters. Tibaijuka (2006) notes that the prevention of disasters can be greatly enhanced through the adoption and enforcement of more appropriate land-use planning and building codes. However, the developing countries face several difficulties in this area. These countries often use regulations formulated elsewhere, under different economic, physical (such as climatic) and social conditions. For example, many of them continue to use planning and building regulations and codes from the former colonial powers. Arimah and Adeagbo (2000: 293) note that, in Nigeria, 'In essence the urban development and planning regulations currently in force are those tailored after the 1932 British Town and Country Act which was adopted in 1946'.

The development and building control regime in Singapore has many special features which were introduced after the collapse of a hotel in 1986. In Singapore, the design should be by Qualified Persons, who are Registered Architects and Engineers. For sizeable buildings, an Accredited Checker must be engaged to scrutinise the submitted calculations and drawings of the structural engineer. Building activity on site only starts after a Building Permit has been obtained. The Client must pay for a full-time Resident Engineer or Clerk of Works on the site to act as the client's representative and to control the works. The law also requires a Safety Officer and an Environmental Officer to be engaged. A Temporary Occupation Permit (TOP) must be obtained after completion. This is given after all the installations and systems are tested. A Certificate of Statutory Completion must be granted. The requirements for this are even more stringent than those for the TOP. An annual certificate of inspection is required for all lifts in buildings. The law also requires five-yearly inspections of structural efficacy for commercial buildings; and ten-yearly inspections for residential buildings.

Researchers have explored ways of putting in place systems and procedures which enable the construction process to deliver products that are resilient against disasters, as well as being ready to effectively address disasters when they occur. Bosher et al. (2007) offer PRE-EMPT, an iteratively developing tool which incorporates disaster risk management, and can guide and improve how buildings and infrastructure are planned, designed, built and retrofitted to cope with disasters. Le Masurier et al. (2007) note that existing regulatory provisions and frameworks that are intended to ensure safety, health and environmental protection in routine land use and development may constrain reconstruction efforts, cause difficulties in multi-agency responsibilities and co-ordination as well as in resource allocation, and result in a lack of pragmatism in disaster practitioners. They suggest that building regulations should provide effective means of reducing and containing vulnerabilities, and facilitate post-disaster reconstruction. The International

Federation of Red Cross and Red Crescent Societies (Gospodinov 2001) notes that there is insufficient understanding of the part disaster preparedness must play in the setting of housing and construction standards. Thus, the Federation has committed itself to advocating for better responses to housing and construction standards in order to mobilise the power of humanity for disaster prevention, preparedness, mitigation and response. There are examples of well-formulated development codes. The International Code Council, a membership association dedicated to building safety and fire prevention, develops the codes used to construct various types of buildings. Most US cities, counties and states have chosen to use the council's codes.

Some change is evident in developing countries. Mozambique has reconstructed and rehabilitated its roads and social infrastructure, which had been devastated by massive floods, to higher standards (Cossa 2001). It is also encouraging the populations to abandon areas of risk and move to more secure ones. In Rwanda, a ministerial decree in 1997 instructs that in the urban areas, construction should be done only in the surveyed plots (Nkusi 2001). Bangladesh has taken measures to protect the natural water bodies and to tighten up its planning and environmental laws and their enforcement (Siddique 2001). The governments of several Caribbean and Pacific countries have adopted legislation strengthening their construction codes as part of a comprehensive mitigation programme following the destruction caused by major hurricanes.

In most developing countries, owing to inadequate executive capacity, the existing urban development and planning regulations are not enforced. Arimah and Adeagbo (2000) found very low levels of compliance with such regulations in Nigeria. This is typical of the developing countries where there are, usually, weak administrative agencies; and lack of clarity in responsibilities for enforcing and administering regulations. Munich Re's annual review of natural catastrophes in 2003 noted that the earthquake that devastated Bam in Iran in December 2003 killed more than 40,000 people in large part because the mud brick houses were not designed to handle a major tremor (World Bank 2004a). The traditional buildings of mud brick and heavy roofing were particularly unsafe when earthquakes struck. Munich Re suggests that disaster risk should be included in the evaluation of investment projects; this would allow for more careful selection and design of projects, and development and implementation of risk management measures to protect the benefits of primary projects. Authors offer proposals for capacity building in the public agencies of developing countries in the area of disaster mitigation and recovery. Keraminiyage et al. (2007) note that developing countries take longer than their industrialised counterparts to recover from disasters because of lack of 'intellectual capacity' in their local governments. They propose good practice in the formulation and implementation of capacity building strategies.

Design and procurement for resilience

The ability of a constructed facility to withstand damage, or to prevent or reduce loss of lives during a disaster can be enhanced through appropriate design and judicious selection of construction materials and methods. This involves the structure, installations, layout and dimensions. The OED (2005) suggests that, when planning new infrastructure, their future maintenance, and the training of the technical personnel for it, should be considered. Thus, the construction industry must have adequate capacity and capability to undertake planning and design which gives due cognisance to the possibility of all forms of disasters in the particular context of the locations of the items.

In addition to designing in compliance with appropriate statutes and codes, the development of suitable concepts for the circumstances of the affected persons is critical. Many building systems have been proposed, for both temporary construction after disasters, or permanent rebuilding. Maiellaro et al. (2007) offer a system of low-cost temporary or semi-permanent housing based on corrugated steel and earth coverings, which they consider suitable for use in the 'second emergency' phase of medium- to large-scale disasters. Johnson (2007: 36) describes temporary housing as 'a crucial but controversial part of disaster recovery' and notes that, in the literature, there is no agreement on its merits and appropriate approaches to it. Such housing is considered to be overly expensive, too late, too long-lasting, and responsible for undesirable impacts on the urban environment. Johnson (2007) suggests that temporary housing should be part of the overall disaster recovery programme. Its organisational and technological design should consider the following factors: (i) temporary housing should be rapidly available for the affected people; (ii) efforts should be made to draw on local suppliers and resources; (iii) the units should meet local living standards in terms of amenities, comfort and location; (iv) the units should be designed for the length of time the units are needed, or incorporate a long-term plan for their usage; and (v) the units should be easy to remove.

Suitable arrangements for the procurement of the building and infrastructure works are also necessary. Wilkinson et al. (2007) note that fast and efficient contractual systems are necessary for the rebuilding process after disasters; those for projects in 'normal' situations may not be suitable. Wilkinson et al. (2007) offer recommendations for the development of contractual systems for disaster reconstruction, with special focus on the procurement approaches. Similarly, Davidson et al. (2007: 114) suggest: '... in a disaster-prone area, the principles governing procurement ... should be established before disaster occurs so that in the chaos following the disaster there is a framework for taking the applicable procurement decisions coherently ...'. The OED (2005) suggests that to avoid delays in reconstruction, streamlined decision making and procedures for contracting civil works should be put in place early. Seed money should be made available to finance small-scale post-disaster actions such as constructing model infrastructure or demonstrating mitigation techniques. In countries prone to disasters, disaster mitigation and recovery should be part of long-term institutional development.

Construction and operation

The actual construction on site determines the final quality of the completed building or item of infrastructure. Thus, the nation should have construction organisations which are able to produce sound construction. The competence and capability of local contractors must be increased. The transfer of knowledge on effective ways and means of enhancing the resilience of constructed items and implementing post-disaster reconstruction programmes is critical if lessons are to be learnt, lives saved, physical and socio-psychological damage reduced, and livelihoods safeguarded. Pathirage *et al.* (2007) propose a knowledge-based framework, developed from research on the post-tsunami reconstruction in Sri Lanka, to facilitate the sharing of information, good strategies and practices, and skills in disaster mitigation.

Construction industry development

Considering the importance of planning, designing and constructing buildings and items of infrastructure to reduce their vulnerability to disasters and to respond effectively to disasters in order to save lives, rehabilitate vital infrastructure, and reinstate economic activities, it is necessary to provide the construction industry in each developing country with the requisite capacity and capability. This can only be achieved through deliberate, planned, strategic, systematic efforts. Ofori (1993) suggests that *construction industry development* has the following components: human resource development; materials development; technology development; corporate development; development of documentation and procedures; institution building; and development of the operating environment of the industry.

For a construction industry in a developing country to play an effective role in disaster mitigation and recovery efforts, it must have certain key features. First, human resource development should equip construction professionals with the knowledge and skills required to undertake appropriate designs and construction, and with the attitude of life-long learning to keep themselves informed of developments in knowledge overseas. It is also necessary to build up the nation's capacity of skilled tradespersons. Moreover, programmes for passing on technical skills quickly to members of the community in order to enable them to participate in the rehabilitation of their own homes should also be developed. Second, a programme of materials development should be instituted in each geographical region to find high-performing disaster-resistant materials which are suited to the local context and are of good quality, durability and affordability. Possible components and systems based on these materials, especially those which lend themselves to rapid deployment, should also be developed. Third, measures should be in place in pursuit of the technological development of the industry to ensure that it has the capability to handle the various projects which will be required to provide protection and resilience against disasters, and those which will be applied in post-disaster reconstruction.

The fourth aspect is corporate development where the companies in the construction industry are provided with the policy and administrative guidance and incentives to progressively upgrade their operations and strengthen their organisations. The fifth consideration is institution building. The professional institutions and trade associations can be a powerful force for change in the construction industry. Thus, these organisations should be strengthened. The final component of an industry development programme in this context is a conducive operating environment. In the public sector, there should be a regime of statutory regulations and codes which guides planners and designers to take preventive action, and contractors to produce items of the requisite quality and durability. There should also be an efficient and effective enforcement framework to give practical effect to the regulations. The operating environment also includes the practices and procedures of the sectors of the economy with which construction firms do business. For example, the contract forms and project procedures should facilitate the relatively complex projects relating to disaster management.

Good practice in implementing recovery programmes

There is a growing volume of literature on appropriate measures to ensure disaster resilience and recovery. The IEG (2006) notes that whereas the consequences of disasters are not completely preventable, it is often technically possible to mitigate them so that fewer lives and less of the constructed environment are lost. Systems of prediction and risk analysis for many types of disasters have been developed and are available for application, and with suitable institutions and adequate resources catastrophe is avoidable. The ProVention report (World Bank 2004b) called for: (a) governments to make natural hazard risk analysis and reduction an integral part of the development process; (b) more independent reviews of recovery efforts after major disasters to ensure lessons were learned; (c) more attention to be paid to supporting the livelihoods of the affected populations; and (d) mechanisms for periodic updating of damage and needs assessments.

The World Bank (2001) observes that international experience from disaster-hit areas suggests that the recovery programme should follow principles including: revival of the economy; empowering individuals and communities; affordability, private sector participation, and equity;

decentralisation; and communication and transparency. Consultation with, and participation by, the affected communities must be at the heart of the recovery programme, including, as far as possible, rebuilding of their own houses by individuals in their original location. Effective community participation is stressed by many authors, such as El-Masri and Kellett (2001) who studied the post-war reconstruction of villages in the Lebanon.

The OED (2005) draws the following conclusions from the earthquake and tsunami disaster of 26 December 2004. First, at the initial stages, providing survivors with income-earning opportunities tied to physical work such as the clearance of rubble and recovery of building materials is often as helpful as grief counselling. Second, it is necessary to assess carefully the technical necessity for relocating people or entire villages before doing so because of the social dislocation and possible loss of livelihood; and the possibility of taking advantage of the existing infrastructure in the original location. Third, reconstruction of damaged infrastructure is imperative but insufficient in itself; it is necessary to identify local vulnerabilities and determine ways of reducing them most effectively in the particular context, such as by offering financial incentives, reviewing land use and management practices and land tenure patterns, upgrading building codes, and training construction craftsmen.

Davidson et al. (2007) note that the literature stresses that community participation is the key to the success of low-cost housing in developing countries. It enables the decentralisation of decision making in urban management and empowerment of the population; it may be a compulsory component of funding agencies, or encouraged by NGOs; it makes use of existing strong community ties, social cohesion and values; it makes the best use of local knowledge; it reduces costs; and can be a strategy of survival as it uses the resources of the informal sector and the owners. However, this premise is not clearly reflected in reconstruction practice. Davidson et al. (2007) suggest that users can be involved in a range of ways, from being the labour force only to playing an active part in decision making, but that their capabilities are often wasted. They suggest that the organisational and technological design of programmes should consider the local social, political and economic contexts to enable the participation of users in decision making during the planning and design stages, as this leads to positive results.

Billing (2006: 14) notes that, after a disaster, every attempt should be made to 'buy local', and that 'starting immediately on household repair and reconstruction using local artisans (who may only need their lost tools replaced) helps re-establish local business and gets the community involved from the outset'. Minervini (2002) observes that after the devastation of the war in Kosovo, the rebuilding process in both rural and urban areas relied mainly on reinforced concrete, and did not use the local method that many of the locals normally utilised to build their own houses, which was adobe reinforced with recurrent horizontal wooden boards. Thus, the bulk of the materials and skilled personnel were imported; only companies skilled in this method took part in the process, the local users were seldom involved in rebuilding, and it was costly (further discussion on this topic is provided in Chapter 4). This led to the creation of a society which depended on external aid and external capital investment.

Davis (2006) suggests there must be continuity between (the often two-stage process of) initial shelter following the collapse of a dwelling to a permanent home. This provides a 'unique opportunity to teach new building, financial management and contract management skills, [and] ... to focus on safe building and disaster preparedness' (p. 8). There should also be: (i) opportunities and mechanisms for full participation in decision making by the affected communities; (ii) an emphasis on 'building back better' (p. 8); (iii) a review of capacities, including recycling building debris, using local builders and craftsmen, consulting leaders and local institutions, finding out the number of unoccupied, undamaged buildings or community buildings that may offer temporary shelter.

Steinberg (2007) notes that the wave of international support for Aceh, Indonesia after the tsunami disaster led to the involvement of several agencies in the construction of housing; many of these agencies had no prior experience. Despite the huge volume of funding which had been pledged, the reconstruction has been slower than expected, and has been affected by numerous bottlenecks. There is also concern about the quality of the finished product, for greater integration of housing with the residential infrastructure, and for livelihood support. Steinberg (2007) observes that 'it is not only habitat which matters but reconstruction of lives and communities' (p. 150). The reconstruction programme in Aceh faces the following obstacles: (i) slow reconstitution of land tenure and titles, hindering confirmation of beneficiaries; (ii) large areas of unbuildable land (now flooded by the sea); (iii) difficulties in the selection and identification of beneficiaries, with absentee relatives appearing, and claims and counter claims; (iv) environmental problems of some sites which are difficult and costly to address; (v) cost increases owing to the need to import materials, labour and other resource and capacity shortages; (vi) logistics problems in importation of materials; (vii) ignorance of new improved anti-seismic building codes among some agencies, and absence of a building control system, means poor quality and lack of earthquake resistance in some completed houses (which some beneficiaries have rejected); (viii) insufficient budget allocations for habitat-related infrastructure which had not been adequately considered at the planning stage; (ix) absence of livelihood reconstitution as a result of uni-sectoral planning; (x) inadequate or no provision for renters of housing; (xi) uncertainties concerning repair and rehabilitation of damaged housing units owing to lack of a clear policy of government and donor support for owners of such units; and (xii) 'housing without village planning' (p. 162) – provision of thousands of housing units without the necessary residential infrastructure, or consideration of the physical environment.

Jorgensen (2006) summarised the priorities of risk mitigation investments as: (i) establish a permanent disaster management department, committee or agency at central, district and local levels and maintain close liaison with other government organisations; (ii) make budgetary provision for disaster and risk mitigation; (iii) undertake mapping and legal surveys of all existing structures, and maintain these assets; (iv) modify the existing bye laws and planning regulations; (v) ensure that structures comply with planning regulations and building by-laws, and institute an inspection and reporting system; and (vi) ensure regular drills and training at grassroots level.

The World Bank's US$ 505 million reconstruction loan to Turkey after the 1999 earthquake included measures to update and enforce building codes. The reconstruction programme will introduce better planning for land-use, and requires compulsory insurance for housing. Emergency response management will also be upgraded. Another initiative in Central America by the World Bank, Inter-American Development Bank and the Japanese government is assisting six governments in assessing disaster risks, setting up emergency warning and response systems, improving building codes and their enforcement, and carrying out studies to identify environmental measures that would reduce the threat of natural hazards.

International knowledge banks and partnerships

A great deal of knowledge and experience has been built up at the international level on how to attain resilience, and undertake disaster recovery programmes. The UNHSP's Risk and Disaster Management Programme endeavours to enhance knowledge and experience in countries on disaster risk management in order to promote sustainability and self-reliance. It provides local government, communities and business organisations with practical strategies for mitigating and recovering from conflicts and disasters. From its experience, it is continually refining its approaches and techniques to reduce the vulnerability of human settlements to disasters. The programme focuses on: protection and rehabilitation of housing, infrastructure and public facilities; resettlement of displaced persons and returnees; restoration of local social structures through settlement development; land and settlements planning and management for disaster prevention; and creation of co-ordination mechanisms for improved disaster risk management. It fields assessment and technical advisory missions to disaster-prone countries to assess demands for support; designs and implements projects at national, regional and global level in collaboration with countries and support agencies; strengthens co-operation, co-ordination and networking among experts, communities, NGOs, governments and external support organisations in performing disaster-related activities; develops techniques and tools for the management of disaster prevention, mitigation and rehabilitation; and designs and implements training programmes.

The World Bank's Disaster Management Facility (DMF) aims to reduce human suffering and economic losses caused by natural and technological hazards, by promoting the integration of disaster prevention and mitigation efforts into development activities. The DMF provides technical support to World Bank operations, promoting capacity building, and establishing partnerships with the international and scientific community working on disaster issues. Its objectives are: (i) to improve the management of disaster risk in member countries and reduce vulnerability in the World Bank portfolio; (ii) to promote sustainable projects and initiatives that incorporate effective prevention and mitigation measures; (iii) to promote the inclusion of risk analysis in World Bank operations, analysis and country assistance strategies; (iv) to promote training in the areas of disaster prevention, mitigation and response; and (v) to identify policy, institutional and physical interventions aimed at reducing catastrophic losses from disasters through structural and non-structural measures, community involvement and partnerships with the private sector.

The *Hyogo Framework for Action (HFA) 2005–2015: Building the Resilience of Nations and Communities to Disasters* was adopted by 168 countries at the UN World Conference on Disaster Reduction in Kobe, Hyogo in 2005 to galvanise international and national efforts to reduce vulnerability to natural hazards (ISDR and World Bank 2007). The principal goal of the HFA, which is the primary international agreement for disaster reduction, is to integrate disaster considerations into sustainable development policies, planning, programming and financing at all levels of government. The HFA offers a strategic framework for multi-stakeholder commitment to, and engagement in, disaster risk reduction. Its priorities are: (i) ensure that disaster risk reduction is a national and a local priority with a strong institutional basis for implementation; (ii) identify, assess and monitor disaster risks and enhance early warning; (iii) use knowledge, innovation and education to build a culture of safety and resilience at all levels; (iv) reduce the underlying risk factors; and (v) strengthen disaster preparedness for effective response at all levels.[2]

There is much scope for partnerships involving the public and private sectors, and across countries. The Global Facility for Disaster Reduction and Recovery is a global partnership of the International Strategy for Disaster Reduction (ISDR) to support the implementation of the HFA. It is supporting joint efforts of the ISDR, the World Economic Forum and the World Bank to foster public-private partnerships, particularly in countries prone to multiple disaster risks. For example, following the 1999 earthquake in Turkey, the governments of Turkey and the UK, in partnership with some British engineering and construction firms operating in the country, developed the British Earthquake Consortium for Turkey. The consortium pooled expertise and resources to identify areas which were relatively more earthquake-resilient and thus better suited for development. It contributed to the efforts of the Marmara Emergency Earthquake Reconstruction Project. As another example, the UNHSP partnered with the German chemical firm, BASF, in tsunami relief and reconstruction (Rollnick 2006).

Concluding remarks

Researchers in construction should contribute by developing means for safeguarding the nations' accomplishments and savings in physical development. Strategies to attain disaster resilience should be built into the planning, design, construction and operation processes for buildings and items of infrastructure.

National governments and local construction industries should work to realise the components of construction industry development discussed in this chapter. In most developing countries, there is a need for the reform of regulations and codes, and their enforcement frameworks to enable them to provide resilience to the built items in the particular local context, while striking a balance between complexity and affordability, as well as enforceability and time considerations. The UNHSP and the World Bank would be able to help in these efforts. This effort would also benefit from financial and technical assistance from the World Bank, such as under the ProVention initiative.

The ongoing procurement reform in many countries should also provide systems for recovery programmes. There is need for more research and development (R&D) in this area. Other areas where R&D is required include: the changing patterns of the causes of disasters and their implications for construction; and appropriate design in response to local knowledge on the impact of disasters on constructed items. International best practices on these issues should also be compiled and disseminated. The UNHSP could lead this effort. To facilitate the R&D, there should be timely and accurate data on disasters. Thus, it would be suitable if such data were included among those which are collected and reported under the Global Urban Observatory (GUO).

The professional institutions and trade associations should enhance the awareness of their members of the need to assess the risks of disasters and take necessary precautions at all stages of the planning, design and construction processes. It would also be useful if they could provide the practitioners with checklists of appropriate actions for various disasters which might happen locally. It is also necessary for the curriculums of professional programmes to be redesigned to cover the relevant aspects of disaster management. There is also commercial merit in some construction organisations developing a rapid-response capability in order to attend to post-disaster reconstruction in developing countries. In the medium-term, the International Council for

Research and Innovation in Building and Construction (CIB) could consider forming a volunteer team of practitioners to help with post-disaster recovery in developing countries.

Many of these recommendations and R&D issues will be discussed in more depth in the forthcoming chapters of this book.

Notes

1 National University of Singapore, Department of Building, National University of Singapore, 4 Architecture Drive, Singapore 117566. Tel: (65) 6874 3421; Fax: (65) 6775 7401. e-mail: bdgofori@nus.edu.sg.
2 http://www.unisdr.org.

References

Arimah, B.C. and Adeagbo, D. (2000) 'Compliance with urban development and planning regulations in Ibadan, Nigeria'. *Habitat International*, 24: 279–94.

Badiane, A. (2001) 'Speech at High Level Segment of Economic and Social Council on the Role of the United Nations System in Supporting the Efforts of African Countries to Achieve Sustainable Development', Geneva, 16–18 July.

Billing, K. (2006) 'Building back better: post-crisis economic recovery and development'. *Habitat Debate*, 12(4): 14.

Bosher, L., Dainty, A., Carrillo, P. and Glass, J. (2007) 'PRE-EMPT: developing a protocol for built-in resilience to disasters'. *Proceedings, CIB World Congress: Construction for Development*, Cape Town, 14–17 May.

Cossa, H. (2001) 'Statement by H.E. Mr. Henrique Cossa, Vice-Minister of Public Works and Housing of the Republic of Mozambique before the 25th Special Session of the United Nations General Assembly for the Overall Review and Appraisal of the Implementation of the Outcome of the United Nations Conference on Human Settlements (Habitat II)', New York, 8 June.

Davidson, C., Johnson, C., Lizarralde, G., Dikmen, N. and Sliwinski, A. (2007) 'Truths and myths about community participation in post-disaster housing projects'. *Habitat International*, 31: 100–15.

Davis, I. (2006) 'Getting the right approach to long-term post-crisis shelter strategies'. *Habitat Debate*, 12(4): 8.

du Plessis, C. (2001) 'Doing more, better, with less: a developing country perspective on sustainability through performance'. Keynote Paper presented at World Building Congress 2001, Wellington, 2–6 April.

El-Masri, S. and Kellett, P. (2001) 'Post-war reconstruction: participatory approaches to the damaged villages of Lebanon. A case study of al-Burjain'. *Habitat International*, 25: 535–57.

Gospodinov, E. (2001) 'Statement by Encho Gospodinov, Head of Delegation, The International Federation of Red Cross and Red Crescent Societies, United Nations General Assembly, Twenty-fifth special session on Human Settlement (HABITAT II)', New York, 8 June.

Gueli, R. (2007) 'Construction for development (but also for security?)'. *Proceedings, CIB World Congress: Construction for Development*, Cape Town, 14–17 May.

Han, S.S. and Ofori, G. (2001) 'Construction industry in China's regional economy'. *Construction Management and Economics*, 19: 189–205.

Hillebrandt, P.M. (2000) *Economic Theory and the Construction Industry*, third edition. Basingstoke: Macmillan

Iglesias, G. (2006) 'Cities at risk: a case for better planning, management and policies'. *Habitat Debate*, 12(4): 7.

Independent Evaluation Group (2006) *Hazards of Nature, Risks to Development.* Washington, DC: World Bank.

International Federation of Red Cross and Red Crescent Societies (2006) *World Disasters Report: living and dying in the shadows*, Geneva: International Federation of Red Cross and Red Crescent Societies.

International Strategy for Disaster Reduction and World Bank (2007) *Global Facility for Disaster Reduction and Recovery: Partnership Charter.* Geneva/Washington DC: ISDR/World Bank.

Johnson, C. (2007) 'Impacts of prefabricated temporary housing after disasters: 1999 earthquakes in Turkey'. *Habitat International*, 31: 36–52.

Jorgensen, A.K. (2006) 'International experience with disaster management operations: lessons learned from the Asian Development Bank', Paper presented at the Regional Hazard Risk Management Conference, New Delhi, India, 19–20 December.

Keraminiyage, K., Haigh, R., Amaratunga, D. and Baldry, D. (2007) 'EURASIA: role of construction education in capacity building for facilities and infrastructure development within a developing country setting'. *Proceedings, CIB World Congress: Construction for Development*, Cape Town, 14–17 May.

Le Masurier, J., Rotimi, J.O. and Wilkinson, S. (2007) 'Regulatory frameworks for post-disaster reconstruction: improving resilience in the process'. *Proceedings, CIB World Congress: Construction for Development*, Cape Town, 14–17 May.

Maiellaro, N., Lassandro, P., Lerario, A., Regina, G. and Zonno, M. (2007) 'Modular building solutions for emergency in developing countries'. *Proceedings, CIB World Congress: Construction for Development*, Cape Town, 14–17 May.

Member Countries of the Commonwealth (2001) 'Statement by Member Countries of the Commonwealth Consultative Group on Human Settlements at the Special Session On Habitat II', New York, 8 June.

Minervini, C. (2002) 'Housing reconstruction in Kosovo'. *Habitat International*, 26: 571–90.

Montoya, L. (2004) 'Book review of Natural Disasters and Development in a Globalizing World'. *Habitat International*, 28: 499–500.

Moor J. (2002) 'Cities at risk'. *Habitat Debate*, 7(4): 4–5.

Munich Re (2006) *Annual Review: Natural Catastrophes 2005.* Munich: Munich Re.

Nkusi, L. (2001) 'Statement by Prof. Laurent Nkusi, Minister of Lands, Human Resettlement and Environmental Protection of Rwanda, to the 25th Special Session of the General Assembly on the Review and Appraisal of the Implementation of the Habitat Agenda', New York, 6–8 June 2001.

Ofori, G. (1993) *Managing Construction Industry Development: Lessons from Singapore's Experience.* Singapore: Singapore University Press.

Ofori, G. (2000) 'Challenges for construction industries in developing countries'. *Proceedings, Second International Conference of the CIB Task Group 29*, Gaborone, Botswana, November, pp. 1–11.

Operations Evaluation Department (2005) *Lessons from Natural Disasters and Emergency Reconstruction*. Washington, DC: World Bank.

Pathirage, C., Amaratunga, D., Haigh, R. and Baldry, D. (2007) 'Knowledge sharing in disaster risk management strategies: Sri Lankan post-tsunami context'. *Proceedings, CIB World Congress: Construction for Development*, Cape Town, 14–17 May.

Pelling, M. (2006) 'Cities are getting more and more vulnerable.' *Habitat Debate*, 12(4): 6.

Rollnick, R. (2006) 'The aftermath of natural disasters and conflict'. *Habitat Debate*, 12(4): 4–5.

Samaraweera, M. (2001) 'Statement by Honourable Mangala Samaraweera, Minister of Urban Development, Construction & Public Utilities at the Twenty-fifth Special Session of the United Nations General Assembly for Overall Review and Appraisal of the Implementation of the Outcome of the United Nations Conference on Human Settlements (Habitat II)', New York, 7 June.

Siddique, Q.I. (2001) 'Statement by Quamrul Islam Siddique, Permanent Secretary, Ministry of Housing and Public Works, Bangladesh at the Special Session on Habitat II', New York, 6 June.

Steinberg, F. (2007) 'Housing reconstruction and rehabilitation in Aceh and Nias, Indonesia – rebuilding lives'. *Habitat International*, 31: 150–66.

Tibaijuka, A.K. (2006) 'A message from the Executive Director'. *Habitat Debate*, 12(4): 2.

United Nations (1996) 'Report of the United Nations Conference on Human Settlements (Habitat II)', Istanbul, 3–14 June. New York.

UNHSP (2001) *Draft Declaration on Cities and other Human Settlements in the New Millennium*. Nairobi: United Nations Centre for Human Settlements.

UNHSP (2006) 'Report shows urban slum dwellers most vulnerable to floods'. November. UN-HABITAT Online. Available: HTTP: (accessed 1 May 2007) http://www.unhsp.org/content.asp?cid=4105&catid=286&typeid=6&subMenuId=0.

UNHSP (2007a) *UN-HABITAT Annual Report 2006*. Nairobi: UN-HABITAT.

UNHSP (2007b) *Vulnerability Reduction and Disaster Mitigation*, UN-HABITAT Online. Available: HTTP: (accessed 1 May 2007) www.unhsp.org/content.asp?typeid=19&catid=286&cid=866&activeid=867.

Wilkinson, S., Zuo, K., le Masurier, J. and van der Zon, J. (2007) 'A tale of two floods: reconstruction after flood damage in New Zealand'. *Proceedings, CIB World Congress: Construction for Development*, Cape Town, 14–17 May.

World Bank (1970) 'East Pakistan – cyclone protection and coastal area rehabilitation project', Press Release, 18 November.

World Bank (2001) *Complete Preliminary Gujarat Earthquake Damage Assessment and Recovery Plan*. Joint report by the World Bank and Asian Development Bank, Washington DC/Bangkok: World Bank/Asian Development Bank.

World Bank (2004a) 'Natural disasters: Counting the cost'. World Bank press release 2 March, Online. Available: HTTP: (accessed 1 May 2007) http://go.worldbank.org/NQ6J5P2D10.

World Bank (2004b) *Natural Disasters: Lessons From the Brink Countries and Donors Urged to Study Past Disaster Recovery Efforts to Avoid Repeating Mistakes.* World Bank press release 2 March, Online. Available: HTTP: (accessed 1 May 2007) http://go.worldbank.org/MVCQK8FE10.

World Bank (2007a) *Global Economic Prospects: Managing the Next Wave of Globalization.* Washington, DC: World Bank.

World Bank (2007b) *Infrastructure in Latin America and the Caribbean: Recent Developments and Key Challenges.* Washington, DC: World Bank.

Chapter 4

More to lose

The case for prevention, loans for strengthening, and 'safe housing' insurance – the case of central Vietnam

John Norton and Guillaume Chantry

Introduction

Setting aside the human 'cost' of injury and fatality, the value of the losses in goods and services and destruction to buildings of all kinds caused by disasters has steadily risen over recent decades. In this perspective, Hurricane Katrina was the most costly disaster ever to hit the United States; the storms that hit Western Europe in December 1999 and again in 2007 (storm Kyrill) were the most costly so far to hit Europe. This pattern of increased *economic* loss is repeated around the world. This does not necessarily suggest, however, that the degree of *material* loss has increased significantly, nor that the violence of 'natural' disasters is greater than before – even if some data suggest that the latter might be the case. What this pattern really highlights is that the social and economic costs of recovery or rebuilding have increased massively.

Certainly, the economic losses caused when natural hazards affect wealthy industrial nations will tend, by their very location and degree of sophistication, to head the tables of economic loss. But in human terms, the level of *vulnerability* to loss and its impact in poor societies is just as significant and if anything more devastating in local terms, even though such losses attract little attention and even less support for recovery. Loss of property, and in particular loss of the home – to many crucial to their economic survival, either as the base of a craft industry or for raising or growing food – is in many poor communities central to increasing vulnerability to economic loss from disasters. This vulnerability stems in large part from the widespread shift from traditional, often locally-resourced techniques and materials used in building, to increasingly monetarised methods of material production and construction deemed to be 'modern' (for more on this see Chapter 5). The result is that families have *more to lose*, because the replacement value of goods and shelter in many fast-modernising poor societies has increased and will continue to do so.

We are talking here of communities with no access to insurance against the impact of disasters, where the consequences of a disaster on a poor family in both social and economic terms can be – literally – catastrophic, even when they are fortunate enough to survive it. Approaches that encourage *preventive* action to reduce risk can and need to be promoted. These are some of the issues that need to be addressed to reduce vulnerability.

This chapter draws on the experience of several poor communities in Asia, where changes in socio-economic circumstances and consequent changes in building practice over recent decades, perceived as an *improvement* in living conditions and building quality, have in reality contributed to *increased vulnerability* – a tragic paradox. The chapter considers the nature and cause of such vulnerability risk, and the steps that can be, and are being, taken to reduce it.

Social and economic aspects of vulnerability

Vulnerability to natural, and indeed human induced, hazards is and has been on the increase for decades. This is essentially the result of social and economic changes, which, when occurring in the context of hazards, have increasingly significant consequences for the afflicted population, and on their ability to resist, cope and recover. Although there are strong indications that certain types of natural hazard – cyclones and floods, for example – are increasing in intensity and frequency, these hazard increases are essentially marginal compared with the consequences of socio-economic and material changes that have taken place and that contribute to increased vulnerabilities (for example see Wisner *et al.* 2004; Bosher 2007). To consider the issue of disaster risk reduction therefore requires placing the idea of vulnerability in its broad context, where diverse non-hazard factors have an effect on the level of exposure of individuals and families to the potential impact of a hazard.

One of the tendencies of recent years, however, has been to disassociate natural hazards – and the short-term responses to these – from the broader context of social development or decline. This has been largely due to policies that, often by the nature of the donor mandates, limit the scope and duration of actions for the prevention and response to disasters, thus reducing the opportunities to both address the root causes of disasters in a comprehensive manner, and to consider these in a long-term context – that of 'development'. Therefore if one does not address the developmental context in which risks develop and disasters take place, then there is a strong probability that the victims will be equally or even more vulnerable next time round, because the root causes of their vulnerability have not been addressed.

We can in particular see a very clear link between poverty and vulnerability, and it is equally clear that we cannot address the latter without addressing the former. Second, we can see a clear issue of declining or inadequate knowledge, incentives and means to build safely in hazardous environments. We need to address these issues if the built environment is to become safer and if the economic losses that are sustained when buildings and their contents are damaged or destroyed are to be reduced.

Housing and the built environment are, in a poverty and hazard context, extremely telling examples of the way in which vulnerability has increased and why. They are also telling examples of the effect on vulnerability of poverty, inadequate investment in housing and lack of knowledge when using new materials, which all combine to increase vulnerability to economic and human loss.

Here, we will principally examine the experience of 20 years' work in Vietnam by Development Workshop France (DWF) and how changing socio-economic circumstances have increased vulnerability to economic loss that is the result of damage or destruction to the home caused by a disaster.

The negative impact of economic development – more to lose

Prior to 1985, central Vietnam was desperately poor and emerging from the ravages of war. Those very few who were sufficiently well off managed to build very solid timber-frame houses with fine detailing and long-understood techniques, producing buildings that were capable of withstanding most of the natural hazards that, then as now, affect the exposed Vietnam coast – floods and typhoons being regular annual events. For the rest of the population, infinitely poorer, their houses were built with materials gathered in the vicinity – rice thatch, bamboo frames, earth, lime and thatch for the walls. These houses could be quickly assembled with neighbour assistance using no-cost materials, and were, inversely, rapidly destroyed by even quite small storms. And whilst injury and mortality levels when storms came were often high, these were so for reasons largely unrelated to the quality of the home.[1] Thus, devastating though the loss of these frail shelters was at the moment of the event, several days later nearly all the damaged buildings in a village would have been rebuilt and at little material cost. Notwithstanding the short-term suffering and inconvenience, in terms of local capacity to recover, vulnerability was actually low because recovery to the previous status was cheap and relatively easy to achieve.

However, in 1986 economic policy changes, known as Đôi mói or renewal, and initiated by liberal socialist party leaders to help Vietnam out of its economic crisis, heralded a set of six major policy changes. For the purposes of this chapter, the most significant was that a peasant farmer could now choose what crops he was going to grow and that the more he grew the more profit he would keep for himself. Overall the farmer was now working for himself rather than for the state. And this translated gradually over the following years to the rural population having small sums of money that they could invest in improving the quality of their home. Over the next decade house construction moved from being 100 per cent based on very local gathered resources, through an initial phase of 'transition' construction – when it was common to see a bamboo frame thatched wall house covered with expensive ceramic roof tiles – to the ultimate replacement by early 2000 of nearly three-quarters of all rural housing by newly built houses. Thus fifteen years after Đôi mói, commune statistics[2] showed that some 70 per cent of provincial and rural housing stock had been replaced by new homes built with bricks, blocks, tiles, tin sheeting and cement, all of which are costly materials.

However, the same commune statistics – and visible evidence – show that many of these new houses have remained, as before, 'semi-solid', meaning that they are vulnerable to damage caused even by relatively small disasters (Table 4.1). Hence the paradox in the context of poverty and natural hazards: millions of families invest their time and hard-earned savings in a house that they believe is much more solid than the house of the past. In reality, their technical knowledge of the new materials, and lack of skills to build well, combined with the incremental nature of building under the constraint of scarce funds, results in houses that have cost a lot of money, but which are essentially utterly exposed to damage caused by floods and storms.

The typical rural family in Central Vietnam, with a monthly income of between US$ 20 and US$ 35, has a main house with a floor area of 35 m^2 (5m × 7m) divided into three bays, not including a covered terrace (2m × 7m) in front, and a kitchen/multipurpose enclosed area to the right of the house. Urban houses tend to be narrower and more elongated. The average house has a construction cost of between US$ 1,000 and US$ 1,500 for the main (three-bay) part of the house. The family may have made their cement tiles and blocks, but they will buy most other materials including cement, steel, ceramic roof tiles or roof sheeting, bricks, shutters, doors and windows. Often, they will borrow from relatives or even take out a loan in order to cover these costs. Since one of the objectives of building a new home is to have a stronger house, why does so much damage still occur? Essentially because neither the materials nor the structure of these new homes have been used in a manner that assures the integrity of the building. Many of the storm-resistant features of traditional housing have been neglected, whether for reasons of misguided faith in new materials, false economy, or simply because the home has remained 'unfinished'. House surveys show that costly roof tiles have not been tied down, shutters have not been installed, the walls are often weak, the roof frame is not held together, and that insufficient

Table 4.1 Proportion of weak and semi-solid houses, central Vietnam

Households	Solid house	Weak/semi-solid house	Wooden frame house	Simple house
12,520	3 per cent	70 per cent	10 per cent	17 per cent

Source: Communes in Thua Thien Hué Province, Central Vietnam.

attention has been paid to supposedly minor details. Worse, despite using more durable and 'modern' materials the quality of construction is often poor and houses consistently have defects that make them liable to damage even from small storms. A critical factor here is that although considerable investment and effort has been made, poverty compels people to skimp on materials and to fail to complete all the details of their home that would make it safe. A common example: when reinforced concrete columns are used in the corners of a house, they tend to only have two or maybe three thin steel bars; therefore in such circumstances both the steel bars and concrete are being wasted.

The overall result is that with the shift from traditional low- or no-cost techniques and materials to 'modern' materials, vulnerability has increased enormously, firstly because the quality of building work is poor, partly because of poverty and inadequate applied knowledge about safe building, and secondly because the damage caused to these new houses now has an economic value and so the recovery cost has huge negative consequences for the family, consequences that reach out to all aspects of their lives.

The issue of the knowledge gap and poor quality building alone is a major factor in increased vulnerability, and certainly not one that is exclusive to Vietnam. After the December 2004 tsunami hit Banda Aceh and many other regions in the Indian Ocean basin, surveys showed that poor quality building work, poor use of the building materials themselves, and an inadequate grasp of safe construction principles themselves all contributed directly to the failure of many buildings (Dixit and Shrestha 2005). There was widespread concern that the reconstruction programmes demonstrated the same level of poor quality construction detailing and application (Boen 2005). The same reports for Banda Aceh suggested that traditional buildings generally have a good record or performance in past earthquakes, and that the shift to often relatively unfamiliar 'modern' building materials carried with it a decline in quality along with an increase in both cost and the misguided conviction that the new buildings were safer (Boen 2005).

Modern non-engineered buildings invariably fail adequately to apply the principles of safe construction, whether for reasons of ignorance or economy, and in far too many cases engineered buildings do not have them applied adequately both because engineers and architects are not sufficiently aware of practicalities of disaster 'resistant' design and because contractors skimp on quality – poor quality control and ignorance again playing a part. Above all, faith in the capacity of modern materials to provide a secure home – or any other sort of building whether public or private – in hazard-prone environments is in effect dangerously deceptive. It is based on the *theoretical* potential quality of materials and techniques, not on the practical reality of how people actually use these.

So we have a compounded situation that generates vulnerability: poor quality construction and lack of application of safe construction techniques contributing directly to vulnerability; lack of collective and professional sensibility to the fact that the issue of safe construction matters and thus should be applied on the ground; and lack of capacity, and indeed motivation and priority, to make safe construction an issue that can save money as well as lives, largely driven by the constraints of poverty and the desire or need to save money.

What can be done?

In the central province of Thua Thien Hué, Vietnam, and building on previous DWF experience in Vietnam dating back to 1989, supported initially by Canadian International Development Aid (CIDA), and since 2003 by the European Commission Humanitarian Aid Office (ECHO), DWF has since 1999 promoted the preventive strengthening of existing houses and public buildings based on ten essentially generic key principles of typhoon-resistant construction. More recently, DWF has also encouraged the application of these same principles to the construction of new buildings by both government and the private sector and in particular as part of the Vietnam Government's 'Temporary House Replacement' programmes, which are designed to provide a more secure environment in which the extremely poor can improve their situation.

Each of the ten key points of typhoon-resistant construction[3] describes a principle that either reduces the risk of damage to the building or reduces the risk of loss of materials (Box 4.1).

For example, the veranda roof (a high-risk item) should be structurally separate from the main roof of the house. The connections between all individual parts of the structure, from the ground to the ridge, have to be strong. Doors and shutters should allow the building to be closed up. All parts of the roof and wall structure must be firmly connected. Roof sheets or tiles must be held or tied down. Trees should be planted to form wind breaks. The ten key points (Figure 4.1) can be applied to almost any type of building in the communes, regardless of the type of structure or the type of materials that have been used. All but the frailest of houses can be strengthened.

In Central Vietnam, the first task for DWF was to convince both the public and local decision-makers that preventive strengthening could even be considered a financially and technically viable option. Local scepticism was considerable. National and provincial decision-makers alike tended at first to view DWF's ideas as both worthless and unworkable. The proposition that families and communities can make a major contribution to reducing their own vulnerability to disasters; that their houses can be retrofitted to resist the effects of storms and floods; that not only can this preventive strengthening be done at a fraction of the cost of rebuilding, but also that families will willingly contribute to the costs; all these ideas met with official resistance when they were first suggested.[4] Interestingly, there was particular scepticism

---

Box 4.1 The ten key points of typhoon-resistant construction

1   Choose the location carefully to avoid the full force of the wind or flood (typhoon and flood).
2   Build a house with a simple shape to avoid negative pressure.
3   Build the roof at an angle of 30° to 45° to prevent it from lifting off.
4   Avoid wide roof overhangs; separate the veranda structure from the house.
5   Make sure the foundations, walls, roof structure and roof covering are all firmly fixed together (typhoon and flood).
6   Reinforce the triangular bracing in the structure, strengthen walls to increase stiffness (typhoon and flood).
7   Make sure the roof covering is attached to the roof structure to prevent it from lifting.
8   Match opposing openings (typhoon and flood).
9   Use doors and windows that can be closed.
10 Plant trees around the house as wind breaks and to reduce the flow of water (typhoon and flood).

Figure 4.1   Poster used in Central Vietnam to illustrate the ten key principles of storm-resistant construction (Source: DWF 2000)

amongst many building technicians – architects, engineers and disaster experts – who rejected the idea that non-engineered buildings built by families for themselves could be rendered safe by preventive strengthening, or indeed that such non-engineered structures merited such strengthening – better to tear them down and start again. Paradoxically these same professionals also considered that their designs provided for safe buildings and were scantly aware that there was a significant gap between the design in the office and the reality on the building site. This reality has been sadly demonstrated by some of the buildings constructed in Thua Thien Hué province by the Vietnam Government 'Temporary House Replacement' programme, which aims to replace the very fragile shelters of the extremely poor with solid new houses as a contribution to alleviating poverty. Alas, a considerable number of these newly built houses have now had their roofs blown off by small whirlwinds. As a result DWF was asked by the Government to assist in making sure that this did not happen again to any future houses that were to be built in the same programme.

The process of achieving public and official provincial acceptance has taken time: firstly and centrally, it has required working with hundreds of families, helping them to strengthen their existing homes, and working on small public buildings such as kindergartens and markets that demonstrate the same principles to a larger public. It is this above all that has shown that preventive strengthening is indeed both possible and viable, and moreover popular. No two houses have the same strengthening requirements, but for a houses with a *reconstruction* value in the order of US$ 1,000 to US$ 1,200, preventive strengthening costs less than 25 per cent of this sum.

In each partner commune, DWF also encouraged the creation of Commune Damage Prevention Committees set up within the Commune's People's Committee – the lowest administrative level in Vietnam – and these became DWF's practical partner in each commune, managing the variety of activities that were and are agreed on for each annual disaster prevention action plan with DWF staff in support.

In addition, DWF has provided rapid two-day training on the issues and practicalities of typhoon-preventive strengthening and construction to all commune builders and community leaders in each commune where DWF operates. This ensures that there is local access to genuinely widespread knowledge of what the issues of safe construction are and how these should be applied. Very few of these community builders have previously had any opportunity to discuss the issue of building to reduce the risk of damage by flooding and storms, and they have welcomed this chance, and applied the results in their work. Training community leaders has been equally important as these are also the people in the community well placed to disseminate practical information to the public in their locality (Figure 4.2).

A fourth major component of the programme has been to make the general public aware of the need for preventive strengthening. To this end

Figure 4.2 DWF staff in Central Vietnam using a model during training for local builders (Source: DWF 2001)

DWF works with the Prevention Committees to organise a very wide variety of publicity events using theatre, song, poetry, sport, traditional community communication tools and local television to get the prevention message across. The programme also works with kindergarten and primary schools, to make the next generation aware of the issue and to encourage children to raise this with their parents. The quality of tales, poetry, paintings and drawings that these children produce on the subject of disaster prevention is ample confirmation of the awareness that children develop around this issue (Figure 4.3).

The strategy of demonstrating preventive strengthening (Figure 4.4), of creating the prevention institutions within the communes, of providing training on strengthening to nearly all the people concerned in building, and of raising public awareness, has paid off. When Typhoon N° 6 Xangsane hit central Vietnam in October 2006, the levels of damage to property and housing was huge. But damage to houses that had been strengthened by applying DWF's ten key points was absolutely minimal, and people took note. The provincial authorities rapidly issued an official recommendation that henceforth local authorities and the population should apply the principles promoted by DWF for storm-resistant construction. Taking this further, and consolidating the example of the DWF programme, the Thua

Figure 4.3 Painting by a child in Thuy Xuan School, Thua Thien Hué province, done during a school competition about disaster prevention and safe houses (Source: DWF 2007)

Figure 4.4 Mrs Mai Thi Lai proudly stands in front of her house in Thuy Thanh Commune, Central Vietnam. This house is one of many strengthened with DWF help and the beneficiary's own contributions (Source: DWF 2005)

Thien Hué province has prepared a project for 'research, demonstration and regulation' to ensure the application of safe construction principles for housing and small public buildings and to integrate safe construction principles and practices into the formal construction sector. These are already important steps in bringing about change in attitude and behaviour at domestic and official levels. They represent real progress after many years – but this is still not enough. There remain several major barriers to getting families and the government to devote scarce resources to preventive action, and these cannot be addressed by animation and demonstration alone.

The most important of these issues in communities as poor as those of central Vietnam is to reach the point where families are ready and able to invest in preventive safety within the framework of existing or future mechanisms, mostly financial. For this, they need help.

Since 1999, DWF has always provided a financial subsidy to cover part of the cost of preventive strengthening. The rest of the cost has been covered directly by the beneficiary family. At the outset in 2000, the subsidy was nearly 60 per cent of the overall cost of strengthening, families providing the rest. Once the principle of preventive strengthening had been established, however, and people's faith in the results was high, the local family contribution rose steadily to reach an average in the order of 65 per cent of the total cost, whilst the 35 per cent contribution from DWF has remained an important incentive for families to commit to making their home safer.

However, early on it was observed that many families resorted to borrowing from private moneylenders at usurious rates to cover their contribution. This posed a real danger that house strengthening might cause additional financial suffering. In 2002, DWF therefore organised a credit system designed to assist families who wished to borrow specifically for the strengthening of their home. No such fund previously existed, and local banks have not thought to provide such form of loans. Small credit funds have normally been reserved for income-generation activities, and common wisdom suggested that high repayment rates depended on this link to income generation.

A key question was therefore whether families would make the same commitment to repay loans for house strengthening, even though this was not in itself an income-generating activity. The DWF hypothesis was that families genuinely consider a safer house to be a fundamental base in which all other activities can develop in a healthy manner. Although the DWF pilot credit programme has been small, with a total of US$ 40,000 loaned over three years (2002–5 figures), high repayment rates have clearly shown that families are just as reliable in making repayment of this type of credit as they have generally been for more classic income-generating activities, and confirms that families make the connection between a safe house and income-generation capacity.

The credit system has been organised at commune level by the Commune Damage Prevention Committee often in collaboration with the Women's union or the Farmers' union which have more general credit management experience. A typical loan is repayable over 18 months, at 0.3 per cent monthly interest (in line with Government-imposed rates) and with a families borrowing an average of about US$ 130 (see Table 4.2).

Surveys by DWF in several communes in Thua Thien Hué province have suggested that about 60 per cent of families are interested in loans for house strengthening and improvement. The challenge now is to get the local banks to take over the role of providing loans for house strengthening, and they already tacitly recognise that many families who borrow do so to make their home better and safer.

Conclusion

If, as has been demonstrated, houses in Central Vietnam can be made significantly more resistant to the effects of typhoons, floods and storms, at a cost affordable by individual families, then an additional incentive for families to take this step is, for these same families, to have access to insurance against future damage to their home, once the strengthening has been done, since these houses represent a greatly reduced risk to insurers. This too, as in the case of the house-strengthening credit system, has to be pitched at a price that people can afford. In a region where natural hazards are part of the almost routine risks that families face, putting in place practical steps to reduce the risk and vulnerability and the creation of incentives for people to take these steps must become a key part of national disaster risk reduction strategy. At present, in Vietnam, there is no private house insurance against disasters, nor the incentive of some form of post-disaster subsidy to cover a much lower risk of damage that could be offered to families who have taken preventive action to strengthen their home against disasters. Just as making credit for preventive strengthening more widely available, developing a system of incentives to families is another important challenge for the future.

So what about poverty reduction as a fundamental issue in reducing vulnerability, since the very poor have difficulty accessing any supporting financial mechanism?

Vietnam Government surveys, of which the key source is the Vietnam Living Standards Survey report in 2002,[5] show that for both the very poor

Table 4.2 Example of a loan

Cost of works	DWF subsidy	Family contribution	Credit	Duration (months)	Interest (by month)	Repayment (by month)
4,600,000 dongs	2,000,000 dongs	600,000 dongs	2,000,000 dongs	18	0.3%	114,000 dongs
300 US$	130 US$	40 US$	130 US$			7.5 US$

and those living on the precarious borderline of poverty, a critical issue in the struggle to emerge from poverty is the impact of household and community wide material and economic shocks on individual families (Bosher 2007). One of the principle economic shocks that a family in Central Vietnam can face on a repeated basis is, as well as losing means of production such as trees and fish ponds, that of having the family home damaged or destroyed. The destabilisation that results, including loss of productive capacity, loss of health amongst the family members, and the costs of recovery and rebuilding, all contribute to keeping families in poverty or to returning them to it.

It follows that a key contribution to reducing poverty and vulnerability is to ensure that the home is safe and secure. Achieving this is therefore a major contribution to reducing the vulnerability to economic shocks cased by loss or damage to the home. In this hazard-prone environment, 'vaccinating' the home against the risk of damage is as important as preventing pandemics.

Notes

1   Vietnam's capacity to reduce human losses and injury has in recent years been excellent, through short-term mobilisation and early warning capacity.
2   Reports provided to DWF by communes in Thua Thien Hué province, Central Vietnam in 2004.
3   These were originally developed and tested in the course of the DW/GRET implementation of UNDP/UNCHS programme VIE /85/019 'Demonstration of typhoon resistant building techniques' 1989–91.
4   The pattern of initial resistance followed by gradual acceptance is familiar: the principle of community-based disaster reduction received scant attention at the 1994 Yokohama World Conference on Disaster Reduction but was widely acknowledged at the Kobe conference in 2005.
5   In 2002 the Ministry of Labour, Invalids and Social Affairs (MOLISA), Vietnam, estimated the poverty line to be VND 100,000 per month in rural plain areas (VND 1.2 million per year), equivalent to $ 5/month.

References

Boen, T. (2005) 'Observed reconstruction of houses in Aceh seven months after the great Sumatra earthquake and Indian Ocean tsunami of December 2004', *Earthquake Spectra*, 22(S3): S803–18.

Bosher, L.S. (2007) *Social and Institutional Elements of Disaster Vulnerability: The Case of South India*, Bethesda, USA: Academica Press.

Development Workshop France (2005) 'Housing and microfinance in Communes – a response to natural disasters?' Thua Thien Hué: Development Workshop France.

Dixit, A.M. and Shrestha, S.N. (2005) *Seismic Safety of Buildings: Where Did We Go Wrong and how Can We Improve*, Kathmandu: National Society for Earthquake Technology (NSET).

Nugent, N. (1996) *Vietnam: The Second Revolution*, Brighton: Brighton In Print.

Wisner, B., Blaikie, P., Cannon, T. and Davis, I. (2004) *At Risk: Natural Hazards, People's Vulnerability, and Disasters*, second edition, London: Routledge.

Chapter 5

Structural adaptation in South Asia

Learning lessons from tradition

Rohit Jigyasu

Introduction

Much of the subcontinent in South Asia is highly vulnerable to earthquakes. In fact, over the last few decades, earthquakes have been one of the main reasons for the heavy loss of life and property in the region. Post-earthquake damage assessments have revealed that much of the damage and destruction was due to the poor quality of built fabric, which was structurally too weak to resist even the mild lateral forces associated with earthquakes.

The increasing vulnerability of structures can be attributed to various factors such as the poor quality of construction, lack of maintenance, and unsympathetic additions and alterations. However in most cases, the blame is squarely placed on 'non-engineered' vernacular structures using traditional construction materials and techniques, which are largely perceived as outdated and weak, thereby resulting in their outright demolition during post-earthquake rehabilitation, only to be subsequently replaced by 'modern' structures using contemporary materials and construction practices.

However, detailed investigations of earthquake-prone regions in the subcontinent has revealed that many so-called old and non-engineered vernacular structures have indeed performed remarkably well during earthquakes due to traditional construction systems that have indigenously developed over long periods of time. Nonetheless, the hard fact is that most of these structures are increasingly vulnerable due to several factors including the gradual loss of traditional knowledge. This chapter will look at the scope and extent of this knowledge and the reasons for its loss and degeneration in the context of earthquake-prone regions of Gujarat and Kashmir in India.

The crucial challenge indeed is how to make traditional knowledge relevant to the contemporary building context. This chapter will therefore address the critical thrust areas for the contemporary structural adaptation of traditional knowledge that is necessary for achieving a safe and sustainable future.

Increasing disaster vulnerability in South Asia

The frequency and intensity of disasters is increasing at a very rapid pace across the South Asian subcontinent. Take the case of two devastating earthquakes that hit the subcontinent during the last decade, namely the Gujarat earthquake in 2001 and more recently the Kashmir earthquake in 2005.[1]

In both cases, most structures whether 'modern' or 'vernacular', suffered heavy damage, causing enormous loss of life and property. Many Reinforced Cement Concrete (RCC) constructions did not follow even the basic rules of construction. For instance, rather than resting directly on the beams, the roof slabs of many structures were cast on two or three brick courses placed on the beams, which in some cases were not even at the same level (Figure 5.1). In other cases, the roof slabs had virtually no reinforcement bars and the successive layers of mud added over time for terracing significantly increased the vertical load. As a result, they simply cracked and collapsed like a 'pack of cards' due to the force of the earthquake (Figure 5.2). In some structures, paper-thin walls laid in poor masonry were unable to withstand even a low-intensity earthquake. There were other instances where RCC beams resting

Figure 5.1 Poor contemporary construction in Indian-administered Kashmir with a RCC slab cast on two or three courses of brick placed over the beams, which are not even at the same level

Figure 5.2 This roof slab had virtually no reinforcement bars and the layers of mud that had accumulated on top increased the vertical load. As a result, the structure simply cracked and collapsed like a pack of cards due to the impact of the earthquake

on the columns made of slender brick piers or poorly-reinforced concrete columns simply gave way due to the lateral impact of earthquake.

Most of the vernacular structures also did not perform well due to the poor quality of stone masonry. Although many stone walls were clad with well laid out courses, their inner core was built of random rubble masonry in low quality mud mortar. Due to improper bonding and absence of through stones, these walls collapsed due to the earthquake (Figure 5.3). Inadequate corner joints between the perpendicular walls were also one of the reasons for the poor behaviour of these buildings. In structures with sloping roofs, free-standing gable walls could not withstand the lateral forces of earthquake and collapsed. One of the major reasons for the extensive damage sustained by most structures was incompatible additions, resulting in the loss of their structural integrity. For example, in the structures built of load-bearing stone walls, the upper floors were added using RCC, which simply collapsed.

Earthquakes have always occurred, however the increasing scale of devastation caused by them in spite of the overall economic growth seen in

Figure 5.3 These random rubble masonry walls collapsed during the earthquake due to improper bonding and the absence of through stones

the region, prompts us to think about the core reasons for the increasingly poor built fabric.

The earthquake survivors – repository of traditional knowledge systems

On close inspection of earthquake-prone regions such as Gujarat and Kashmir, we discover several examples of 'good quality' vernacular constructions that did survive these devastating earthquakes, owing to their earthquake-safe construction systems/features.

During the Kashmir earthquake, the vernacular structures built using local Kashmiri building techniques of *Taq* (timber laced masonry bearing wall) and *Dhajji Dewari* (timber frame with masonry infill) in part or in whole, performed much better than many poorly built 'modern' structures. Good examples of *Taq* and *Dhajji Dewari* constructions are mainly found in the urban areas of Kashmir valley, notably Srinagar. Although there were cracks in the masonry infill, most of these structures did not collapse, thereby preventing the loss of life (Figure 5.4).

Figure 5.4 Vernacular house in Poonch town using local Kashmiri building techniques of Dhajji Dewari (timber frame with masonry infill). This type of house performed much better during the earthquake than many of the poorly built 'modern' structures

Also several vernacular constructions such as wooden log houses in which the logs are laid alternately, with the end joints dovetailed to one another[2] and those employing the use of well-laid masonry with through stones, well-designed arches, retaining walls or bastions around the corners, performed well against the earthquake. Other earthquake-safe features found in several vernacular constructions in the earthquake-affected region in Kashmir include ceilings with joists resting on wooden bands running all along the walls, well-designed trusses, 'tongue and groove' joinery and balconies resting on projecting wooden joists. In some constructions, extensive use of wood on the upper floor in the form of wall panelling, balconies, staircases etc. significantly reduced the weight, thereby enhancing the earthquake performance of the structures.

Earthquake-resistant construction systems are also found in Gujarat. The typical traditional dwellings of the Kutch region, the *bhungas*, have withstood the test of time for centuries and have also withstood the earthquakes, thanks to their circular form, which is very good in resisting earthquakes. Moreover, wattle and daub constructions, especially where wood is used as reinforcement for the wall, have proved to be very effective. It is worth mentioning that *bhungas* are not only earthquake safe, they also demonstrate sensitive understanding of the locally available resources, climatic conditions and the spatial requirements of people (Figure 5.5). In fact, all these factors play a significant role in the evolution of vernacular architecture at any given place.

In Gujarat, many structures built prior to the 1950s had floor joists extending through the rubble stonewalls to support the balconies. These type of structures were more successful in stabilising the walls than those where joists terminated in pockets and therefore performed much better during the 2001 earthquake (Langenbach 2001). In fact, in the severely affected town of Anjar, this kind of structure was one of the rare ones found standing amidst the debris of collapsed houses (Jigyasu 2002) (Figure 5.6).

Some vernacular constructions employing wooden frames with masonry infill also performed well against the earthquake due to their capacity to dissipate the energy. In other constructions, earthquake-safe features like tie beams, knee bracing, and tongue and groove joinery are also found (Figure 5.7). In fact, due to these features, the structures located in the old part of Ahmedabad city performed remarkably well compared to the 'modern' multistoried constructions, which suffered heavy earthquake damage.

Figure 5.5 Bunga in a village near Bhuj in Gujarat, India (October 2001). The walls are made of mud applied to an inner wooden trellis, a comparatively ductile and light system. The mud walls and the thatched roof provide reasonable cooling and respond to the climate conditions in the arid Rann of Kutch (image courtesy of Matthias Beckh)

Figure 5.6 One of the very few surviving houses in Anjar was built before the 1950s

Figure 5.7 Several earthquake-safe features can also be found in many traditional constructions, such as knee bracing in this case

Based on the above findings, we can conclude that traditional knowledge systems for earthquake mitigation do exist in some structures located in the earthquake-prone regions of Kashmir and Gujarat, although their percentage is very small.

Nature and scope of traditional knowledge for disaster mitigation

In the light of the above findings, it is important to discuss the nature and scope of traditional knowledge. Such knowledge develops locally and demonstrates optimum levels of technology evolved by skillfully balancing other factors. Moreover it is experience laden, practice oriented and culturally embedded, thus more holistically oriented. Paul Sillitoe (1998: 223) describes this knowledge as 'by definition interdisciplinary; local people think of and manage their general environment as a whole system'. The tangible manifestation of this knowledge is seen in the built fabric, while skills and cultural practices form the intangible dimensions. Moreover, there is an inseparable link with the local social and cultural practices, which act as subconscious carriers of this knowledge.[3]

This knowledge is continuously evolving, assimilating and adapting to changing needs, constraints and aspirations.[4] Therefore it often gets tested over a long period of time and integrated into a larger cultural context (Peter Schroder 1995, translated from German by Schmuck-Widmann 2001). In this process, new materials, techniques and architectural vocabulary are assimilated in the ever-evolving vernacular form.[5] Since this knowledge is in continuous evolution, it is not possible to draw rigid boundaries around traditional knowledge by freezing it in time.

The physical manifestation of traditional knowledge is mostly seen in vernacular structures.[6] It is important to recognise subtle variations/adjustments in the traditional knowledge systems according to various typologies of the vernacular fabric in which they find application. These typologies evolve on the basis of localised factors such as the geographical location, social and economic context, which result in very different sets of variables and corresponding structural and architectural features.[7] Therefore it would not be appropriate to assume that hazard mitigation is the only criteria for the evolution of a specific traditional knowledge, which may be the conscious or subconscious result of multiple factors.

Last but not least, it is important to make a clear distinction between the 'pure traditional construction systems' and the 'hybrid vernacular form'. The former might be part of the latter but not vice versa, since the latter is the result of complex local variables as stated before. Again, let us take the case of Kashmir, where the majority of the vernacular fabric is made of stone and load-bearing brick walls. Traditional construction systems like *Dhajji Dewari* are just one of the many systems, used only in part as a partition wall/side wall in a construction that is largely made of random rubble masonry.

The increasing vulnerability of the 'vernacular': critical issues

Loss or degeneration of traditional knowledge

The key issue is the loss or degeneration of traditional building systems over the last few decades, which has made the buildings vulnerable to disasters in the first place and in some cases has even increased existing vulnerability through inappropriate post-disaster reconstruction. The underlying reasons for this loss or degeneration therefore need to be explored.

First, economics tend to influence the owner's choice of materials, potentially lowering the specifications below the desired optimum levels before and after the disaster. For example, wood was one of the primary building materials for housing in several earthquake-prone regions and its skilful combination with stone masonry helped in better seismic performance. However, wood slowly became unaffordable and people started making alterations to their structures, which in many cases made them more vulnerable to earthquakes. For example, in the Kutch region of Gujarat as well as in Kashmir, the walls were extended up to over 15 feet in unbraced height. This was simply undertaken as a measure to support the ridge of the roof and thereby avoid the use of the large amounts of wood that would be required to build a roof truss (Langenbach 2001). Also in many instances, sophisticated tongue and groove joints were replaced with simple nailing of wooden members, which could easily give way in the event of an earthquake (Jigyasu 2002).

The degeneration of traditional knowledge is also the result of several factors that are linked to underdevelopment, such as poverty resulting in the lack of maintenance, loss of traditional livelihoods and general misperceptions. Some of these are discussed in detail in the subsequent subsections.

Misperceptions against the use of traditional materials

Overriding perceptions today favour the use of new materials like cement in place of traditional materials such as adobe and stone, which are largely perceived as 'weak', 'unsafe' and 'outdated'. Needless to say, the specifications needed with the introduction of new materials and technology are not feasible in many earthquake-prone regions owing to the local unavailability of resources, for example appropriate curing of concrete is virtually impossible in the drought-prone regions of Kutch in Gujarat. Moreover poor economy also forces people to make compromises in their constructions.

After the Gujarat earthquake, wrong perceptions were also evident in the way traditional structures were pulled down to make way for 'modern' structures, even where they were still standing, as seen in the affected towns of Anjar, Bhuj and Morbi. Ironically, in most cases, the new structures were no better, thanks to the poor workmanship (Jigyasu 2002).

Similar trends were seen after the Kashmir earthquake, where many traditional constructions that had in fact performed better against the earthquakes were abandoned by their owners due to the widely prevalent perception that traditional buildings were 'old' and 'outdated' and therefore 'unsafe' and 'unliveable'. Many of these structures were also on the verge of demolition and proposed to be replaced by 'modern' reconstructions. In the absence of any proper technical assistance, people started rebuilding on their own using the limited resources available at their disposal, including the compensation money being provided by the Government. Not many people realised that the main problem did not lie with the use of stone as a building material but the way it was being used.

Difficulties are often experienced in implementing repair and retrofitting programmes during post-earthquake rehabilitation because of the prevailing misperceptions against vernacular structures, which discourage people from undertaking these measures. Ironically, in most cases, the new constructions after the earthquakes are found to be even poorer than before because, with no technical assistance forthcoming, people are typically left with no option but to provide a roof above their heads in the quickest time possible. Moreover, these constructions are observed to be unsustainable in terms of the available skills and resources. Stone is the locally available material and is part of the local building culture. Replacing it with concrete would prove to be economically unaffordable for the far-flung villages located in a difficult terrain. As a result, over time people start reconstructing by themselves in stone without employing earthquake-safe practices[8] (Jigyasu 2002) (Figures 5.8 and 5.9). This also brings out the importance of long-term considerations for sustainability during post-disaster rehabilitation because short-term fears slowly give way to long-term issues of livelihoods, social networks, and cultural and religious associations.

The challenge of integrating traditional and modern materials and technology

Due to the introduction of new materials such as concrete, the original strength of the traditional materials tends not to be utilised effectively, thereby affecting the capacity of the vernacular structures to withstand earthquakes. Materials such as brick and concrete, which were introduced later in some earthquake-prone regions, are combined randomly with traditional materials of stone and wood even in post-earthquake reconstructions, thereby adversely affecting their structural integrity and seismic performance.[9]

In other instances, partial replacement of traditional materials with modern ones, notably concrete, has not only reduced the inherent capacity

Figures 5.8 and 5.9 Reconstruction of rural houses undertaken by the owners following the 2005 earthquake in Kashmir. The use of composite materials and discontinuous concrete bands show total disregard for basic techniques for earthquake resistance

of these structures but also increased their earthquake vulnerability to a great extent. Take the case of Majuli in Assam, a large river island with unique local ecology; the vernacular housing in the area using bamboo constructed on stilts has evolved as a sensitive response to local factors, notably floods that inundate the island on a regular basis. The light bamboo structure enables easy dismantling and relocation, in the event that the area is affected by floods. Moreover, the nature of the material and joinery allows for the flexibility of structures, especially useful in the event of an earthquake to which this region is highly prone (Figure 5.10). This case demonstrates the remarkable capacity of traditional technology to adapt to the nature of hazards and develop a harmonising living relationship with them, rather than merely resisting them.

Gradually however, the structural framework in bamboo, including the stilts, is being partially replaced with very poorly constructed concrete piers with little or no reinforcement resting directly on the soft soil, thereby creating extremely vulnerable soft-storey structures (Figure 5.11).

Misperceptions and lack of proper knowledge have also resulted in inappropriate repairs to the damaged 'modern' as well as 'vernacular' houses, especially following major disasters in the region. For example after the Gujarat and Kashmir earthquakes, people returned to their damaged

Figure 5.10 Vernacular bamboo houses constructed on stilts in Majuli Island, Assam, India demonstrate good understanding of local hazards: flood and earthquake

Figure 5.11 Replacement of bamboo posts with poorly constructed concrete piers resting directly on the soft soil is increasing the vulnerability of vernacular houses in Majuli

houses after undertaking repairs by merely filling up 'through cracks' with cement grout without achieving structural soundness and the compatibility of materials (Jigyasu 2001b).

Considering the above trends, it won't be wrong to opine that the contemporary building character in the earthquake-prone regions such as Gujarat and Kashmir is pretty confusing since neither is it able to draw upon the traditional knowledge and capacity, nor is it able to fully adapt modern materials and construction systems, thanks to many factors that have been outlined in the previous subsections.

Let us now consider the appreciable efforts on behalf of some organisations to develop earthquake-resistant technology during the post-earthquake rehabilitation process. After the Gujarat earthquake, a consortium of NGOs promoted alternative technology using pre-cast 'compressed soil blocks' with or without an interlocking dry stacked masonry system, ring reinforcement and wooden rafters. It also set up a laboratory to experiment and test 'new' technologies.

However, such alternative methods also required strict quality control and proper curing. During the construction phase, the concerned NGOs took care of this but since these technologies were not based on traditional knowledge and required proper curing (a difficult proposition in a drought-prone area), there were challenges regarding 'internalising' these among the local community, once these organisations withdrew from the scene. Considering this trend, it is highly doubtful whether such technologies can take root within the building culture of the area (Jigyasu 2002).

Let us consider another example to substantiate the issue of appropriate technology for post-earthquake rehabilitation. After the 1993 Marathwada earthquake in India, as many as ten 'building centres' were established in the region with the support of HUDCO (Housing and Urban Development Corporation) and the Maharashtra State Government. These centres were supposed to promote construction activity and generate employment through training programmes for construction artisans, unskilled labour and unemployed youth. The centres supplied building materials to construction sites and educated people with respect to earthquake-resistant technology. This was a very good idea and would have ensured sustainability. Unfortunately, only five years after they were conceived, all these centres had shut down. Today, they appear like ruins, with unfinished concrete blocks, dry tanks and rusted machines. Why did this happen? There are several reasons. Firstly, the technology which was supposed to be inculcated was alien and unsustainable. Secondly, these centres were established through outside financial resources without a proper plan for internalising the whole process with the local community. Thirdly, there was considerably less involvement of traditional artisans, who were made to neglect their existing skills and learn something totally outside their previous experience. 'Earthquake-resistant technology' was taught as rigid design packages, without any scope for experimentation (Jigyasu 2001a).

In the context of the above discussion, the importance of 'appropriate building technology' has been addressed over the recent decades, thanks to the pioneering efforts of several professionals and NGOs. However, it is important to recognise that 'low cost' is not the only criteria determining the appropriateness of technology. Factors such as environmental impact, climatic suitability, social adaptability and cultural compatibility, all of which contribute towards the sustainability of technology, also need to be considered. Moreover, it is important to consider both 'technology' and 'design' aspects while seeking to improve the resilience of the vernacular fabric. Traditional knowledge systems, if conceived as 'packaged products' for fast duplication and transfer, are also in danger of falling into the same trap if perceived in a static manner.

The 'traditional' as well as the 'modern' technology on disaster mitigation both have their own merits, when considered with respect to the specific objectives and different contexts within which both kinds of technology are produced. Although with the fast pace of globalisation, one cannot and should not overlook the impact of 'modern' on 'traditional', the real challenge lies in marrying the two by building on the vast local experience on the one hand and rapid strides in technological development on the other. This indeed is a challenge that needs to be addressed through innovative participatory approaches.

Lack of utilisation of the local capacity

With changes in materials and technology, traditional craftsmen find themselves increasingly incapable of using their skills, for example, local masons who are skilled in shaping and laying stones are not trained to handle brick and concrete constructions. While they find themselves incapable of using new materials, their traditional knowledge of stone masonry has degenerated to a considerable extent, primarily because of the lack of demand (again linked to the general misperceptions about traditional constructions) over the last few decades, forcing them to move to other occupations. As a result, the successive generations are unable to learn the skills from their masters. Even those who can afford modern RCC constructions, are unable to afford the level of workmanship required for these types of constructions due to the unavailability of a skilled workforce. The increased demand, before and after an earthquake, for outside craftsmen who tend to be unfamiliar with local traditional construction practices, has made matters more complicated. For example, it has been found in Indian administered Kashmir that one of the reasons for poor construction practices after the earthquake was the lack of involvement of local masons. In fact, most of the reconstruction is being undertaken by masons from Bihar, who are not well conversant with the use of local materials and building techniques.

Here it would be worthwhile to mention the sad plight of building craftsmen in the region. In the absence of any official recognition, favourable policies and general apathy, most of them are unable to reap the benefits of general economic growth and development process and have become entrapped in an endless cycle of poverty, forcing them to leave their traditional sources of livelihood.[10]

Mainstreaming traditional knowledge into contemporary building culture

Recovering 'scientific aspects' of traditional knowledge and vice versa

A large part of the writing on local knowledge attempts to 'package' and 'market' traditional knowledge as something complete in itself by marking an artificial boundary between it and formalised, scientific knowledge (Schmuck-Widmann 2001). However, Richards (1985) rightly emphasises experimentation as an important aspect of traditional knowledge, and thus makes a claim that traditional knowledge is scientific. According to him, traditional knowledge is knowledge that is in conformity with general scientific principles, but which, because it embodies place-specific experience, allows better assessment of risk factors in production decision. This kind of knowledge arises where local people undertake their own experimentation, or where they are able to draw inferences from experience and natural experiments (Richards 1985).

Similar emphasis is given by Flavier *et al*. (1995), who state that traditional information systems are dynamic, and continually influenced by internal creativity and experimentation as well as contact with external systems. This continuous process of experimentation, innovation and adaptation enables traditional knowledge to blend with science and technology as well.

Therefore, rather than categorising traditional and scientific knowledge into mutually exclusive domains, attempts should be made to recover the 'scientific' aspects of traditional knowledge and the 'traditional' aspects of scientific knowledge. While the former will enable traditional knowledge systems to be easily understood by professionals, the latter would demand that larger scientific concepts get translated into modes of communication that are locally understood. This process of rediscovering, recovering, encoding and decoding is an organised scientific activity in itself.

Before we can make any generalised statement/theory on the earthquake performance of traditional construction techniques, we need to conduct 'scientific experiments' to evaluate their specific performance with respect to the vernacular structures in which they are incorporated. This is crucial because even within a single technique, say *Dhajji Dewari*, there might be different performance, depending on the spacing of panels, type of wood, infill, bonding, mortar mix etc. as well as the specific location of the technique within the structure. Moreover, since earthquake risk is a complex result of many factors that dynamically influence each other, the structure needs to be assessed in its entirety and not in a piecemeal manner.[11]

Without this rigorous process, merely believing that the traditional techniques are effective against earthquake can be very dangerous as people start using these techniques without really understanding the basic principles behind their good performance. So ironically, we may end up with the same problem as faced by 'modern structures' that are perceived to be strong but in reality are not, due to low understanding of the basic construction principles and necessary specifications.

This scientific approach for traditional knowledge ought to have profound implications on engineering education in the region, which is heavily focused on modern materials and construction techniques. In fact, the civil engineering curriculum of the premier institutes of engineering and technology in the region is almost entirely devoted to brick, concrete and steel constructions. The research agenda is also predominantly focused on these materials. In fact very little research is seen on the behavioural and design aspects of stone and wood constructions, which ironically form the bulk of the rural built fabric today.

There is an urgent need to expand the scope of research and education to include traditional materials and techniques, especially stone and wood, to cater to the needs of majority of the built fabric. This will not only help in developing innovative solutions to reduce disaster vulnerability but also remove the misperceptions of engineers against the effectiveness of traditional materials.

Replacing, restoring or evolving?

Critical choices need to be made regarding the basic philosophy governing interventions for improving the resilience of vernacular structures and the role of traditional knowledge in formulating these. Should we restore traditional knowledge systems by recovering and reusing them as when they existed in their pristine glory? Or should we attempt to restore their essence by bringing back the creative process of evolution by responding to changing needs, constraints as well as aspirations but at the same time building on the accumulated experience of the past and maintaining the local sense of identity? The latter seems to be an obvious choice if we wish traditional knowledge to play a proactive role in disaster mitigation and recovery.[12]

No matter how strongly we feel about these traditional constructions, the hard reality is that no one will buy our argument until or unless we are able to rationally debate the logic of our advocacy for the role of traditional construction techniques in earthquake mitigation and their potential in the light of changing needs, aspirations and affordability.

This requires us to address the improvement in construction practices during disaster preparedness as well as in the recovery phase in two ways. Firstly, by developing innovative and sustainable technological ideas for new constructions by building on traditional knowledge, and secondly by developing workable alternatives for repair and retrofitting of traditional and historic structures, instead of focusing on standard engineering recipes and design packages for the new constructions. The latter will require us to consider the following questions:

- How best can we repair and retrofit poor quality/damaged traditional construction systems? Here we need to discuss/debate appropriate technology/interventions that can improve the resilience and also maintain the heritage values to the best possible extent.
- How can we repair and retrofit 'hybrid vernacular structures' by using traditional construction techniques?[13]
- To what extent (if at all) and how best can we use modern materials/techniques for repairing and retrofitting hybrid constructions without compromising on their safety as well as heritage values?

Based on the analysis of empirical data from decades of research on the issue, there are a few general patterns in the way buildings behave during earthquakes that can guide repair and retrofitting interventions to address the above-mentioned questions.

However these 'ideal' principles are bound to be tempered by context-specific issues such as poor economy, knowledge limitation and lack of material availability that are far more controlling than any abstract notion of engineering perfection. Therefore, practical testing and application holds much more relevance here than purely academic views. This may also require lowering the thresholds for earthquake safety by establishing optimum acceptable standards for managing risks in response to local constraints and opportunities.[14]

Last but not least, this would require engagement with the community for mutual sharing and development of knowledge through dialogue between the local craftsmen and trained professionals and not merely through the transfer of technology.[15]

Balancing safety considerations with heritage values

While proposing interventions in the historic and/or vernacular fabric (especially in the former case), we encounter the challenge of balancing safety considerations with the heritage values embedded in the structures. While it is critically important to take all possible measures to improve the performance of structures against natural hazards, such as earthquakes, to protect lives and property, it is also important that the material and structural authenticity and integrity as well as architectural, associational and other values embedded in the fabric are retained to the best possible extent. Retaining these values is vital, not least for the sake of protecting these cultural resources that are the source of identity and psychological recovery after disasters.

However the crucial challenge is about overcoming highly defensive positions taken by heritage professionals on the one hand and engineers on the other, when neither of them are willing to make any compromise on their rigid professional standpoints, resulting more often in a deadlock. Instead, both sides need to strive for optimum risk-reduction measures by balancing the safety aspects with the protection of significant values in order to achieve safe and sustainable built fabric that assimilates and builds upon the rich knowledge from past experience but at the same time engages in a creative process of adaptation to current needs and realities.

After all, we do not wish that traditional knowledge merely decorates books, exclusively meant for the elites and intellectuals. Rather, we want this knowledge to evolve and benefit future generations and at the same time secure precious lives and properties from disasters.

Notes

1   According to the official figures, the Northern Kashmir Earthquake on 8 October 2005 killed around 73,000 and injured more than 70,000 people in Pakistan (ERRA 2006) and killed around 1,300 people and injured 7,510 in India (Arya and Agarwal 2005). The earthquake that struck the Kutch region of Gujarat in India on 26 January 2001 killed 18,253 people and injured 166,836 (Mistry *et al.* 2001).

2   The wooden log houses of Kashmir are mostly located in the northern part of the valley, especially in the picturesque Lolab valley. These generally follow a simple plan with a small square room on the ground floor and a similar room on the first floor. Access to the upper floor is from a wooden staircase aligned along the walls. The houses are provided with a steeply pitched sloping roof consisting of layers of hay laid over the roof purlins directly, or in certain cases the hay is placed on top of wooden beams in turn supported on a wooden truss. In most of the older wooden houses, the joinery detailing is very rudimentary, simply comprising notches, for holding the logs together (INTACH J&K 2006).

3   In the Marathwada region in the western Indian state of Maharashtra, a special technique of adobe brick preparation seen in Khel village is linked with the local cultural practice of ramming the earth by feet while singing devotion to the local saint. Such brick is found to be so light that it floats on water. Unfortunately, the original technique of making this brick is now lost.

4   Globalisation is not a new phenomenon although its pace has now greatly increased, thereby not allowing gradual assimilation. For example, *Dhajji Dewari* is not a purely indigenous construction technique of Kashmir as is wrongly perceived by many. It has developed over the last two centuries evidently due to external contacts/influences. The *Dhajji* technique has been used in England as well as in other European countries since the eighteenth century. Therefore it would be inappropriate to brand traditional knowledge as purely indigenous or localised.

5 In Kashmir, the traditional sloping roof using shingles is now almost entirely replaced by corrugated iron sheets because of various factors such as the availability of material, cost effectiveness and easy application. As such, corrugated iron roofing can now be considered very much a part of the Kashmiri vernacular.

6 It is important to mention here that the vernacular fabric is not necessarily traditional although it might well be the case in a few instances. The former evolves with time and therefore is dynamic while the latter focuses on the continuity aspect. A problem exists in some places because 'vernacular' and its translations into other languages, is understood more pejoratively than is true in Britain and the USA.

7 The vernacular architecture of Kashmir has distinct typologies depending on the geographical and geological regions (valley, low or high mountain ranges). Moreover the vernacular architecture of urban areas especially in and around Srinagar is very distinct from the non-engineered rural constructions. All these subtle variations much be explicitly considered without mixing them all under the banner of 'Kashmiri vernacular' (Desai amd Desai 2006).

8 This trend has been observed by the author during the course of his research on the long-term impacts (after seven years) of contractor-driven reconstruction in Marathwada, India following the 1993 earthquake. In the immediate aftermath of the disaster, local people preferred to move to the modern 'concrete block' constructions provided by the Government. However over time, they started making additions and alterations to these houses using stones, since it was the most easily available and affordable material in the region.

9 We must draw clear boundaries between the 'vernacular' and the 'modern'. Modern materials and construction techniques are being increasingly used these days to varying degrees. However, these cannot be qualified as 'vernacular' because the latter essentially undergoes a gradual process of assimilation. In many earthquakes, those buildings that mixed old and new construction methods and materials (steel or concrete with adobe or stone masonry, etc.) have proved to be far more lethal than the truly traditional ones, or the genuinely modern ones (that is, the modern ones constructed to an acceptable standard). In Bam, Iran, this was profoundly the case. It also applies to the structures in Kashmir with rubble stone walls and reinforced concrete floors and roofs.

10 The plight of the building craftsmen in the region is demonstrated by the fact that they are treated on par with a labourer and paid many times less than a junior architect or engineer. Compare this with Japan, where a young craftsman gets the same (if not higher) wages than a senior architect. However, their situation is dramatically different from that of other craftsmen, such as those involved in handicrafts. The latter have found favourable policies and opportunities to improve their economic condition, mainly through tourism.

11 In many cases, it is observed that the recommendations by 'experts' are primarily based on random observations and studies, pieced together to address specific objectives, which the expert identifies for 'proving a point' rather than undertaking vigorous 'scientific testing and verification'. Instead, a comprehensive approach for understanding the structure and its systems including the traditional ones is necessary to arrive at good design interventions that can be illustrated by showing the nature of the damages to different structures, and describing how one can respond to these with a coordinated set of improvements and repairs.

12 In fact, one of the main reasons for the loss or degeneration of traditional knowledge is the disruption of its process of evolution, thereby putting a stop

to the 'creative' search for solutions through continuous trial and error. This evolutionary process is what defines the true essence of traditional knowledge.

13 The proposal for armature crosswalls is an interesting example to illustrate this. Armature crosswalls, proposed to be used as earthquake hazard mitigation for reinforced concrete and masonry infill-wall buildings vulnerable to collapse, are based on the flexibility and energy dissipation properties found to have existed in traditional constructions with non-bearing infill walls located in earthquake-prone regions around the world, such as Portugal, Turkey and Kashmir (Langenbach *et al*. 2006).

14 By lowering the thresholds, we may achieve optimum levels of safety for far greater number of structures compared with the case where these standards are maximised. However these thresholds must ensure that there is no danger to lives, even if there is slightly more damage to the structures.

15 Knowledge sharing between the traditional craftsmen and the professional architects and engineers can only be realised if the former undergo formal education and the latter are formally trained in traditional materials and construction systems besides modern concrete and steel constructions. The formal education of craftsmen will surely enhance their capability to comprehend the 'professional language' and communicate with the professionals, thereby improving their ability to understand and adapt to changing needs and directly deal with the market at large. Of course this also needs to be supported by adequate public policies.

References

Arya, A. and Agarwal, A. (2005) *Guidelines for Earthquake Resistant Reconstruction and New Construction of Masonry Buildings in Jammu & Kashmir State*, National Disaster Management Division, Ministry of Home Affairs, Delhi: Government of India.

Desai, R. and Desai, R. (2006) 'Draft of the manual on repair and retrofitting of rural structures in rural Kashmir', Unpublished report.

ERRA (Earthquake Rehabilitation and Reconstruction Authority) (2006), *Earthquake Emergency Recovery Project, Livelihood Support Cash Grant; Umbrella Project Document*, Government of Pakistan, Online. Available HTTP: http://www.erra. gov.pk/Reports (accessed 24 June 2007).

Flavier, J.M, de Jesus, A. and Navarro, C. (1995) 'The regional program for the promotion of indigenous knowledge in Asia', in Warren, D.M., Slikkerveer, L.J. and Brokensha, D. (eds) *The Cultural Dimension of Development: Indigenous Knowledge Systems*, London: Intermediate Technology Publications, pp. 479–87.

INTACH J&K (Indian National Trust for Art and Cultural Heritage – Jammu and Kashmir Chapter) (2006) *Research Report on Disaster Risk Management of Vernacular Heritage in Kashmir*, Kyoto: Research Center for Disaster Mitigation of Urban Cultural Heritage, Ritsumeikan University.

Jigyasu, R. (2001a) 'From "natural" to "cultural" disaster: consequences of post-earthquake rehabilitation process on cultural heritage in Marathwada region, India', *Bulletin of the New Zealand Society for Earthquake Engineering*, 33(3): 237–42.

Jigyasu, R. (2001b) 'Post-earthquake rehabilitation in Gujarat: nine months after', Online. Available HTTP: <http://www.radixonline.org/gujarat4.htm> (accessed 28 April 2007).

Jigyasu, R. (2002) 'Reducing disaster vulnerability through local knowledge and capacity', Dr Eng thesis, Trondheim: Norwegian University of Science and Technology.

Langenbach, R. (2001) 'A rich heritage lost, the Bhuj, India, earthquake', *Cultural Resource Management Magazine*, 24(8), US National Park Service.

Langenbach, R., Mosalam, K., Akarsu, S. and Dusi, A. (2006), 'Armature crosswalls: a proposed methodology to improve the seismic performance of non-ductile reinforced concrete infill frame structures', *8th US National Conference on Earthquake Engineering (8NCEE)*, San Francisco 1906 anniversary, Online. Available HTTP: <http://www.conservationtech.com> (accessed 2 May 2007).

Mistry, R., Dong, W. and Shah, H. (eds) (2001) *Interdisciplinary Observations on the January 2001 Bhuj, Gujarat Earthquake*, World Seismic Safety Initiative & Earthquakes and Megacities Initiative.

Richards, P. (1985), *Indigenous Agricultural Revolution; Ecology and Food Production in West Africa*, London: Hutchinson.

Schmuck-Widmann, H. (2001) *Facing the Jamuna River – Indigenous and Engineering Knowledge in Bangladesh*, Dhaka: Bangladesh Resource Centre for Indigenous Knowledge.

Sillitoe, P. (1998), 'The development of indigenous knowledge', *Current Anthropology*, 39(2): 223–53.

Chapter 6

Developments in seismic design and retrofit of structures

Modern technology built on 'ancient wisdom'

Stefano Pampanin

Introduction

Significant advances have been accomplished in the last decades in the seismic protection of structures thanks to the development of new technologies and advanced materials as well as to the introduction and refinement of innovative design approaches for earthquake resistance.

Following the worldwide recognised expectation and ideal aim to provide a modern society with structures able to sustain a design-level earthquake with limited or negligible damage, emerging solutions have been developed for high-performance, still cost-effective, seismic resisting systems, based on the combination of traditional materials and more recently available technology.

In parallel, remarkable accomplishments have been achieved in terms of a better understanding of the behaviour of the existing structures under earthquake ground motions, leading to the development of alternative and efficient strengthening/retrofit solutions.

The crucial need for a prompt and wide implementation of such techniques has been further emphasised by the catastrophic effects of recent earthquake events (Turkey, Greece, Colombia and Taiwan, 1999; India, 2001; Pakistan/ India, 2005). A broad consensus between public, politicians and engineers or scientists communities seems to be achieved when stating that excessively severe socio-economical losses due to earthquake events, as still observed in recent years in seismic-prone countries, should be nowadays considered unacceptable, at least for 'well-developed' modern countries.

It could in fact be argued (Bertero 1997) that the rapid increase in population, urbanisation and economical development of our urban areas would naturally result in a generally higher seismic risk, defined as 'the probability that social and/or economical consequences of earthquake events will equal or exceed specific values at a site, at various sites, or in an area during a specific exposure time'. Even though the seismicity remains constant, the implications of business interruption or downtime due to the damage to the built environment are continuously increasing and should be assigned an adequate relevance in the whole picture of seismic risk, typically defined as a combination of seismic hazard and vulnerability.

As a result, in an attempt to develop adequate seismic risk mitigation strategies, more emphasis needs to be given to a damage-control design approach after having assured that life safety and the collapse of the structure are under control (in probabilistic terms).

Such an approach should apply to the design of new constructions as well as to the strengthening/upgrading of existing ones, regardless (in principle) of the constitutive material, with the clear aim of achieving more resilient earthquake-resisting structures.

In this chapter, an overview of recently developed seismic design philosophies, supported by the use of modern technology and advanced materials, for either the design of new structures or the assessment and retrofit of existing ones will be presented. Although most of the concepts can be referred to different construction materials, particular attention will be given to multi-storey reinforced concrete buildings, considering (a) their predominance in the existing built environment as a result of the post-World War II reconstruction in the 1950s and 1960s in most of the Mediterranean area, as well as many other seismic-prone countries and (b) their inherently high seismic vulnerability as confirmed by recent catastrophic earthquake events.

As a support to this overview, interesting and fascinating suggestions and lessons can be obtained from the less-recent past, by comparing the current trends in modern 'innovative' solutions for the future generation of buildings systems with the major achievements in the architectural solutions used by the ancients in the attempt to preserve glorious example of their culture for posterity, in spite of the regular earthquake ground motions (Figure 6.1).

When walking this 'bridge of knowledge' of our cultural heritage with the critical eyes of a curious and passionate observer, surprising commonality in

Figure 6.1 Dramatic representation of the destruction of Sparta in 464 BC, frame by Egisto (Courtesy of NISEE Image Collection, EERC Library Berkeley)

engineering problems and their successful (or just attempted) solutions can be appreciated. It will appear evident how by understanding and, sometimes unconsciously, implementing these lessons with the support of our modern technology, major breakthroughs in the development of 'innovative' solutions for earthquake-resisting structures can be achieved.

Towards performance-based seismic design and retrofit

In response to a recognised urgent need to design, construct and maintain facilities with a limited level of damage following an earthquake event, an unprecedented effort has been dedicated in the last decade to the preparation of a platform for ad-hoc guidelines involving the whole building process, from the concept and design to the construction aspects.

In the comprehensive document prepared by the SEAOC Vision 2000 Committee (1995), Performance Based Seismic Engineering (PBSE) was given a comprehensive definition, as 'consisting of a set of engineering procedures for design and construction of structures to achieve predictable levels of performance in response to specified levels of earthquake, within definable levels of reliability and interim recommendations have been provided to actuate it'. Within this proposed framework, expected or desired performance levels are coupled with levels of seismic hazard by performance design objectives as illustrated by the Performance Design Objective Matrix shown in Figure 6.2.

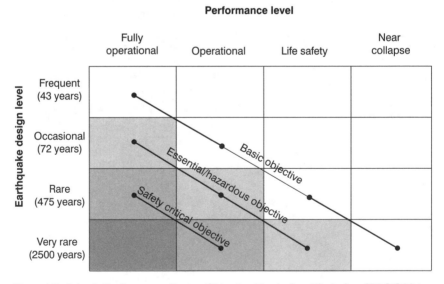

Figure 6.2 Seismic Performance Design Objective Matrix (modified after SEAOC Vision 2000, 1995)

Performance levels are an expression of the maximum acceptable extent of damage under a given level of seismic ground motion, thus representing losses and repair costs due to both structural and non-structural damage. As a further and fundamental step in the development of practical PBSE guidelines, the actual conditions of the building as a whole should be expressed not only in qualitative terms, intended to be meaningful to the general public, using general terminology and concepts describing the status of the facility (i.e. Fully Operational, Operational, Life Safety and Near Collapse) but also, more importantly, through appropriate technically-sound engineering terms and parameters, able to assess the extent of damage (varying from negligible to minor, moderate and severe) for single structural or non-structural elements as well as the whole system.

The choice of appropriate engineering parameter(s), damage indicator(s) or index(es), able to uniquely characterise the status of the structure after the earthquake as well as the definition of appropriate values for lower and upper bounds of each performance level, represents the most critical and controversial phase of a reliable performance-based design or assessment approach.

Recent developments in performance-based design and assessment concepts (Pampanin et al. 2002; Christopoulos and Pampanin 2004) have highlighted the limitations and inconsistencies related to current PBSE approaches, whereby the performance of a structure is typically assessed using one or more structural response indices, as the maximum deformation (i.e. interstorey drift or ductility) and/or cumulative inelastic energy absorbed during the earthquake. The role of residual (permanent) deformations, typically sustained by a structure after a seismic event even when designed according to current codes, has instead been emphasised as a major additional and complementary damage indicator. As noted by the authors and observed after real earthquake events (i.e. Kobe, 1995), residual (permanent) deformations can result in the partial or total loss of a building if static incipient collapse is reached, if the structure appears unsafe to occupants or if the response of the system to a subsequent earthquake is impaired by the new at-rest position of the structure. Furthermore, they can also result in increased cost of repair or replacement of non-structural elements as the new at-rest position of the building is altered. These aspects have not been properly reflected in current performance design and assessment approaches.

Emerging trends in the design of high performance seismic resisting systems

Recognising the economic disadvantages of designing buildings to withstand earthquakes elastically as well as the associated severe consequences following an earthquake event with a higher-than-expected intensity (e.g. as occurred with the Great Hanshin event, Kobe, 1995), current seismic design philosophies favour the design of 'ductile' structural systems able to undergo inelastic reverse cycles while sustaining their integrity. According to the aforementioned original PBSE concepts, different levels of structural damage and, consequently, repairing costs should thus be expected and, depending on the seismic intensity, should be typically accepted as an unavoidable result of the inelastic behaviour. The basics of this design philosophy, referred to as capacity design or hierarchy of strength and developed in the mid-1960s by Professors Park and Paulay from University of Canterbury in New Zealand, is to ensure that the 'weakest link of the chain' within the structural system is located where the design, behaving as a ductile 'fuse', thus protecting the structure from more undesired brittle failure mechanisms (i.e. beam sway vs. soft-storey mechanism, flexural yielding vs. shear failure).

Regardless of the structural material adopted (i.e. concrete, steel, timber), traditional solutions capable of providing adequate global and local ductile structural behaviour rely on the inelastic behaviour of the material, allowing for plastic deformations to occur within selected discrete and sacrificial regions (typically referred to as 'plastic hinge regions').

Alternatively, seismic protection can rely on the use of energy dissipation devices or base isolation systems, designed with passive, active or semi-active control to protect the main structural skeleton from inelastic mechanisms and damage. A recent comprehensive overview of passive supplemental damping and base isolation can be found in Christopoulos and Filiatrault (2007).

A revolutionary alternative design approach has been achieved by recent advanced solutions, referred to as jointed ductile connections, originally developed for seismic-resistant precast/prestressed concrete buildings under the US-PRESSS (PREcast Seismic Structural System) Program coordinated at the University of California, San Diego (Priestley 1991, 1996; Priestley *et al.* 1999; Pampanin 2005).

In 'dry' jointed ductile solutions, contrary to the traditional emulation of cast-in-place solutions, thus opposite to wet and strong connection approaches (Figure 6.3), precast elements are jointed together through unbonded post-tensioning tendons/strands or bars. The inelastic demand is accommodated within the connection itself (beam-to-column, column-to-foundation or wall-to-foundation critical interface, Figure 6.4), through the opening and closing of an existing gap (a sort of 'controlled rocking' motion).

A reduced level of damage, when compared with equivalent cast-in-place solutions (Figure 6.5), is expected in the structural precast elements, which are basically maintained in the elastic range. Moreover, the self-centering contribution due to the unbonded tendons can lead to negligible residual (or permanent) deformations, which, as mentioned before, should be adequately considered as a complementary damage indicator within a performance-based design or assessment procedure. A particularly promising and efficient solution within the family of jointed ductile connections is given by the 'hybrid' systems (Figure 6.3), where self-centering and energy-dissipating properties are combined through the use of unbonded post-tensioning

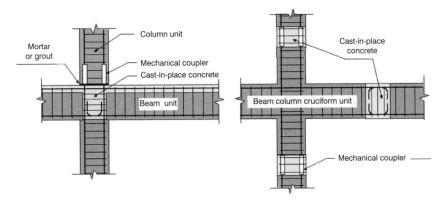

Figure 6.3 Typical arrangements of precast units in the emulation of the cast-in-place approach that is widely used in Japan and New Zealand (*FIB* 2003) as well as, to a lesser extent, in Mediterranean countries

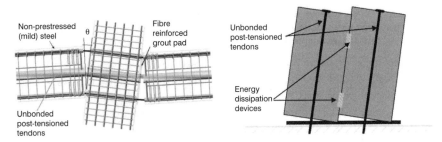

Figure 6.4 Jointed precast 'hybrid' frame and wall systems, implementing a controlled rocking motion, as developed in the US-PRESSS Program (courtesy of Suzanne Nakaki)

tendons/bars and longitudinal non-prestressed (mild) steel or additional external dissipation devices (Stanton *et al.* 1997).

As a result, extremely efficient structural systems are obtained, which can undergo inelastic displacements similar to their traditional counterparts, while limiting the damage to the structural system and assuring full re-centering capability. A damage control limit state can thus be achieved according to either the traditional or the more recent definitions of performance levels, leading to an intrinsically high-seismic-performance system almost regardless of the seismic intensity.

Dissemination of knowledge on new technology: combination of top-down and bottom-up approaches

Several on-site applications adopting PRESSS-type technology have been implemented in the last few years in seismic countries around the world including the US, Europe, Central and South America, Japan, and New

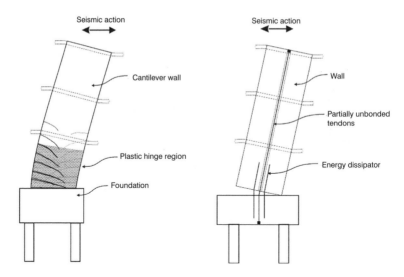

Figure 6.5 Comparative response of a traditional monolithic system (damage in the plastic hinge region at the base) and a jointed precast (hybrid) solution (damage limited to the fuses and negligible residual deformations) (*fib* 2003)

Zealand. Major seismic codes or design guidelines (*fib* 2003; ACI T1.2-03 2003; NZS3101:2006) have incorporated the possibility of using such technological solutions. An overview of recent developments of hybrid solutions in precast concrete construction including research outcomes, modelling and design aspects, code provisions and guidelines as well as practical applications can be found in Pampanin (2005).

Following the rapidly increasing interest from the industry or, generally speaking, from the market, a programme for wider dissemination of this free technology has been for example promoted by the New Zealand Concrete Society in collaboration with the University of Canterbury by means of: (a) seminars/workshops to practitioner engineers, precasters, contractors, local authorities' representative (NZCS 2005), as well as (b) courses for undergraduate and graduate students currently delivered in the form of intensive short courses also outside New Zealand (e.g. Europe and South America).

A combination of top-down and bottom-up approaches is thus confirmed to be a successful strategy to introduce innovative technologies and disseminate the required know-how in the market with the aim to create safer and more resilient earthquake-resisting constructions.

Extension of the hybrid system concept to bridge construction and other materials

The concept of hybrid systems has been extended in the last few years to bridge construction (either cast-in-situ or segmental precast), with several contributions in the literature studying the behaviour of single bridge piers or whole bridge systems. An example is the concept of controlled rocking systems for use in bridge construction introduced by Palermo et al. 2005 (Figure 6.6), following and integrating previous work on unbonded post-tensioned solution for bridge piers (e.g. Mander and Chen 1997; Hewes and Priestley 2001; Kwan and Billington 2003).

Similarly, the basic concepts of PRESS technology originally developed for concrete construction, has recently been proposed and experimentally tested for multi-storey seismic resisting solutions in steel (Christopoulos et al. 2002) as well as timber (Laminated Veneer Lumber, LVL; Palermo et al. 2006; Figure 6.7).

Very satisfactory and promising results have been obtained in terms of higher performance of these solutions when compared with their traditional monolithic counterparts, i.e. welded or bolted connections in steel construction; multi-nailed, bolted or steel dowel connections in timber construction.

Dealing with the existing built environment: retrofit strategies and solutions

The crucial need for strengthening or retrofitting existing modern structures designed with substandard details, in order to withstand seismic loads without collapsing or with relatively moderate damage, has been highlighted by the catastrophic effects of earthquake events in the recent past. Extensive experimental–analytical investigations on the seismic performance of existing reinforced concrete frame buildings, mainly (if not only) designed for gravity loads, as typically found in most seismic-prone countries before

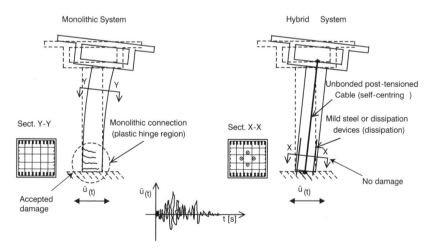

Figure 6.6 Concept of hybrid system applied to bridge piers and systems (modified from Palermo et al. 2005)

Figure 6.7 Experimental testing under simulated seismic loading of multi-storey buildings or bridge pier subassemblies (Canterbury University, New Zealand, Palermo *et al.* 2006)

the introduction of adequate seismic design code provisions in the 1970s (Aycardi *et al.* 1994; Beres *et al.* 1996; Hakuto *et al.* 2000; Pampanin *et al.* 2002; Park 2002) have confirmed the expected inherent vulnerability of these systems observed in past earthquakes (Figure 6.8).

As mentioned above, the revolutionary capacity design approach, capable of protecting a structure from sudden brittle collapse, was developed in New Zealand by Professors Park and Paulay in the mid-1960s and introduced in major international codes only in the 1970s. Inevitably, most of the existing buildings designed and constructed prior to that period (1930–70) would be inadequately detailed to sustain significant lateral loads due to an

Figure 6.8 Observed damage in pre-1970s designed existing reinforced concrete frame buildings with masonry infills (Turkey, left, and Greece, right, earthquakes, 1999; from NISEE Image Collection, EERC Library Berkeley)

earthquake event. This does not exclude the fact that more recently designed and constructed buildings might have not properly implemented the code prescriptions. An example of the typical features of a pre-1970s reinforced concrete building is shown in Figure 6.9.

Plain round bars, instead of deformed bars, were generally adopted until the mid-1960s. Very few stirrups were used in columns and beams, as typical of gravity load design. No requirement for transverse reinforcement (stirrups) inside the joint region was in place. Typically the minimum required amount of longitudinal and transverse reinforcement (as low as 0.8–1 per cent of the gross section area in some cases) was placed in the columns, mostly designed to sustain vertical loads, not bending moments or shear due to lateral earthquake-induced loads. The columns were often tapered along the elevation in a multi-storey frame construction, again as natural and economically-wise in a design for gravity-loads-only (or -mainly).

As a result of this inadequate reinforcement detailing, of the lack of transverse reinforcement in the joint region, as well as of the absence of any capacity design principles, brittle failure mechanisms are expected either at local level (e.g. shear failure in the joints, columns or beams) or global level (e.g. soft-storey mechanism leading to full collapse of the building). Moreover, the presence of infills (e.g. typically un-reinforced masonry bricks) can lead to unexpected and controversial effects due to the interaction with the bare frame (Crisafulli *et al*. 2000; Magenes and Pampanin 2004).

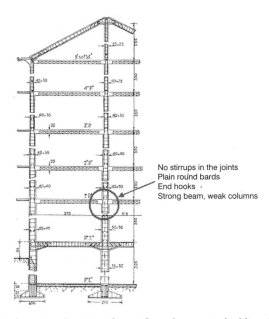

No stirrups in the joints
Plain round bards
End hooks ·
Strong beam, weak columns

Figure 6.9 Typical section elevation of a reinforced concrete building designed in Italy in the late 1950s (Santarella 1957)

It is worth remembering that this inherent vulnerability is common in pre-1970s designed buildings worldwide, as demonstrated by extensive research (both experimental and analytical) carried out in the past decades confirming the observed results of catastrophic collapses in past earthquakes (Northridge 1994, Kobe 1995, Turkey 1999, and Taiwan 2001).

The main reasons for such numerous collapses are thus clearly and 'simply' related to:

a   obsolete design principles;
b   the general lack of a more comprehensive knowledge in the field of earthquake engineering as available today, regarding: the intensity and recurrence of a design earthquake, the behaviour of structures under seismic conditions by means of experimental testing and more refined numerical non-linear analyses tools, the introduction of a hierarchy of strength and capacity design principles, a better understanding of the importance of the detailing of the reinforcement as lap splices, bending of the bars, etc;
c   the lack of current technology and quality controlled materials.

It is worth noting that bad workmanship or wrong prescriptions from the design engineers, still occurring in spite of clearly written and updated seismic code design provisions, and identifiable as 'corrupted behaviour' as discussed in Chapter 12, can certainly aggravate the problem of the inherent vulnerability of the existing built environment. However, for the technical weaknesses discussed before, this latter cultural-related issue cannot and should not be said to be the main source of post-earthquake collapses, as is sometimes done by non-technical articles presented by the media with the clear aim of capturing the attention of the general reader.

Experimental testing in several universities or research laboratories around the world have demonstrated that pre-1970 buildings, even when designed with the best attention to workmanship (laboratory controlled) and in full accordance with the older design code specifications available in those decades (in terms of overall design, material strength and structural detailing), would still be very likely to suffer major damage, if not a complete collapse, under a moderate-to-strong earthquake event.

The political repercussions of disclosing this technically-sound awareness to a general public (i.e. outside the scientific community) are clearly significant. The responsibility for seismically upgrading existing multi-storey buildings might for example be claimed to belong to local or central authorities, particularly when referring to government-owned buildings for public use (schools, hospitals, theatres, museums, churches and monuments). Similar expectations could be extended to the numerous government-funded high-rise buildings designed between 1950 and 1960. An example of proactive initiative in this sense, is given by the financial support provided by the World Bank in dedicated projects related to seismic risk analysis and mitigation, as for the feasibility studies of seismic strengthening in the city of Istanbul (as part of the Marmara Earthquake Emergency Reconstruction Project, Metropolitan Municipality of Istanbul, 2003), considered to be at a very high seismic risk.

Available retrofit solutions and controversial issues

Alternative seismic retrofit and strengthening solutions have been studied in the past and adopted in practical applications ranging from conventional techniques, which utilise braces, jacketing or infills, to more recent approaches including base isolation, supplemental damping devices or advanced materials (e.g. Fibre Reinforced Polymers, FRP (*fib* 2001) or Shape Memory Alloys, SMA (Dolce *et al.* 2000)). Most of these retrofit techniques have evolved in viable upgrades. However, issues of costs, invasiveness, and practical implementation still remain the most challenging aspects of any intervention. Also, a proper understanding (assessment) of the seismic vulnerability of the structure is a delicate and crucial step before selecting the most appropriate retrofit strategy.

Based on recent lessons learned from earthquake damage and extensive experimental and analytical investigation in the past few decades, it is becoming more and more evident that major controversial issues can arise when, for example: (a) deciding whether the retrofit is needed and in what proportions; (b) assessing and predicting the expected seismic response with alternative analytical/numerical tools and methods; (c) evaluating the effects of presence of infills, partitions or 'non-structural' elements on the seismic response of the overall structure; (d) deciding to weaken a part of the structure in order to strengthen the whole structure; (e) adopting a selective upgrading to independently modify strength, stiffness or ductility capacity; (f) relying on the deformation capacity of an under-designed member to comply with the displacement compatibility issues imposed by the overall structure; (g) defining a desired or acceptable level of damage the retrofit structural should sustain after a given seismic event, i.e. targeting a specific performance level after the retrofit.

Due to improper assumptions or approaches during either the assessment or the retrofit phase, dramatic consequences could occur in a ground shaking. Structures assessed to be sufficiently earthquake-'resistant' might end up suffering major damage if not a total collapse. On the other hand, an inadequate retrofit or strengthening intervention could condemn an 'earthquake-risk' structure to a secure collapse.

Finally, consideration on cost-effectiveness, invasiveness, architectural and aesthetic aspects further complicate such a complex decision-making process, along with issues related to the socio-economical consequences of excessive damage and related downtime due to a limited or interrupted functionality of the structures after the seismic event.

Assessment of seismic vulnerability: the fundamental and delicate role of an appropriate diagnosis

Prior to, and irrespective of, the technical solution adopted, the efficiency of a retrofit strategy strongly depends on a proper assessment of the internal hierarchy of strength of beam-column joints as well as on the expected sequence of events within a beam-column system (i.e. shear hinges in the joints or plastic hinges in beam and column elements). The effects of the expected damage mechanism on the local and global response should also be adequately considered. As in any diagnosis phase, a simple but reliable analysis procedure should be able to provide useful preliminary information as support for any retrofit intervention, maybe temporary due to lack of time, knowledge, technology or funds as well as due to the anticipation of possible changes due to expected loads, boundary conditions or state of the structure. Incorrect and non-conservative assessment of the sequence of events can otherwise result, leading to inadequate, and not necessarily conservative, design of the retrofit intervention.

Understanding the seismic behaviour of beam-column joints in frame systems: the devil is in the details

The complexity of the behaviour of beam-column joints has been well recognised in the past, as confirmed by the major effort dedicated in the past thirty to forty years to the development of appropriate design guidelines. In particular, when capacity design principles were introduced in the standard seismic design philosophy, the idea of dealing with 'the weakest link of the chain' naturally promoted a higher sensibility to the need to properly protecting this part of a frame structure. However, when dealing with an existing structure, the lack of information on the structural detailing, on the material properties and on the original design could be crucial for a correct assessment of its vulnerability under a seismic event and the subsequent definition of an adequate retrofit strategy.

Minor differences in the structural details could improve or degrade the expected behaviour under seismic loading, significantly enough to lead to a marginal survival when not to a sudden collapse of the whole structure. Different damage or failure modes are in fact expected to occur in beam-column joints depending on the typology (exterior or interior joint) and on the adopted structural details (i.e. total lack or presence of a minimum amount of transverse reinforcement in the joint; use of plain round or deformed bars; alternative bar anchorage solutions), as shown in Figure 6.10

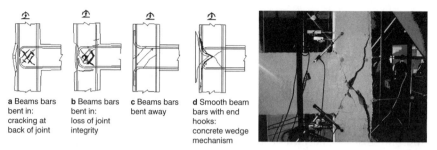

a Beams bars bent in: cracking at back of joint

b Beams bars bent in: loss of joint integrity

c Beams bars bent away

d Smooth beam bars with end hooks: concrete wedge mechanism

Figure 6.10 Alternative damage mechanisms expected in exterior tee-joints depending on the structural detailing (a), (b) beam bars bent inside the joint region; (c) beam bars bent outside the joint region; (d) plain round beam bars with end-hooks 'concrete wedge' mechanism. Right side observed damage in an exterior beam-column joint subassembly subjected to laboratory-simulated seismic loading (Pampanin *et al.* 2002)

for an exterior tee-joint with no transverse reinforcement (typical in older construction).

After diagonal cracking, the shear transfer mechanism in the joint region has to basically rely on a compression strut mechanism, whose efficiency is critically related to the anchorage solution adopted for the longitudinal beam reinforcement. When beam bars are bent into the joint (Figure 6.10a,b), they can provide a limited resistance against the horizontally expansion of the joint, until the hook opens under the combined action of the diagonal strut and the pulling tension force in the beam reinforcement, leading to a rapid joint degradation. When beam bars are bent away from the joint (Figure 6.10c), as typical of older practice in New Zealand and Japan, no effective node point is provided for the development of an efficient compression strut mechanism, unless a significant number of transverse column hoops are placed immediately above the joint core. A rapid joint strength degradation after joint diagonal cracking is expected. The worst scenario is however provided by the solution shown in Figure 6.10d, typical of Mediterranean construction practice, where plain round bars with end hook anchorage were used. As recently shown by experimental tests on beam-column joint specimens and a three-storey frame system (Pampanin *et al.* 2002; Calvi *et al.* 2002), the combination of strut action and of a concentrated compression force at the end-hook anchorage, due to slippage of the longitudinal beam bars, can lead to the expulsion of a 'concrete wedge' (Figure 6.10, right side), with rapid loss of load-bearing capacity.

The controversial effects of masonry infills on the seismic response: an open debate

The effects of infills on the response of the overall structure still represents an open topic, with a critical need of further investigations for the seismic vulnerability assessment of extensive classes of existing buildings. Controversial effects on the global inelastic mechanism can be expected depending on the infill properties (mechanical characteristic and distribution) and on the joint damage mechanism.

In recent studies (Magenes and Pampanin 2004; Pampanin *et al.* 2006), the interaction between un-reinforced masonry infills and reinforced concrete (RC) frame systems, when appropriately considering the joint zone non-linear behaviour, has been investigated through non-linear pushover and time-history analyses on 2-D and 3-D multi-storey frame systems under uni-directional or bi-directional input motions.

The presence of infills can guarantee higher stiffness and strength, reducing the inter-storey drift demand (ratio between the relative floor displacements and the interstorey height), while increasing the maximum floor accelerations. A further positive influence of the infills can be recognised in the reduction of column interstorey shear contribution as well as in the possible delay of a soft-storey mechanism which might instead develop in a bare frame solution. On the other hand, the interaction between un-reinforced masonry infills and RC frames can lead to unexpected or peculiar effects when compared with the response of the bare frame, either at a local level (e.g. shear failure in columns; damage to joint region) or on the global seismic response (e.g. soft-storey mechanism). Moreover, the sudden reduction of storey stiffness due to the damage of the infills can lead to the formation of an unexpected soft-storey mechanism, independently of the regular or irregular distribution of the infills along the elevation.

Similarly, when investigating the response of 3-D frames under either uni-directional or bi-directional earthquake input excitation, inelastic torsion mechanisms can occur due to the irregular distribution of damage to the infills. Interestingly, or unfortunately, enough such problems seem to be likely to occur in some (fewer) cases even when considering more recently constructed buildings designed according to current code provisions.

Multi-level retrofit strategy: a rational compromise with reality

According to a multi-level retrofit strategy approach (as suggested by Pampanin *et al.* 2007), alternative objectives can be targeted in terms of hierarchy of strength within the beam–column–joint system, depending on the joint typology (interior or exterior) and on the structural details adopted. As discussed by the authors, the ideal retrofit strategy would in fact not only protect the beam-to-column joints that were identified as the major deficiency in these frames, but would further upgrade the structure to exhibit the desired weak-beam strong-column behaviour (beam-sway mechanism) which is at the basis of the design of new seismic-resistant RC frames. However, due to the disproportionate flexural capacity of the beams when compared with the columns this is difficult to achieve in all cases and for all beam-to-column connections without major interventions. This is especially true for interior beam-to-column connections where the moment imposed on interior columns from the two framing beams is significantly larger than for exterior columns. As mentioned before, interior joints are less vulnerable than exterior joints and exhibit much more stable hysteresis behaviour with hardening after first cracking. It is thus conceivable, in a bid to protect the interior columns from excessive curvature demand, to tolerate some joint damage prior to hinging of the columns.

Two levels of retrofits can therefore be considered, depending on whether or not interior joints can be fully upgraded. A complete retrofit would thus consist of a full upgrade by protecting all joint panel zones and developing plastic hinges in beams while columns are protected according to capacity design principles. A partial retrofit would consist of protecting exterior joints, forming plastic hinges in beams framing into exterior columns, while permitting hinging in interior columns or limited damage to interior joints, where a full reversal of the strength hierarchy is not possible.

Ultimately, the viability of the partial retrofit strategy must be investigated on a case-by-case basis to ensure that the localised damage to interior joints does not severely degrade the overall response of the structure or jeopardise the ability of the interior columns to safely carry gravity loads.

Implementation of the multi-level retrofit strategy using alternative solutions

The feasibility and efficiency of two alternative retrofit solutions, following the proposed multi-level retrofit strategy approach and relying either on the use of FRP composite materials (Fibre-Reinforced Polymer) or on a low-invasive metallic haunch connection, has recently been presented in the literature (Pampanin 2005; Pampanin *et al.* 2006, 2007), based on experimental and analytical investigations carried out on a series of as-built and retrofitted 2-D beam-column joint subassemblies and on two three-storey frame systems. As shown in Figure 6.11, the occurrence of brittle mechanisms at local or global level was adequately protected and a more desirable hierarchy of strengths and sequence of events achieved (i.e. plastic hinge in the beam), with both the selected interventions leading to more ductile and dissipating hysteresis behaviour.

Feasibility, low invasiveness and reversibility of the intervention

It is worth recalling that, once a satisfactorily structural performance is guaranteed, issues related to practicality as well as, in general, to cost-benefit analysis should be evaluated. Accessibility of the joint region, invasiveness

Figure 6.11 Use of Fibre Reinforced Polymers (left) or a metallic haunch retrofit solution (right) to protect the joint region from a sudden shear failure and relocate the damage in the beam, to activate a more favourable global mechanism using a FRP or haunch retrofit solution (Pampanin *et al.* 2006, 2007)

of the intervention, maintenance and reversibility will also have to be considered in real applications based on a case-by-case evaluation. It is worth noting however that a typical geometrical and plan configuration of existing buildings designed for gravity load only in the 1950s to 1970s period might consist of frames running in one direction only and lightly reinforced slab in the orthogonal direction, the latter being quite typical of the construction practice in Mediterranean countries. In these cases, the adoption of the proposed retrofit intervention can be somehow facilitated, when compared with more recently designed buildings with frames in both directions and cast-in-situ concrete slabs providing flange effects. Examples of application in situ of FRP on an exterior (corner) beam-column joint are shown in Figure 6.12. Experimental testing is on going to further investigate the efficiency of FRP retrofit solutions on 3-D exterior beam column joints when minimising the invasiveness of the intervention (Figure 6.12, right side).

Suggestions for advanced retrofit solutions

The aforementioned recent emphasis on residual deformations, re-centering capability as well as limited level of damage thanks to a controlled rocking system, has resulted in the development and proposal of an advanced seismic retrofit strategy and technology able to provide a higher performance with a limited level of damage and permanent deformations.

Use of SMA devices or post-tensioning systems

Unbonded braces systems have been developed using traditional solutions, while the unique properties of Shape Memory Alloys (SMA) have been exploited to develop a self-centring and dissipating device (with a typical flag-shape behaviour) to be used either in series with a typical brace (Figure 6.13)

Figure 6.12 FRP retrofit of the exterior face of a corner beam-column joint. Left: in situ application using carbon fibres (courtesy of Interbau s.r.l., www.interbau-srl. it). Right: experimental testing using glass fibres (Pampanin *et al*. 2007)

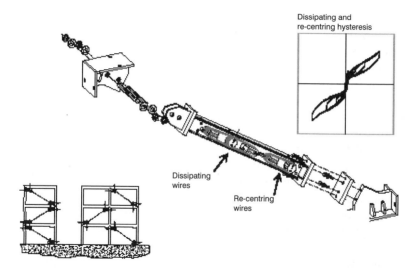

Figure 6.13 Brace system based on re-centring and dissipating Shape Memory Alloys wires (after Dolce *et al*. 2000)

or as a dissipater and re-centring system in a base isolation intervention (Dolce *et al*. 2000; Dolce and Cardone 2003).

SMA have also been used in the form of supplemental reinforcement or post-tensioning systems for the seismic rehabilitation of monuments and historical buildings as in the cases of the Basilica di San Francesco in Assisi (Croci *et al*. 2000), damaged after two consecutive earthquake shocks in the 1997 Umbria-Marche earthquake or the Bell Tower of the S. Giorgio Church in Trignano (Figure 6.14), damaged after the 1996 Modena and Reggio earthquake in Emilia, Italy (Indirli 2000; DesRoches and Smith 2004).

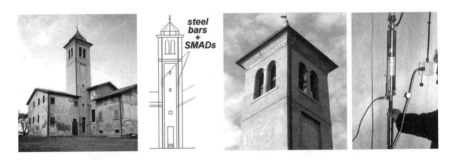

Figure 6.14 Bell Tower of S. Giorgio Church in Trignano; retrofit intervention with vertical SMA wires in series with prestressed steel tendons and detail of the device (Photo courtesy of FIP Industriale)

Strengthening or weakening as a basis of seismic retrofit?

The concepts of strengthening and retrofit are often improperly associated. According to advanced retrofit techniques proposed by Elnashai and Pinho (1998) the structure could be upgraded without needing any strengthening (increase of strength). Stiffness-only, strength-only or ductility-only intervention can be carried out on a single element as part of a selective retrofit approach.

By further developing the concept and referring to the significant advantages of hybrid or controlled rocking systems (in terms of limited level of damage, control of the stress level acting as a fuse) presented in the previous paragraphs, it could actually be argued that a weakening, not strengthening approach could be a more valuable, though clearly counterintuitive, approach to protecting an existing structure from an expected earthquake ground motion. The concept of a *selective weakening* approach has recently been presented in the literature (Pampanin 2006) and implemented with experimental validation on shear wall systems (Ireland *et al.* 2006).

The concept is explained in Figure 6.15. By saw-cutting the longitudinal bottom reinforcement of a gravity load dominated beam or of a shear-dominated wall a better control of the overall mechanism can be achieved, according to hierarchy of strength principles. A flexure-dominated rocking mechanism can be activated, which is able to guarantee a limited level of damage in the structural member as well as the upper limit of the level of stress (fuse action) directed to the beam–column joint panel zone or to the existing foundation protecting weak links of the fuse. Moreover, shear walls with low aspect ratio in existing buildings could be suggested to be split into two adjacent rocking coupled walls, with significant reduction of shear failure concerns as well as overturning demand on the foundation.

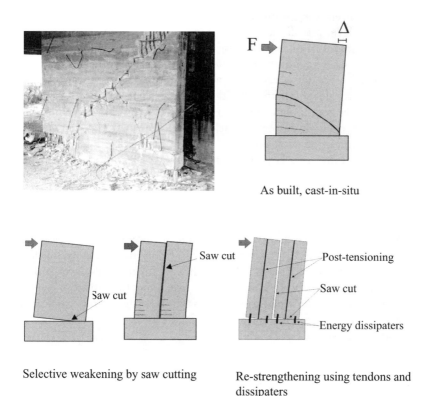

As built, cast-in-situ

Selective weakening by saw cutting

Re-strengthening using tendons and dissipaters

Figure 6.15 Typical shear failure mechanism of an existing under-designed RC wall (Izmit-Kocaeli, Turkey, 1999) and conceptual application of the proposed selective weakening technique (after Ireland *et al.* 2006)

Further enhancement of the behaviour could also be achieved at this stage by using advanced energy dissipation devices (e.g. viscous-elastic, friction, SMA, combined in advanced flag-shaped systems; Kam *et al.* 2006). Acknowledging that monolithic behaviour might imply damage (if resulting from a cast-in-situ technique), a low-damage retrofit or new design approach could be used to appropriately 'isolate' the floor system from the lateral load resisting frame or wall systems (Jensen *et al.* 2006).

Seismic risk analysis for mitigation strategies based on retrofit solutions at territorial scale

The selection of the most appropriate retrofit solution to be implemented is typically accomplished by targeting, on a case-by-case basis, an acceptable weighted balance between the benefits due to the improved seismic performance and the direct (and sometimes indirect) costs associated to the intervention. Recent developments of viable and low-cost retrofit solutions within a multi-level retrofit approach, suggest the possibility to implement 'standardised' solutions on an urban or territorial scale.

However, when expanding the scale of the intervention (and analysis) to a territorial level (city, region, country), more complex criteria and intervention strategies should be considered and evaluated in order to define the most effective action plan to minimise the overall risk. In particular, the actual limits of available resources, including budget, material, human and technical resources, logistics and supporting infrastructures, can represent the critical constraint for a large-scale intervention.

Mitigation analyses, although not yet codified, are expected to become in the near future a fundamental decision-making tool for the allocation of funds by local authorities, as already observed in some part of the world. As an example, in the USA the Federal Emergency Management Agency (FEMA) act of 2000 mandates that states and local governments conduct mitigations analyses as a fundamental condition to receiving Hazard Mitigation Grant Programme (HMGP) funds. Mitigation analyses should describe: (1) the prioritising mitigation actions and (2) how the overall mitigation strategy is cost-effective and maximises the overall wealth; in other words which one amongst the possible alternatives should be funded to minimise the overall risk.

The efficiency of alternative structural mitigation strategies is under investigation within the framework of a seismic risk analysis approach (Giovinazzi and Pampanin 2007; Figure 6.16). Single- or multi-criteria approaches, implemented within either a deterministic or a stochastic evaluation and applied to a case study area, has confirmed this as a valuable support for decision-making. In particular, targeting partial or total retrofit interventions with non-uniform spatial distributions, has been demonstrated on a case-study area to be an efficient approach to mitigate the overall seismic risk (or expected consequences) when, due to resources constraints, a massive global intervention is not feasible.

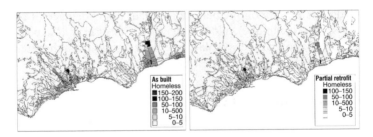

Figure 6.16 Example of GIS-based representation of efficiency of a partial retrofit intervention at territorial scale (Giovinazzi and Pampanin 2007)

Historical developments in seismic engineering: understanding and implementing lessons from our ancient heritage

The introduction of jointed ductile systems, assembled by post-tensioning and able to undergo a severe seismic event with minor structural damage, certainly represent a major achievement in seismic engineering in the last decade if not a fundamental milestone in the historical development in the field, as shown in Figure 6.17. The conceptual innovation introduced by capacity design (or hierarchy of strength) principles as part of the design approach for ductile systems led in the mid-1970s to a revolutionary implication in seismic design philosophy. Similarly the development started in the early 1990s of ductile connections able to accommodate high inelastic demand without suffering extensive material damage, appears to be a promising step forward for the next generation of high-performance seismic-resisting systems, based on the use of conventional materials and techniques.

Interestingly enough, as shown in the following figures, such a breakthrough has been inspired by ancient architecture, thus representing a clear example of use of modern technology to further develop and refine very valuable solutions built in our ancient heritage. In the following concluding pages, a few self-explanatory examples of lessons from the ancient architecture will be shown, looking at traditional construction techniques in seismic regions as well as at the original development of the aforementioned concepts for high seismic performance of either newly designed or retrofitted/repaired structures.

Intentionally, only a minimum amount of text is used, allowing the reader to enjoy a naturally complex reaction and a possible mix of different emotions: (a) fascinating surprise for the achievement of our ancestors two thousand years ago without the advanced technology available nowadays;

Figure 6.17 Evolution of seismic-resisting connections (Pampanin 2005) performance of beam-column joints designed according to pre-1970 codes (shear damage in the joint); following capacity design principles as per the New Zealand concrete standard (beam plastic hinge) and adopting hybrid jointed ductile connections as per the special provisions in Appendix B of the NZS3101:2006 (controlled rocking system)

(b) sincere curiosity and a desire to learn more about those times; (c) hidden pride for the confirmation of the efficiency of a brilliant and independent idea of our modern times; (d) genuine enthusiasm for a new idea for further development and study; (e) ultimate recognition that science and research might be really based on a 'cyclic process' with periods of 20–30 years for 'frequent' or 'rare' events, but much longer periods up to 2000–3000 years, for 'very rare' events or breakthroughs.

The real origin of rocking systems, self-centring and limited damage response under earthquake loading is illustrated in Figure 6.18.

The use of post-tensioning for retrofit strategies throughout the centuries is illustrated in Figures 6.19 to 6.22.

Figure 6.18 Typical use of rocking systems and segmental (multi-block) construction in the practice of ancient Greek and Roman temples axial load for re-centring; use of a lead element as shear and torsion key

Figure 6.19 Top: seismic-resisting solutions for timber buildings used in the ninteenth century that performed well during the 29 June 1873 Belluno earthquake (Barbisan and Laner 2001). Note the use of wooden rafters and beams passing though the full thickness of the walls and well anchored at the edge. Example of a timber beam post-tensioned as a chain. Bottom: example of retrofit intervention on churches or ancient building facades using post-tensioned chains in the nineteenth and twentieth centuries (Pavia, Gubbio; Italy)

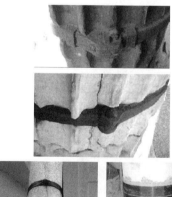

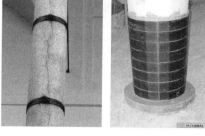

Figure 6.20 Confinement on columns for retrofit intervention in past centuries (e.g. top: by architect Giuseppe Valadier in nineteenth century, Vesta Temple; bottom left, Collegio Borromeo, Pavia) through iron or steel chains and advanced solution using FRP composite materials on circular columns (bottom right: courtesy of Interbau s.r.l., www. interbau-srl.it)

Figure 6.21 Typical representations of collapse of towers during historical seismic events (Napoli, 1805; Tuscany, 1896; courtesy of Jan T. Kozak Collection, EERC Library Berkeley) awareness of their intrinsic vulnerability. Bottom right: modern use of a half-collapsed medieval tower as residential private house (Pavia, Italy)

Figure 6.22 Application of a temporary retrofit intervention based on transverse post-tensioning and confinement of the corners to a medieval tower (Pavia, Italy) after the collapse of the Cathedral Bell Tower in 1989. Implementation of similar intervention by using FRP composite materials, on a square column (top right:courtesy of Interbau s.r.l., www.interbau-srl.it).

References

ACI T1.2-03 (2003) 'Innovation Task group 1 and collaborators, special hybrid moment frames composed of discretely jointed precast and post-tensioned concrete members (ACI T1.2-03) and commentary (ACI T1.2R-03)', American Concrete Institute Farmington Hills, MI.

Aycardi, L.E., Mander, J.B. and Reinhorn, A.M. (1994) 'Seismic resistance of R.C. frame structures designed only for gravity loads: experimental performance of subassemblages', *ACI Structural Journal*, 91(5): 552–63.

Barbisan, U. and Laner, F. (2001) 'Historical antiseismic building techniques: wooden contribution', *Tecnologos Publisher*, 1(1), published online at www.tecnologos.it.

Beres, A., Pessiki, S., White, R. and Gergel,y P. (1996) 'Implications of experimental on the seismic behaviour of gravity load designed RC beam-column connections', *Earthquake Spectra*, 12(2): 185–98.

Bertero, V. (1997) 'Performance-based seismic engineering: a critical review of proposed guidelines', in Fajfar, P. and Krawinkler, H. (eds) *Seismic Design Methodology for the Next Generation of Code*, Rotterdam: Balkema.

Calvi, G.M., Magenes, G. and Pampanin, S. (2002) 'Experimental test on a three storey reinforced concrete frame designed for gravity only', *Proceedings of the 12th European Conference on Earthquake Engineering*, London, September, paper no. 727.

Christopoulos, C. and Filiatrault, A. (2007) *Principle of Passive Supplemental Damping and Base Isolataion*, Pavia, Italy: IUSS Press.

Christopoulos, C. and Pampanin, S. (2004) 'Towards performance-based design of MDOF structures with explicit consideration of residual deformations', *ISET Journal of Structural Engineering*, Special Issue 41(1): 53–73.

Christopoulos, C., Filiatrault, A., Uang, C.M. and Folz, B. (2002) 'Post-tensioned energy dissipating connections for moment resisting steel frames', *ASCE Journal of Structural Engineering*, 128(9): 1111–20.

Crisafulli, F.J., Carr, A.J., and Park, R. (2000) 'Analytical modelling of infilled frame structures: a general review', *Bulletin of the New Zealand Society of Earthquake Engineering*, 33(1): 30–47.

Croci, G., Bonci, A., and Viskovic, A. (2000) 'Use of shape memory alloy devices in the Basilica of St Francis of Assisi', *ISTECH Project-Proc. of Final Workshop*, pp. 110–33.

DesRoches, R. and Smith, B. (2004) 'Shape memory alloys in seismic resistant design and retrofit: a critical review of their potential and limitations', *Journal of Earthquake Engineering*, 8(3): 415–29.

Dolce, M. and Cardone, D. (2003) 'Seismic protection of light secondary systems using different base isolation systems', *Journal of Earthquake Engineering*, 7(2): 223–50.

Dolce, M., Cardone, D. and Marnetto, R. (2000) 'Implementation and testing of passive control devices based on shape memory alloys', *EESD*, 29(7): 945–68.

Elnashai, A.S. and Pinho, P. (1998) 'Repair and retrofitting of RC walls using selective techniques', *Journal of Earthquake Engineering*, 2(4): 525–68.

fib (2001) 'Externally bonded FRP reinforcement for RC structures', *fib Bulletin*, 14, Lausanne: Federation International du Beton.

fib (2003) 'Seismic design of precast concrete building structures', *fib Bulletin* 27, Lausanne: Federation International du Beton.

Giovinazzi, S. and Pampanin, S. (2007) 'Multi-criteria approaches for regional earthquake retrofit strategies', *Proceedings of the 8th Pacific Conference on Earthquake Engineering*, Singapore, 8–10 Dec.

Hakuto, S., Park, R. and Tanaka, H. (2000) 'Seismic load tests on interior and exterior beam-column joints with substandard reinforcing details', *ACI Structural Journal*, 97(1): 11–25.

Hewes, J.T. and Priestley, M.J.N. (2001) 'Experimental testing of unbonded post-tensioned precast concrete segmental bridge columns', *Proc. of the 6th Caltrans Seismic Research Workshop Program*, Sacramento, CA, 12–13 June.

Indirli, M. (2000) 'The demo-intervention of the ISTECH Project: the Bell-Tower of S. Giorgio in Trignano (Italy)'. *ISTECH Project- Shape Memory Alloy Devices for Seismic Protection of Cultural Heritage Structures Proceedings of Final Workshop*, 23 June, 134–46.

Ireland, M., Pampanin S. and Bull, D.K. (2006) 'Concept and implementation of a selective weakening approach for the seismic retrofit of R.C. buildings 2006', *Proceedings of the Annual NZSEE Conference*, Napier, New Zealand, 8–10 March.

Jensen, J., Bull, D.K., and Pampanin, S. (2006) 'Conceptual retrofit strategy for existing hollowcore seating connections', *Proceedings of the NZ Society of Earthquake Engineering Conference*, Napier, New Zealand, 8–10 March.

Kam, W., Pampanin, S., Palermo, A. and Carr, A. (2006) 'Advanced flag-shape systems for high-seismic performance', *Proceedings of the 1st ECEES*, Geneva, Switzerland, Sept.

Kwan, W.-P. and Billington, S. (2003) 'Unbonded post-tensioned bridge piers II: seismic analyses', *ASCE Journal of Bridge Engineering*, 8(2): 102–11.

Magenes, G. and Pampanin, S. (2004) 'Seismic response of gravity-load designed frame systems withj masonry infills', *Proceedings of the 12th World Conference on Earthquake Engineering*, Vancouver, August, paper no. 4004.

Mander, J.B. and Chen, C.T. (1997) 'Seismic resistance of bridge piers based on damage avoidance design', *Technical Report NCEER-97-0014* (National Centre for Earth. Eng. Research), Buffalo: State University of New York.

Metropolitan Municipality of Istanbul (2003) *Earthquake Master Plan for Istanbul*, Report by Bogazici University, Istanbil Technical University, Middle East Technical University, Yildiz Technical University.

NZCS (2005) 'Introduction to the design of jointed ductile connections', Seminar Notes, New Zealand Concrete Society. Christchurch, Wellington, Auckland, 1–8 March.

NZS3101:2006 (2006) *Concrete Structures Standard*, Wellington: Standards New Zealand.

Palermo, A., Pampanin, S. and Calvi, G.M. (2005) 'Concept and development of hybrid systems for seismic-resistant bridges', *Journal of Earthquake Engineering*, 9(6): 899–921.

Palermo, A., Pampanin, S. and Buchanan A. (2006) 'Experimental investigations on LVL seismic resistant wall and frame subassemblies', *Proceedings of the 1st ECEES*, Geneva, Switzerland, 3–8 September, Paper no. 983.

Pampanin, S. (2005) 'Emerging solutions for high seismic performance of precast/prestressed concrete buildings', *Journal of Advanced Concrete Technology (ACT)*, 3(2): 207–23.

Pampanin, S. (2006) 'Controversial aspects in seismic assessment and retrofit of structures in modern times: understanding and implementing lessons from ancient heritage' *Bulletin of NZ Society of Earthquake Engineering*, 39(2): 120–33.

Pampanin, S., Calvi, G.M. and Moratti, M. (2002) 'Seismic behaviour of R.C. beam-column joints designed for gravity loads', *Proceedings of the 12th ECEE*, 9–13 September, London, Paper no. 726.

Pampanin, S., Christopoulos, C. and Priestley, M.J.N. (2002) 'Residual deformations in the performance-based seismic assessment of frame systems', *ROSE Research Report 2002/02*, Pavia, Italy: European School on Advanced Studies on Reduction of Seismic Risk.

Pampanin, S., Christopoulos, C. and Chen, T.-H. (2006) 'Development and validation of a haunch metallic seismic retrofit solution for existing under-designed RC frame buildings', *Earthquake Engineering and Structural Dynamics*, 35(14): 1739–66.

Pampanin, S., Bolognini, D., and Pavese, A. (2007) 'Performance-based seismic retrofit strategy for existing reinforced concrete frame systems using FRP composites', *ASCE Journal of Composites for Construction*, Special Issue on 'Recent International Advances in FRP Research and Application in Construction', 11(2): 211–26.

Park, R. (2002) 'A summary of results of simulated seismic load tests on reinforced concrete beam-column joints, beams and columns with substandard reinforcing details'. *Journal of Earthquake Engineering*, 6(2): 1–27.

Priestley, M.J.N. (1991) 'Overview of the PRESSS research programme', *PCI Journal*, 36(4): 50–7.

Priestley, M.J.N. (1996), 'The PRESSS program current status and proposed plans for phase III', *PCI Journal*, 41(2): 22–40.

Priestley, M.J.N., Sritharan, S., Conley, J.R. and Pampanin, S. (1999) 'Preliminary results and conclusions from the PRESSS five-storey precast concrete test building', *PCI Journal*, 44(6): 42–67.

Santarella, L. (1957) *Il cemento armato*, Vol.2, Ed. U. Hoepli, Milan (in Italian).

SEAOC Vision 2000 Committee (1995) *Performance-Based Seismic Engineering*, Sacramento, CA: Structural Engineers Association of California.

Stanton, J.F., Stone, W.C. and Cheok, G.S. (1997) 'A hybrid reinforced precast frame for seismic regions', *PCI Journal*, 42(2): 20–32.

Chapter 7

Residential properties in England and Wales

An evaluation of repair strategies towards attaining flood resilience

Robby Soetanto, David Proverbs, Jessica Lamond and Victor Samwinga

Introduction

Flooding is a global and somewhat inevitable phenomenon, which currently impacts in excess of 520 million people per year worldwide (United Nations University 2004). Some estimates show that the number of people living at risk of devastating floods worldwide is set to double from one billion in 2004 to two billion by 2050 unless more attention is paid to prevention and prediction (United Nations University 2004). In England and Wales, two million homes are in risk areas, although many of these areas are managed adequately. However, 570,000 homes and businesses face a high risk of flooding, twice as many as 2002 estimates (ABI 2006). These figures could rise further if climate change results in more frequent extreme weather events or if properties continue to be constructed on floodplains.

Recent developments in planning legislation (e.g. Planning Policy Statement 25) now require developers to undertake a flood risk assessment if proposing to build on or near to the floodplain and to incorporate appropriate measures to avert or minimise the risk of flooding within the planning application. While this has restricted development in flood risk zones, there remain a considerable number of properties in England and Wales that are exposed to various levels of flood risk. Local Authorities on the other hand are confronted with the political and social pressures of providing new and affordable housing on a scale not experienced since the end of the Second World War. Combined with constantly changing weather patterns and extreme weather events, new and existing properties face the prospects of increased exposure to flood events such as those in Carlisle and large areas of the North East of England (e.g. Sheffield) during the summer of 2007.

In the UK, most standard home insurance provides cover for the repair and reinstatement of damage caused by flooding. Hence, in the event of an inundation, homeowners will turn to their insurance company to assist in dealing with the immediate aftermath and in managing the reinstatement process. The insurance policy normally provides for the reinstatement of the property to (or as near as is practically possible to) its pre-incident condition. As such, insurers have been reluctant to reinstate property with flood-resilient measures for fear of 'betterment'. However, the Association of British Insurers (ABI) is now actively promoting the uptake of such resilient measures during the repair process which must be seen as an appropriate development in the insurance regime. The adoption of resilient measures should reduce the damage (and therefore costs) caused by subsequent flood events, thereby reducing the impact on property owners and lowering the financial risks carried by insurance companies, for the benefit of the wider community.

This chapter utilises data from previous research studies to investigate the extent to which resilient techniques are currently being employed in the repair of flood-damaged property in England and Wales. It discusses the wider implications of the effect of flooding on property values and seeks to offer some conclusions regarding the long-term benefits derived from adopting flood-resilient reinstatement techniques and methods.

The repair of flood-damaged property

The appropriate repair and reinstatement of flood-damaged properties holds the key to incorporating resilient features to better cope with subsequent flooding for properties built on a flood-plain. Making properties flood resilient during the repair process represents a viable and cost-effective strategy (ABI 2004). Following a serious flood event and on being notified by the (insured) homeowner, insurance companies arrange for a qualified loss adjuster, building surveyor and/or damage reinstatement company to visit the affected property. These professionals must determine the extent of damage caused, decide on an appropriate method of drying and agree the scope of works involved in the repair and reinstatement of the property. This can be a complicated process dependent on many factors including the pre-incident condition of the house, the characteristics of the flood (e.g. depth, source, contaminant content) and the characteristics of the property (e.g. construction materials, construction form, age).

As there is little guidance to these professionals when undertaking such scoping works and in determining repair strategies for individual elements within the property, the assessment process is characterised by a high degree of subjectivity, relying on the experience and preferences of the individuals, which ultimately might trigger conflict amongst interested parties (e.g. professionals and homeowners). Recent years have seen the publication of guidance for the repair of flooded domestic properties, such as DTLR (2002), Bowker (2002), Proverbs and Soetanto (2004), Garvin *et al.* (2005), Flood Repairs Forum (2006), and Bowker *et al.* (2007). Apart from helping to 'standardise' the repair work undertaken and removing some of the subjectivity involved in assessment of flood damage, they each recommend the need to incorporate resilient features to reduce damage from future flooding. The recommendation of appropriate repair works in the anticipation of future flooding may also involve redesigning and adapting the property, and possibly changing the behaviour of its occupants (e.g. moving a ground floor kitchen to the first floor).

Given the complexity of repair processes and inter-relationships between stakeholders involved, there is, surprisingly, very scant information on the barriers of incorporating resilient features in flood-damaged properties. Using data from the 2002 flood in Saxony, Germany, Kreibich et al. (2005) investigated the effectiveness of 'precautionary measures' to reduce flood damage by evaluating the costs of incorporating resilient measures against more traditional techniques of reinstatement. However, this study failed to elaborate the impact of incorporating these measures on outturn performance in the repair process. Crichton (2005b) has highlighted increased cost as a significant constraint to the adoption of flood-resilient reinstatement, as there remains some dispute as to which parties should bear the additional costs incurred. Nevertheless, it remains wholly sensible to enhance the resilience of property during the repair process in order to lower costs of reinstatement work following future flooding, as highlighted by Garvin et al. (2005). The ABI (2003, 2004) provides indicative costs of installing flood-resilient measures compared with standard repair techniques, but does not take into account other implications such as the time required to carry out such resilient repairs. Indeed, the ABI and the National Flood Forum (ABI and NFF 2006) have encouraged the uptake of flood resilience by providing a framework for stakeholders to communicate and negotiate terms and to identify other implications of this.

Preliminary evidence of the implications arising from the adoption of flood-resilient repair strategies in terms of cost, time, quality, satisfaction, and overall performance is presented in this chapter. The aim is towards developing an increased understanding of such implications and as a result, further uptake of resilient reinstatement in the future.

Previous research undertaken by two of the authors investigated repair methods recommended by damage-management assessors for flood-damaged properties (Proverbs and Soetanto 2004). This earlier research explored 'present' and 'ideal' repair strategies and examined their performance in terms of perceived cost, quality, time and expected client satisfaction. Here, 'present' repair strategies are remedial solutions which are currently used in practice by damage-management professionals. These 'present' repair strategies are influenced by external factors such as the need to consider time required to undertake repair work or indeed the overall costs of the work. Therefore, the research sought 'ideal' repair strategies which were recommended after the removal of such constraints (such as time). Hence, these 'present' and 'ideal' strategies were found to differ.

A comparison of the perceived performance of present and ideal strategies is presented in this chapter. These ideal strategies were also compared with 'resilient' strategies derived from DTLR (2002), Bowker (2002), Garvin *et al.* (2005), Flood Repairs Forum (2006) and Bowker *et al.* (2007). The following section describes the research methodology employed as part of this study.

Methodology

In an earlier investigation of repair standards by two of the authors (Proverbs and Soetanto 2004), questionnaires were distributed to 1,800 chartered surveyors (identified from the Royal Institution of Chartered Surveyors database) and over 5,000 members of the Chartered Institute of Loss Adjusters. Overall 289 questionnaires were completed, representing a 4.2 per cent response rate. It was evident from the survey that many of those targeted possessed no experience of dealing with flood-damaged properties and were therefore unable to respond. The respondents (60 per cent of whom were loss adjusters and 23 per cent chartered surveyors) possessed a wide range of experience and were deemed as flood-damage assessors for the purpose of the study. The sample was dominated by respondents working throughout England and Wales.

The survey instrument incorporated the use of digital photos (i.e. scenarios) taken from real footage of flood-damaged properties (refer to example in Figure 7.1). Respondents were asked to indicate (i.e. tick) their present and ideal repair strategies for each flood-damage scenario (i.e. each digital image) by choosing from a list of commonly employed strategies identified from the literature and in consultation with damage-management experts, insurers and loss adjusters. Refer to Table 7.1 for repair strategies identified for quarry tiled floors.

Respondents were then asked to assess the perceived performance of both their present and ideal strategies in terms of (i) cost incurred, (ii) quality, (iii) time and (iv) expected percentage of satisfied clients (called 'satisfaction' throughout this chapter for the purpose of brevity), on a five-point scale ranging from 1 to 5 (refer to Table 7.2). Additionally, an overall performance measure was derived from the average of these performance criteria allowing a comparison of overall performance between the strategies. Overall, 31 scenarios were presented, including 6 flood-damaged floors, 11 flood-damaged walls, 4 flood-damaged doors and windows, and 10 flood-damaged utilities.

Statistical tests were employed to investigate whether performance differences between strategies were significant. In this research, the average performance of present and ideal strategies were compared using the Mann–Whitney test. If the difference was significant at 5 per cent levels, it could be concluded that the difference could also be found in the population of

Figure 7.1   Scenario: 'The dwelling has a quarry tiled floor which has been submerged by floodwater' (courtesy of Rameses Associates Ltd)

Table 7.1   List of repair strategies for scenario presented in Figure 7.1

Option	Repair strategy
1	Recommend replacement of floor tiles
2	Recommend replacement of floor tiles if they have been damaged by the floodwater
3	Recommend the floor tiles be carefully removed and cleaned and then relaid
4	Recommend the floor tiles be cleaned in place
5	Recommend the floor tiles be cleaned in place and replaced only where damaged by the floodwater

flood-damage assessors with some (95 per cent) level of confidence. The comparison allowed understanding of the respondents' intention to opt for another strategy (i.e. their ideal strategy). Tables 7.3 to 7.6 present flood-damaged scenarios, together with present, ideal and resilient strategies. The first columns of the tables present descriptions of the flood-damage scenarios. The second, third and fourth columns provide the present, ideal and resilient strategies respectively. The resilient strategies includes due consideration of the risk of future floods (based on Garvin *et al.* 2005);

Table 7.2 Likert scales used for the performance criteria

Level	Performance criteria			
	Cost of repair works	Appropriateness of the repair	Time to conduct the repairs	Expected % of satisfied clients
I	very expensive	very low	very slow	0 to 20
2	expensive	low	slow	21 to 40
3	economically acceptable	acceptable	acceptable speed	41 to 60
4	inexpensive	good	fast	61 to 80
5	very inexpensive	very good	very fast	81 to 100

standard of repair Level A should be used if there is little or no risk of future flooding and therefore is not considered in the following discussion. Level B, should be used if the likelihood of future flood is low to medium (i.e. flood risk is sufficiently high to recommend repairs to increase the resilience of the property above the original specification); and standard of repair Level C, if the risk of future flooding is high (i.e. to recommend repairs that increase the resilience of the property significantly). Table 7.7 provides a summary of the statistically significant differences in perceived cost incurred, quality of repair, time needed, expected client satisfaction and overall performance, between present and ideal strategies for each scenario.

Present, ideal and resilient repair strategies

Present, ideal and resilient strategies for flood-damaged floors, walls, doors and windows, and utilities are discussed in the following sections.

Floors

Six scenarios were considered for different type of floor. For vinyl floor tiles, the present approach was found to be recommending replacement of unbonded floor tiles, compared with the ideal method of replacing all floor tiles. Vinyl is generally water-resistant, but the substrate should be suitably dry before the tiles are put back in place. Difficulties in replacing unbonded floor tiles include finding a correct colour match between new and existing tiles, and the complete drying of the floor substrate (Crichton 2005b). After the flood has subsided, water may be trapped underneath the bonded tiles, and be absorbed by the substrate. Thus, the complete drying of the substrate and replacement of all vinyl floor tiles is probably the most practical solution. The same situation may be encountered when repairing quarry tiled floors. Although quarry tiles do not deteriorate in a flood, problems of stains, matching colour and trapped water make isolated repair quite problematic.

Table 7.3 Present, ideal and resilient repair strategies for flood-damaged floors

Flood damage scenario	Present repair strategy	Ideal repair strategy	Resilient strategy[1]
1. 'The dwelling has vinyl floor tiles installed that have been submerged by floodwater'	Replacement of floor tiles that have become unbonded	Replacement of all floor tiles	Vinyl is generally water-resistant, but the substrate should be dry before the tiles are put in place
2. 'The dwelling has a quarry tiled floor which has been submerged by floodwater'	Replacement of floor tiles if they have been damaged by the floodwater	Replacement of floor tiles	These tiles usually do not deteriorate in a flood, but stains due to floodwater are difficult to remove. Use a full bedding of tile adhesives (and water-resistant grout) to fix the tiles to the (dry) substrate
3. 'The dwelling has a solid concrete floor which has been submerged by floodwater'	The floor is cleaned and allowed to dry	The floor screed be removed, the floor allowed to dry and then the screed replaced	Replace screeds using cement-rich screeds for flood resilience, although drying could take a long time (B)
4. 'The dwelling has a suspended timber (chipboard) floor which has been submerged by floodwater'	Replacement of chipboard	Removal and replacement of all timber components	Replace all chipboard and damaged timber components (with preservative-treated timber joists and floorboards) (B) or Replace suspended floors with solid floors (C)
5. 'When the floorboards are removed, it is discovered that the sleeper walls are constructed directly off the ground (i.e. no concrete slab has been included)'	The sleeper walls are left alone	A damp-proof course layer is installed into the present sleeper walls	A damp-proof course layer needs to be installed
6. 'The dwelling has a concrete floor which has been covered with solid oak blocks'	Replace sections of blocks that have become unfixed from concrete floor	Replace all floor covering (i.e. the oak blocks)	The blocks could be retained if they have not suffered deterioration, warping or cracking, although best consider replacement with more resilient materials (such as tiles)

Note:
[1] B and C (in brackets) denote two standards of repair based on the risk assessment of future floods (defined by Garvin et al. 2005): Standard of repair Level B, if the likelihood of future flood is low to medium (i.e. flood risk is sufficiently high to recommend repairs to increase the resilience of the property above the original specification); and Standard of repair Level C, if the risk of future flooding is high (i.e. to recommend repairs that increase the resilience of the property significantly).

Table 7.4 Present, ideal and resilient repair strategies for flood-damaged walls

Flood damage scenario	Present repair strategy	Ideal repair strategy	Resilient strategy[1]
7. 'The external wall of the property is brickwork with cement mortar joints'	Clean and repoint the wall	The walls be sandblasted to remove any flood debris	Repoint mortar joints; if technically possible and no aesthetic objections, render the external wall, or apply water-resistant paints and coatings (or tanking), or alternatively use flood protection products, such as flood protection skirt (C)
8. 'The external wall of the property has a rendered finish'	Areas of the render that have become unbonded from the wall substrate be replaced	All the render be removed and replaced	Apply a propriety render finish (e.g. polymer-modified system) to reduce water penetration (C)
9. 'The external wall of the property has a pebbledash finish'	Areas of the pebbledash render that have become unbonded from the wall substrate be replaced	All the pebbledash render be removed and replaced	No specific guidance, but apply impermeable render mix (as above)
10. 'An internal wall of the flood-damaged property is constructed of brickwork with a paint finish applied directly to it'	The wall be cleaned and repainted	The wall be cleaned, plastered and decorated	Apply lime-based plaster (or wet-applied plaster) or tiles (C)
11. 'An internal wall of the flood-damaged property has been covered with ceramic tiles'	Only 'loose' tiles be replaced	Replace all tiles	Ensure that wall substrate is completely dry before replacing the tiles, and a good coverage of waterproof tile adhesive on the wall and use water resistant grout (C)
12. 'An internal wall of the flood-damaged property has been covered with a wood veneer on timber grounds'	The veneer in contact with floodwater be replaced	Replace the wood veneer	Replace damaged veneer with treated timber (B). Consider using more resilient materials, such as cement or lime-based plaster, or even tiles (C)
13. 'An external wall of a flood-damaged property has evidence of a rising damp problem'	The client be approached to pay for curing the rising damp problem and the plaster be replaced	Inject the wall with a dpc and replace the plaster	Use of chemical injection dpc (B), and replace the plaster with cement or lime-based (C)

continued…

Table 7.4 continued

Flood damage scenario	Present repair strategy	Ideal repair strategy	Resilient strategy[1]
14. 'Following removal of the wall's plaster, it is found that the wall is incorrectly constructed'	The client be approached and asked to pay for the wall to be correctly constructed	The wall be demolished and reconstructed	Reconstruct the wall with strong engineering brick (C)
15. 'Floodwater has been in contact with an internal block wall which has a gypsum plaster finish'	The plaster be replaced up to the floodwater line (or 15–30 cm above)	Replace all the wall's plaster	Replace the plaster with resilient plaster, such as cement or lime-based (B), or consider using tiles (C)
16. 'Floodwater has been in contact with an internal brick wall which is finished with a lime/ox-hair mix and a lime putty finish'	The plaster be replaced up to floodwater line (or 15–30 cm above)	Replace all the wall's plaster	The existing plaster will provide good resilience; replace of damaged parts with the same (or even better mix or tiles) (C)
17. 'Floodwater has been in contact with an internal timber partition wall'	Replace plasterboard that has been in contact with floodwater	Replace the timber components and the plasterboard	Replace damaged timbers with treated timber; mineral wool insulation with closed-cell type insulation; plasterboard with cement or lime-based plaster and plasterboard, seal the junctions between walls/ partitions and floors using good-quality sealants (B). Strengthen timber frame to resist lateral force and also consider flood protection product on the outside of the building to prevent water from entering (C)

Note:
[1] See note in Table 7.3.

Table 7.5 Present, ideal and resilient repair strategies for flood-damaged doors and windows

Flood damage scenario	Present repair strategy	Ideal repair strategy	Resilient strategy[1]
18. 'A flood-damaged property has a softwood wooden front door that has been in contact with floodwater'	Allow the door to dry out and then assess the damage	Replace the door	Replace the door with hardwood, seal: door frame into building, door into its frame. Maintain a good surface finish (e.g. water-borne paints), replace damaged or corroded hardware with non-corrosive components (B). Consider use of sealed PVC door and/or demountable flood protection (C)
19. 'A flood-damaged property has double-glazed hardwood patio doors that have been in contact with floodwater'	Allow the door to dry out and then assess the damage	Replace the door	Assess the timber components, seal: door frame into building, door into its frame. Maintain a good surface finish (e.g. water-borne paints), replace damaged or corroded hardware with non-corrosive components. Use glazing sealants and gaskets, also sealant capping for external glazing seal (B), replace with toughened glass (B). Consider use of sealed PVC door and/or demountable flood protection (C)
20. 'A flood-damaged property has hollow cellular type infill wooden doors that have been in contact with floodwater'	Allow the door to dry out and then assess the damage	Replace the door	Replace the doors with resistant types, e.g. solid timber doors. Finish the doors properly with a high-built paint system (water-resistant paint). Paint the doors before hanging so that the sides and bottom are fully covered
21. 'A flood-damaged property has wooden window frames that have been in contact with floodwater'	Allow the windows to dry out and then assess the damage	Replace the windows	Assess the timber components, seal: window frame into building, window into its frame. Maintain a good surface finish (e.g. water-borne paints), replace damaged or corroded hardware with non-corrosive components. Use glazing sealants and gaskets, also sealant capping for external glazing seal, if broken, replace with toughened glass (B). Consider use of sealed PVC window (C)

Note:
[1] See note in Table 7.3.

Table 7.6 Present, ideal and resilient repair strategies for flood-damaged utilities

Flood damage scenario	Present repair strategy	Ideal repair strategy	Resilient strategy¹
22. 'A flood-damaged property has steel panel radiators installed that have been in contact with floodwater'	Replace the radiator valves	Replace the radiator and valves	Qualified engineers (e.g. CORGI certified) to inspect the radiators and valves before re-used; floodwater ingress should be resisted by using e.g. polyethylene sheeting to seal them fully (B)
23. 'A flood-damaged property has a gas fired heater that has been in contact with floodwater'	The heater be serviced	The heater be replaced	Qualified engineers (e.g. CORGI certified) to inspect the heater before re-used; gas fittings should be checked for leaks (B)
24. 'A flood-damaged property has a gas meter that has been in contact with floodwater'	The meter be checked for leaks	The meter be replaced	Qualified engineers (e.g. CORGI certified) to inspect the gas meter before re-used; move gas meters to at least 1 m above floor level or above the expected flood level (B)
25. 'A flood-damaged property has a gas wall-hung fire that has been in contact with floodwater'	The fire be serviced	The fire be replaced	Qualified engineers (e.g. CORGI certified) to inspect the fire before re-used; gas fittings should be checked for leaks (B)
26. 'The dwelling has an electrical circuit containing sockets and cables which have been partly submerged by floodwater'	Electrical circuit be checked by an electrician and any faults rectified	Completely replace this installation	Replace this installation and move to a higher level in the structure so that cables drop from first-floor level down to sockets (B)
27. 'The dwelling has an electrical wall-hung heater that has been submerged by floodwater'	The heater be checked by an electrician	The heater be replaced	Qualified engineers (e.g. CORGI certified) to inspect the heater before re-used; move at least 1 m above floor level, depending on the predicted flood depth (B)
28. 'The dwelling has timber skirting boards'	Replace skirting boards that are damaged by floodwater	Replace all skirting boards	Replace skirting boards with more resilient materials, such as ceramic tiles and PVC
29. 'The dwelling has a staircase constructed from timber'	The stairs be allowed to dry and then assess any damage caused	Complete replacement of the staircase	Use timber staircase of solid timber construction; prevent water infiltration through careful maintenance (e.g. painting); check stability and replace any loose treads

Flood damage scenario	Present repair strategy	Ideal repair strategy	Resilient strategy[1]
30. 'The dwelling has built-in wall cupboards'	The cupboards be allowed to dry and then assess any damage caused	Completely replace the cupboards	Consider repositioning cupboards above predicted future flood level, e.g. build off floor with plastic legs concealed behind a removable plinth, and use more resilient materials, such as PVC
31. 'The dwelling has a 'fitted' kitchen that has been partially submerged above the plinths by floodwater'	Replacement of those units that have been in contact with floodwater	Completely replace the kitchen	Replace the kitchen units with proprietary plastic or water-resistant alternatives (PVC) and build off floor (B), or consider moving kitchen to first-floor level (C)

Note:
[1] See note in Table 7.3.

Table 7.7 Summary of statistically significant differences in perceived cost incurred (C), quality of repair (Q), time needed (T), expected client satisfaction (S), and overall (O) performance between present and ideal repair strategies

Flood damage scenario[1]	No. data	Significant differences[2]
1	30	C,Q,S,O
2	12	C,Q,S
3	45	C,Q,T,S
4	19	C,Q,S
5	18	C,Q,T,S
6	22	C,Q,T,S
7	11	C,Q,T
8	16	C,Q,S,O
9	16	C,T,S
10	15	C,Q,S,O
11	54	C,Q,T,S,O
12	23	C,Q,S,O
13	44	C,Q,S,O
14	37	C,Q,S,O
15	13	C,Q,S
16	18	C,Q,S
17	15	C,Q,S
18	54	C,Q,T,S,O
19	28	C,Q,T,S,O
20	16	C,Q,T,S,O
21	24	C,Q,T,S,O
22	11	C,Q,S,O
23	38	C,Q,S,O
24	19	C,Q,S,O
25	34	C,Q,T,S,O
26	14	C,S
27	29	C,Q,T,S,O
28	11	C,Q,S
29	15	C,S
30	31	C,Q,S,O
31	39	C,Q,S,O

Notes
1 Refer to Tables 7.3–7.6 for descriptions of scenario.
2 Statistical significance of difference at 95% confidence level for cost incurred (C), quality (Q), time needed (T), expected client satisfaction (S).

Properly constructed solid concrete floors are particularly resilient to flooding (ABI 2004). For a low or medium risk (i.e. Level B standard of repair; Garvin *et al.* 2005), the screed should be replaced with cement-rich screeds which are impermeable, and the floor should be completely dry prior to the installation of finishes. Trapped floodwater underneath the existing screed could cause long-term instability and unpleasant odours (Crichton 2005b).

For damaged suspended timber floors, replacement of both chipboard and timber components with treated timber joists and floorboards will provide sufficient resilience for medium flood risk. Higher flood risk would necessitate the replacement of the floor with a solid concrete floor. Concrete floors covered with solid oak blocks are quite resilient to flooding, especially if the blocks are painted with water-resistant varnish. However, problems with trapped water and deterioration suggest replacement with more resilient materials, such as tiles. Otherwise, the oak blocks may need to be replaced after flooding (as an 'ideal' repair strategy). Generally, ideal strategies were found to be consistent with resilient strategies, indicating an appreciation of resilient techniques on behalf of damage-management assessors.

Walls

In the UK, walls are commonly constructed from porous materials, such as brick with mortar joints. Installing resilient strategies (in scenarios 7–9) would mean enhancing the impermeable property of these materials, which for external walls, would also improve the flood resistance of houses, by preventing and/or minimising water entering the property. The work could involve repointing mortar joints, rendering with impermeable mix, such as a polymer-modified system, applying water-resistant paints and coating/ tanking. However, before applying these measures, the external wall has to be checked for any damp problems (as in scenario 13). Chemical injection damp-proofing is recommended for this. Homeowners are normally asked to contribute to the high costs incurred with this (as in the 'present' strategy). For properties in high-risk zones, the installation of a flood-protection skirt should ensure better resistance, given adequate warning time and where floodwater is below 1 m.

Improving the resilience of internal walls (scenarios 10, 11, 12, 15, 17) can involve applying lime-based plaster or using tiles (if aesthetically desired). Tiles are resilient to floodwater, but repair work should ensure no trapped floodwater (for an existing tiled wall) and good coverage of waterproof tile adhesive incorporating water-resistant grout. The wood veneer of damaged internal walls (scenario 12) should be replaced with pre-treated timber or, if aesthetically desired, replaced with more resilient materials, such as cement or lime-based plaster, or tile, especially for properties with high flood risk. Gypsum plaster finish (scenario 15) is not resilient to floodwater, hence it should be replaced with cement or lime-based plaster. An internal brick wall with a lime/ox-hair mix and a lime putty finish (scenario 16) is quite common for older buildings and generally provides good resilience (Crichton 2005b). However, the repair of such a finish would require specialist tradesmen and hence increase the cost (Crichton 2005b). Alternatively, homeowners in high-risk flood areas might consider replacing this with more resilient materials, such as tiles (if aesthetically desired).

Several measures may be used to improve the resilience of internal timber partitioning (scenario 17). For properties with medium risk of flooding, this can include replacing damaged timbers with treated timber; mineral wool insulation with closed-cell type insulation; plasterboard with cement or lime-based plaster and plasterboard, and sealing the junctions between walls/partitions and floors using good quality sealants. In higher flood risk areas, additional measures can be strengthening the timber frame to resist lateral forces and the use of flood-protection products, such as flood skirts, to prevent water from entering the building. Here again, 'ideal' and resilient repair strategies are generally compatible (with the exception of scenario 7).

Doors and windows

The repair of flood-damaged doors and windows (scenarios 18–21) currently involve 'wait and see' strategies before an appropriate solution is recommended. Ideal strategies involve the replacement of doors and windows with flood-resilient components. Hardwood doors are resilient to floodwater, but their surfaces need to be maintained with water-resistant paints. Sealing the door/window will prevent or minimise water entering the property. Here, the use of demountable flood protection systems and 'sealed' PVC should be considered for properties with high risk of future flooding. Hollow PVC sections, as normally used in modern double-glazed windows and patio doors, can trap floodwater, which can be difficult to drain resulting in bad odours.

Utilities

For reinstatement of radiators, fires, heaters and gas meters (scenarios 22, 23, 24, 25, 27), qualified engineers, such as those who are CORGI certified, should be consulted before they are re-used (DTLR 2002). Gas fittings should be carefully checked for leaks. Water ingress should be prevented by fully sealing the joints of radiator systems. For properties with medium to high risk of flooding, electrical and gas appliances, meters and sockets should be moved to higher levels, beyond the reach of the predicted flood depth. Specifically, electrical circuits should be moved to the first floor, from where cables can be dropped down to sockets (scenario 26). Generally, present repair strategies tend to recommend checking and servicing the electrical and gas appliances, before complete replacement (as in ideal strategies). However, complete replacement would better assure the homeowner's safety, and allow appropriate resilient measures to be incorporated, at little or no extra cost.

In scenario 28, the present strategy recommends replacing skirting boards that are damaged, whereas the ideal strategy suggests replacing all skirting boards. Skirting boards normally need to be temporarily removed to allow the walls to dry out (DTLR 2002). If the skirting boards are not made from resilient materials, they may distort and need to be replaced. Fully painted solid timber skirting boards are quite resilient to floodwater (DTLR 2002), although the use of more resilient materials, such as ceramic tiles and PVC, is recommended for high-risk properties.

Staircases of solid timber construction are generally resilient to floodwater; water-resistant paint should also be applied to prevent water infiltration. After flooding, the stability of the staircase has to be checked and any loose treads should be replaced (DTLR 2002). Complete replacement of the staircase (as an ideal strategy) should be considered if the staircase was not made flood resilient, especially for properties at risk of future flooding.

Cupboards and kitchen units, especially those made from chipboard or MDF, need to be replaced following flooding. Solid hardwood units may be re-usable, but warping and discolouration of the surface finish may still result in replacement (DTLR 2002). A resilient approach would include the repositioning of the units above predicted future flood levels and replacement with resilient materials, such as PVC. For high-risk properties, the kitchen should also be moved to first-floor level.

Comparisons between present and ideal repair strategies

Figures 7.2 to 7.6 depict the comparisons of present and ideal repair strategies in terms of their cost, quality, time, expected client satisfaction, and overall performance.

Observation of the general pattern of the graphs suggests that there is a large gap between the perceived costs of present and ideal strategies with present strategies being less costly. This large gap is consistent with the satisfaction level achieved, with ideal strategies yielding improved client satisfaction. It is worth noting that clients are deemed reasonably satisfied with present strategies. Observation of the repair quality reveals the same tendency, with ideal strategies yielding higher quality, even though the gap is not as wide as that in the satisfaction measure.

The analysis of time for repair suggests a narrower distinction between present and ideal strategies, and some inconsistency in regard to which are speedier. Further, only 12 of the 31 scenarios were found to have a significant

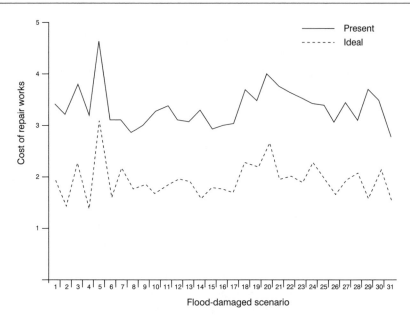

Figure 7.2 Comparison of cost incurred of present and ideal strategies

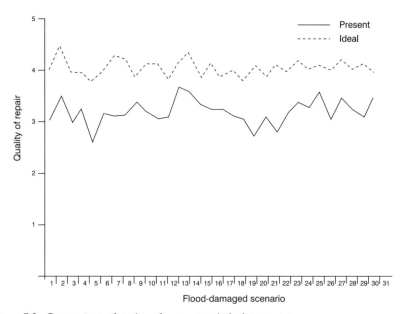

Figure 7.3 Comparison of quality of present and ideal strategies

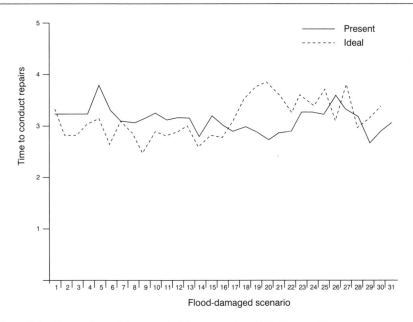

Figure 7.4 Comparison of time needed for present and ideal strategies

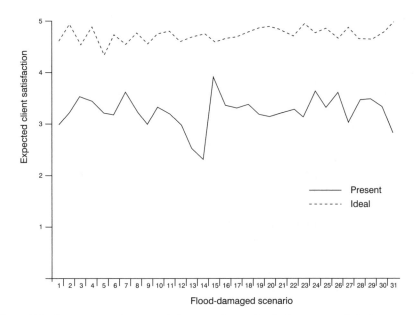

Figure 7.5 Comparison of expected client satisfaction level of present and ideal strategies

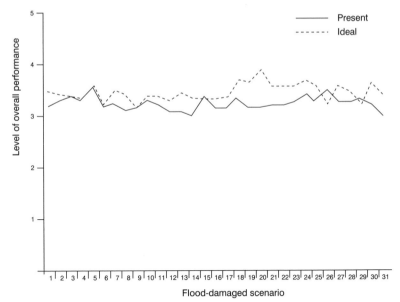

Figure 7.6   Comparison of overall performance of present and ideal strategies

difference in time performance for present and ideal repair strategies (see Table 7.7, column 3). Ideal strategies are considered quicker solutions for those involving replacement of modular packages such as doors, windows and radiators.

Analysis of overall performance (refer to Figure 7.6) reveals a narrow gap between present and ideal strategies, albeit ideal strategies generally result in improved performance. However, this marginal difference is attributed to the cost effectiveness of present strategies which were generally found to be less costly. This needs to be considered against the superior performance in quality and satisfaction measures as achieved by the ideal strategies (and more resilient strategies).

These results indicate some of the challenges faced by the damage-management sector when dealing with flood-damaged domestic properties. Mutual exclusivity was found to exist between the performance of repair strategies. The tendency is that repair methods which derive higher quality and/or satisfaction are perceived to be more costly and vice versa. These results show that resilient reinstatement is currently constrained by the perceived additional costs of the methods involved. While it is logical that the damage-management sector considers the costs of such resilient measures, in view of whole-life costs and making properties more resilient, such measures could help limit the damage caused by subsequent flood events and reduce the amount of time the occupiers have to vacate the property (ABI and NFF 2006). As the frequency and severity of floods are predicted to increase (OST 2004; ABI 2006, 2007) and due to the significant number of properties built on flood plains (Crichton 2005b), it is prudent that these stakeholders look towards solutions which enhance the long-term resilience of properties.

These findings reveal that perceived additional costs currently represent a barrier to flood-resilient reinstatement. However, the literature suggests little evidence of such increased costs and therefore more should be done to develop awareness and understanding among the damage-management sector. Organisations such as the Association of British Insurers, British Damage Management Association, Council of Mortgage Lenders and Royal Institution of Chartered Surveyors, should do more to promote this to their members. Furthermore, in consideration of whole-life costs and the sustainability of proposed repairs, initial cost should not be the overriding criterion, but considered alongside other factors such as maintenance, enhanced resilience, post-incident property value, and user-related criteria (such as minimum disruption due to built-in resilience).

For properties with a high risk of flooding, installing resilient measures may also impact on the sale value and cost of future insurance. This is further considered and discussed in the following section.

The long-term financial implications of flooding for domestic property owners

Short-term financial concerns for owners of property damaged by flooding centre on the costs of reinstatement but homeowners also worry about the potential devaluation of their property and the increased cost of future insurance cover (Watkins and Welsh Consumer Council 1992; EA/DEFRA 2005). In England and Wales, property ownership is the norm and the price of home purchase is one of the most significant financial commitments most individuals will make. The majority of home purchases will be financed via loans in the form of mortgages taken out against the value of the property. Worries about the security of that investment have the possibility of affecting the tendency to claim on insurance and the standard of repair deemed cost effective. For the mortgage lenders flood risk reduces the certainty that they will be able to recoup their money should the mortgagee run into financial difficulty.

As will be described below, most property owners can have confidence that value will be largely maintained in the long term and there is some evidence that the reinstatement process can contribute to ensuring the maintenance of value. For the owner and mortgage lender therefore reinstatement to a good quality ensures the value of their investment and makes sound financial sense. This is particularly the case because the costs are largely borne by insurers. Insurers of the property on the other hand are concerned about repeated flood claims removing the profitability of their domestic insurance portfolio and in the light of anticipated increases in flooding and flood claims must increase premiums to match. This could in itself reduce the value of property in the floodplain because purchasers would regard this extra cost of insurance as a disincentive to purchase. The estimation of the actual increases in insurance cost can be highly problematic as competition between insurers ensures that patterns of increased costs are difficult to discern.

Previous studies of flooding impact on property value

Studies of property value impacts have been carried out in the US, Canada, New Zealand and the UK. The different motivations behind previous studies demonstrate the wide importance of property value, not just in England and Wales but internationally. Apart from the natural concerns of direct stakeholders outlined above, property value is used to inform investment decisions of property developers and building professionals. Real estate valuers have the need to make good estimates of the value of floodplain property (BFRG 2004) and policy makers will require information about the financial implications of mitigation strategies (Chao *et al.* 1998). When interpreting the results from the research to date, and particularly its relevance to England and Wales, it must be borne in mind that different disclosure regimes exist across the world and this seems to be partially reflected in different observed impacts. Potential purchasers may become aware of flood risk status at various stages of the buying process. Analysis of these studies is contained in Chao *et al.* (1998) and in Lamond *et al.* (2005). Some general lessons can be drawn from the literature.

Typically, flooded property retains most of its value once it has been reinstated. The largest average impact observed was 30 per cent, measured by Tobin and Montz (1994), in the immediate aftermath of an event in Linda and Olivehurst, California, and on property flooded to a depth of greater than 10 feet. The majority of studies found average impacts below 15 per cent on transacted property. In some cases, prices were observed to increase after flooding and reinstatement in an area with a low expectation of future flooding, perhaps due to reinstatement resulting in improved building condition (Tobin and Montz 1994).

Individual properties may suffer large discounts, particularly if reinstatement is inadequate or the risk of return of flooding is very high. In rare cases, a property will become almost unsaleable due to financing issues. Examples of this were quoted by the Building Flood Research Group (BFRG 2004), but these comprised a small minority of property transactions. However, for parties with a financial interest in a property, even a small discount can still be highly problematic and leave property owners with mortgage debts exceeding the market value of the property. This could be a particularly acute problem if this meant that restoration work could not be achieved via extending borrowing against the property.

As time elapses after a flood event, the memory of the event fades. The tendency of the community to forget about flood risk can be reflected in the value impact of flood as examples in the literature demonstrate. BFRG (2004) in their questioning of surveyors enquired about the length of time for the recovery in value of a flooded property. There was very little consistency in the responses, with some suggesting under a year and others anticipating longer than an eight-year impact. For one example, in Barlby in England, which is at moderate risk of flood and was subject to flooding during the 2000 event, property price recovery was observed within 18 months (Lamond and Proverbs 2006). This corresponded to an Australian study by Lambley and Cordery (1997). Tobin and Montz (1994) however have studied multiple flood sites and observed different rates of recovery. In one example, Linda and Olivehurst in California, the most severely affected properties had not recovered completely after 10 years. It is interesting to note that in this instance some houses had not been reinstated and served as a visual reminder of the flood.

Theoretical framework for predicting future property value

In studying the impact of flooding in England and Wales, it is imperative to consider a temporal dimension. Theoretical utility profiles are presented in the literature (Tobin and Montz 1994; Lamond and Proverbs 2006). Measurement methodology must take account of this temporal variability as suggested by Eves (2004) and Lamond et al. (2007). Transactions in property tend to cease for about six months following a flood event (Lamond et al. 2007), approximately the average time to reinstatement for flooded property, suggesting that the majority of property owners will not consider selling in the immediate aftermath of a flood without reinstatement.

The impact of insurance cost and availability

Many authors have speculated about the impact of insurance cost on saleability (Clark et al. 2002; Eves 2004; BFRG 2004; Wordsworth and Bithell 2004; Crichton 2005a). In the US, studies have concluded that the cost of insurance can affect sales prices (Shilling et al. 1989; Chao et al. 1998). In England and Wales however, flood insurance is unusual in being packaged with general household insurance. It is therefore not a straightforward matter to disentangle the element of premium attributable to flood risk. In the past, insurers and insured alike have tended to ignore risk during premium setting. Recent events are starting to affect the views of insurers (Crichton 2005a; Lamond et al. 2006) but their policies on pricing are not uniform and so the analysis of impacts is complex. A recent survey of four sites at risk of flooding in England into the effect of flood risk and flooding on the price and availability of insurance, has revealed that neighbours with similar flood risk and flood history can be paying vastly different premiums depending on their insurer (Lamond and Proverbs 2007). At present it seems insurance is available for the majority of floodplain residents at a reasonable cost but this may change in the future as insurers' information about risk improves.

The impact of reinstatement on property value and on cost of insurance

Current research provides weak evidence that the standard of reinstatement can impact on future property value (Tobin and Montz 1994). In most studies, the condition of property prior to and after a flood has not been considered. However, insight into the magnitude of the potential effect of property condition is provided by Bin and Polasky (2003) who included a property condition term in their study of the impact of floodplain location on property prices in California and measured a positive effect of 'good condition' which was larger than the negative impact of floodplain location. Similarly, the impact on resale price of individual properties installing resistant or resilient measures has hardly been tested. A single piece of evidence from Bin *et al.* (2006) showed that coastal property in North Carolina constructed after the flood mapping, therefore likely to be elevated or otherwise protected, was valued no differently from the rest of the market at risk.

Initial indications from a recent survey of flood-risk households suggest that the incidence of resilient and resistant installation in four flood-risk locations in England is very small and that insurers are not acknowledging these installations with premium reductions (Lamond and Proverbs 2007). The low incidence of resilient adaptation makes the statistical measurement of any impacts on insurance cost and property value problematic. However, residents of resilient properties will make fewer or smaller insurance claims on average (ABI 2003) and this will in the long term affect their negotiating position with their insurers (Crichton 2005a). There was evidence in the comments provided by respondents to the insurance survey (Lamond and Proverbs 2007) that some individuals believe that failure to claim helps to keep premiums low.

This is an area where future research could have a real impact on increasing the motivation for resilient reinstatement and thus reduce future flood-damage costs. While the impact of resilient reinstatement on property value will be left up to the market, the inclusion of flood risk assessment and resilience gradings in standard property searches would provide the necessary information for the market to function efficiently in this matter. Insurers have a more direct opportunity to affect the cost benefit of resilient reinstatement for their policyholders via premium discounts. It is a further and even more thorny issue to decide who should pay for mitigation measures for privately owned property. In England and Wales, historically many purchasers were unaware of the flood risk status of their property at the time of acquisition, and it is still possible to buy in ignorance of flood status today. Penalising these homeowners by requiring them to finance flood mitigation seems inequitable and may be beyond the financial capacity of many. Alternatively, in a free and competitive insurance market, insurers will be reluctant to fund mitigation measures which may then be transferred with the policyholder to a competitor's long-term advantage. Full or partial government grants may provide the answer but for timely implementation would require advance agreement on best practice resilience and resistance measures. Despite significant recent debate around this issue there is not a consensus about the relative cost effectiveness of resilience and resistance measures. Further work in the area of the financial justification of resistant and resilient measures is needed.

Conclusions

It has been found that ideal strategies as perceived by flood-damage assessors are comparable to resilient strategies as identified in the literature. This suggests that professionals are appreciative of the need to incorporate resilient measures into the repair of flood-damaged property.

Further, the findings indicate that the perceived additional costs, on the behalf of these professionals, represents a barrier towards the up-take of resilient reinstatement. Recent survey findings indicate limited implementation of resilient reinstatement in flood-affected areas. This is further exacerbated by scant evidence of long-term financial benefits of installing resilient measures and lack of acknowledgment of these installations with premium reductions. Hence, it is difficult to state how this might affect property value and cost/availability of insurance. Organisations representing the professionals involved in flood repair should do more to promote awareness of the benefits and associated costs of flood-resilient reinstatement. Further, professionals should be prepared to consider the long-term analysis (i.e. whole-life costs), including a full range of factors (both tangible and intangible). Importantly, all stakeholders in the reinstatement process should embrace longer-term orientation and attitude, and be prepared to act together for their mutual benefit.

References

ABI (2003) *Assessment of the cost and effect on future claims of installing flood damage resistant measures*. London: Association of British Insurers.

ABI (2004) *Flood resilient homes: what homeowners can do to reduce flood damage*. London: Association of British Insurers.

ABI (2006) *A future for the floodplains*. London: Association of British Insurers.

ABI (2007) *Adapting to our changing climate: a manifesto for business, government and the public*. London: Association of British Insurers.

ABI and NFF (2006) *Repairing your home or business after a flood – how to limit damage and disruption in the future*. Pamphlet prepared jointly by the Association of British Insurers (London) and the National Flood Forum (Bewdley, Worcestershire).

Bin, O. and Polasky, S. (2003) Effects of flood hazards on property values: evidence before and after hurricane Floyd. *Land Economics*, 80: 490–500.

Bin, O., Kruse, J.B. and Landry, C.E. (2006) Flood hazards, insurance rates and amenities: evidence from the coastal housing market. ECU Economics Electronic Working Paper, October 2006. Greenville: East Carolina University.

Bowker, P. (2002) Making properties more resistant to floods. *Proceedings of the Institution of Civil Engineers, Municipal Engineer*, 151(3): 197–205.

Bowker, P., Escarameia, M. and Tagg, A. (2007) *Improving the flood performance of new buildings: flood resilient construction*. Guidance produced by a consortium of CIRIA, HR Wallingford Ltd, Leeds Metropolitan University, WRc and Waterman Group. London: RIBA Publishing.

BFRG (Building Flood Research Group) (2004) *The impact of flooding on residential property values*. RICS foundation report, London: RICS.

Chao, P.T., Floyd, J.L. and Holliday, W. (1998) *Empirical studies of the effect of flood risk on housing prices*. IWR report, United Stated Army Corps of Engineers.

Clark, M.J., Priest, S.J., Treby, E.J. and Crichton, D. (2002) *Insurance and UK floods, a strategic reassessment: report of the tsunami project*. Universities of Southampton, Bournemouth and Middlesex, UK.

Crichton, D. (2005a) *Flood risk and insurance in England and Wales: are there lessons to be learned from Scotland?* Technical report, London: Benfield Hazard Research Centre, University College London.

Crichton, D. (2005b) Toward an integrated approach to managing flood damage. *Building Research and Information*, 33(3): 293–9.

DTLR (2002) *Preparing for floods: interim guidance for improving the flood resistance of domestic and small business properties*. London: Department for Transport, Local Government and the Regions, now part of the Office of the Deputy Prime Minister.

EA/DEFRA (2005) *The appraisal of human related intangible impacts of flooding*. Technical report. London: Environment Agency / Department of Food and Rural Affairs.

Eves, C. (2004) The impact of flooding on residential property buyer behaviour: an England and Australian comparison of flood affected property. *Structural Survey*, 22: 84–94.

Flood Repairs Forum (2006) Repairing flooded buildings: an insurance industry guide to investigation and repair. Watford: BRE Press.

Garvin, S., Reid, J. and Scott, M. (2005) *Standards for the repair of buildings following flooding*. CIRIA publication C623, London: CIRIA.

Kreibich, H., Thieken, A.H., Merz, B. and Muller, M. (2005) Precautionary measures reduce flood losses of households and companies – insights from the 2002 flood in Saxony, Germany. In J. van Alphen, E. van Beek and M. Taal (eds) *Floods, from Defence to Management*. London: Taylor & Francis Group.

Lambley, D. and Cordery, I. (1997) The effects of catastrophic flooding at Nyngan and some implications for emergency management. *Australian Journal of Emergency Management*, 12: 5–9.

Lamond, J. and Proverbs, D. (2006) Does the price impact of flooding fade away? *Structural Survey*, 24: 363–77.

Lamond, J. and Proverbs, D. (2007) *Measuring the long term financial impact of flooding on homeowners: data, issues and opportunities*. European Geosciences Union General Assembly, 15–20 April, Vienna.

Lamond, J., Proverbs, D. and Antwi, A. (2005) The effect of floods and floodplain designation on the value of property: an analysis of past studies. *2nd Probe Conference*, 16–17 November, Glasgow.

Lamond, J., Proverbs, D. and Antwi, A. (2007) Measuring the impact of flooding on UK house prices: a new framework for small sample problems. *Property Management*, 25(4), 344–59.

Lamond, J., Proverbs, D., Antwi, A. and Gameson, R. (2006) Flood insurance in the UK: how does the changing market affect the availability and cost to homeowners. *Building Education and Research Conference*, 10–13 April, Hong Kong.

OST (2004) Future flooding: executive summary. Report produced by Flood and Coastal Defence project as part of Foresight Programme. London: Office of Science and Technology.

Proverbs, D.G. and Soetanto, R. (2004) *Flood damaged property*. Oxford: Blackwell Publishing.

Shilling, J.D., Sirmans, C.F. and Benjamin, J.D. (1989) Flood insurance, wealth redistribution and urban property values. *Journal of Urban Economics*, 26: 43–53.

Tobin, G.A. and Montz, B.E. (1994) The flood hazard and dynamics of the urban residential land market. *Water Resources Bulletin*, 30: 673–85.

United Nations University (2004) *Two billion people vulnerable to floods by 2050*. News Release, 13 June 2004. Online. Available: HTTP: http://www.unu.edu/news/ehs/floods.doc (accessed: 7 July 2004).

Watkins, T. and Welsh Consumer Council (1992) *In deep water: a study of consumer problems in Towyn and Kinmel Bay after the 1990 floods*. Cardiff: Welsh Consumer Council.

Wordsworth, P. and Bithell, D. (2004) Flooding in buildings: assessment, limitation and rehabilitation. *Structural Survey*, 22: 105–9.

Chapter 8

Public attitudes to 'community-based' small-scale flood risk reduction measures in England

A case study in the Lower Thames catchment

Simon McCarthy, Edmund Penning-Rowsell and Sylvia Tunstall

Introduction

This chapter is predominantly a case study, analysing small-scale flood risk reduction measures and the reactions of those in the affected communities. Our survey results show the extent, in this location, of responses from the public in terms of this type of self-help action. The results may be case specific, but we relate our results to a simple model of decision making, and conclude with some generalisations which we hope will be valid elsewhere.

These small-scale flood risk reduction measures are the subject of considerable debate and policy advocacy (Defra 2004) but very little is known about how readily they can be implemented. Demountable flood defences, and enhanced building resilience and resistance measures, have attracted considerable recent attention since the British Government's Department for Environment, Food and Rural Affairs (Defra) approved a £6.5m strategy for flood defences in Shrewsbury and Bewdley on the River Severn in England during 2001, following severe flooding there (Environment Agency 2001). Subsequently, Defra's *Making Space for Water* flood risk management policy explicitly incorporates what it terms 'novel' flood defence approaches such as these (Defra 2005a, 2005b).

Community-based flood risk reduction approaches (CBFRR) include different forms of measures that can be applied in different combinations depending on the requirements of a locality. For at-risk properties already in the floodplain, these can include all the measures listed in Table 8.1. In our view a CBFRR approach also incorporates giving residents and others guidance on adopting such measures and linking this to advice and information about flood warning systems, and flood emergency response and evacuation.

There are strong reasons for exploring the potential of CBFRR approaches in England and Wales. Among these reasons are floodplain congestion (i.e.

Table 8.1 A classification of Community-Based Flood Risk Reduction (CBFRR) approaches

Temporary barriers: used to protect groups of properties where such measures would be more appropriate or cost-effective than local defences. Systems include the Pallet System ('wooden pallets') or Water Filled Dam types ('tube' designs) that, importantly, require minimal ground preparation works.

Demountable barriers: could also be used to protect groups of properties but would require more preparatory works than temporary defences in the form of a foundation beam ('steel slat' designs). Both the temporary and demountable barriers can be erected and dismantled around a flood event.

Communal measures: comprised of local flood walls, earth bunds, local raising of kerbs, roads or footpaths etc. Additional works may be required as part of the defences to maintain local drainage or intercept and dispose of seepage water. This could be achieved through the use of ground reprofiling, culverts, interceptor drains, small pumps, etc.

Individual measures: comprising individual building protection options (or dry proofing) which come in a variety of forms. These can be summarised as:

- Placing gates or covers across flood access points into buildings, including doors and airbricks; some preparation work is usually required for these options.

- Flood skirts around buildings.

- One-way sewer valves to houses.

- Raising of floor levels of existing buildings by jacking the building up; this is most suitable for wooden buildings.

Flood warning and preparation: this can involve both information systems and advice.

development intensity which often limits other flood-defence options or renders them technically infeasible), environmental and social considerations (which might mitigate against other options) and making occupation of floodplains sustainable into the future (where flood damage potential associated with existing development will only rise).

These CBFRR approaches also have the potential to draw members of the community into managing the risk of floods that they face, serving to raise flood awareness, and reinforcing a self-help approach and informed decision-making about future floodplain use. These options therefore could reinforce a resilience and resistance philosophy which we believe has an important role to play in future sustainable floodplain management.

The research context and methods

Our study formed the initial social research component with at-risk residents (a 'preconsultation survey') as part of a strategic overview of options for the management of flood risk in Reach 4 on the Lower Thames, defined as the stretch between Walton Bridge and the tidal limit at Teddington (Figure 8.1)

(Environment Agency 2007). Here there are a number of flood-prone properties along the river banks and on a series of islands in the Thames. Funded under the Environment Agency strategy development exercise, the aim of the study was to assess public attitudes to risk mitigation and to inform decisions regarding the 'social feasibility' of introducing CBFRR as part of a risk-reduction approach in the absence of (or complementary to) justifiable major engineering works. We sought to gauge current levels of resident engagement with flood risk and the acceptability of CBFRR approaches, prior to further consultation on those measures which appeared most publicly acceptable.

The study used a quantitative social survey with a supplementary qualitative research component (McCarthy *et al.* 2006). From an initial list of 39 possible localities, 13 were chosen for the quantitative survey (Figure 8.1). The choice was informed by, first, residents' level of flood risk (targeting those within the 1 in 50 year return period envelope), and secondly, the feasibility of providing CBFRR approaches there. We also sought, thirdly, a spread of localities to represent the length of Reach 4, and, finally, to provide a sufficiently large enough number of households to provide a representative sample and a robust basis for our analysis. The interviewing was undertaken by Market & Opinion Research International (IPSOS/MORI) and a target sample of 200 interviews was set with an upper limit of half the households to be interviewed in each locality; 206 interviews were actually completed. The survey was undertaken in November 2005, preceded by a letter to all

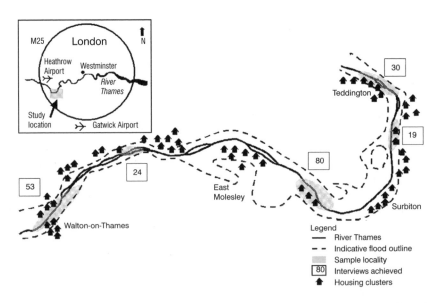

Figure 8.1   The study location, showing the 100-year floodplain (+20% for climate change) with the sampled localities shown circled, with their sample sizes

500 potential respondents to explain the purpose of the interview and to encourage participation.

Following the survey a supplementary qualitative group discussion was also undertaken with residents of Thames Ditton Island to draw out the views of the island communities. A self-selected group of six male and female residents participated at a resident's home in a 75-minute semi-structured discussion. The discussion was steered with an interview guide of issues similar to that of the quantitative questionnaire. However, the qualitative approach allowed for more in-depth questioning and greater flexibility in the pursuit of issues, and the discussion also has informed a clearer understanding of the results emerging from the quantitative survey. The discussion was audio recorded and later a grounded analysis approach undertaken (Maxwell 1996).

This is a prosperous area of south-east England (McCarthy *et al.* 2006). The samples and the localities could be described as comprising affluent, working and retired householders, with a good proportion having been long-term residents at their current address. Respondents were mainly in the 35+ year age group; a quarter were in the 65+ year age group, showing the high representation of retired respondents in the samples, as in the area. There was a strong AB/C1 social group bias (i.e. middle to high income earners) and 67 per cent of those interviewed in the total sample were the chief income earners. Just under half of the total sample were in full-time work. A third of the sample had children in the household, and was equal proportions of both genders. The majority lived in houses rather than flats (apartments), reflecting our sampling strategy, and owned or had a mortgage on their property.

A third of the total sample lived immediately on the river bank. Just under half the sample had lived at their address for up to 10 years, and almost a third of the total sample for 20+ years. In this chapter, statistical analysis of sub-groups has been limited to samples no smaller than 50 respondents, to avoid the potential loss of reliability of the findings when using smaller sample sizes. Any analysis reported as percentage differences or similarities was tested with an appropriate statistical test with a confidence level of 95 per cent or greater.

'Engagement' with the flood risk: public awareness and anxiety

Research elsewhere in the UK suggests that members of the public will only participate meaningfully in discussions and decisions about risk mitigation if they relate clearly to that issue and recognise the threat that they or others face (Fordham 2000; Tunstall 2006).

Building on this, our case study implicitly used a model of human behaviour that posits that the acceptance and adoption of CBFRR approaches depends in some complex way on awareness, anxiety, the evaluation of risk, and the absence of significant constraints, within a context that affects all these variables (Figure 8.2). In essence, this is a simplified version of Mitchell *et al.*'s (1989) contextual hazard decision model, and we recognise that its simplicity masks the importance of many important factors including individual cognitive constraints on taking action (Harries 2008). Contextual constraints might include the incentives that individuals are faced with for adopting individual-scale risk-reduction measures, the role of opinion leaders and information campaigns, the importance of individuals' and combined residents' experience of floods and other hazards, and the ability of households to afford the necessary expenditure or engage in community-level political action to press others to resource these measures on their behalf.

Residents' engagement with the flood risk was measured in our research as a combination of residents' evaluations of the future risk of flooding, their anxiety or 'worry' about that risk, and their current levels of preparation for a flood. In this respect, it should be noted that serious property flooding has not occurred in this location since 1947 although some flooding occurred some 15 km upstream in 2003.

Our results show that just over half our respondents (58 per cent) rated the risk to them as 'not very much' or 'not at all' (Figure 8.3). Taken in the context of other surveys we have conducted (Tunstall 2006), and the fact that our sample covered the whole of the 50-year floodplain (where, by definition, there is a range of risk levels and this risk is low in many areas), we consider this to be a moderate level of acknowledgement of the risk. When a time period was associated with that risk (Table 8.2) a quarter of residents rated as 'likely' or 'certain' that their home would be flooded in the next five years and that proportion rose to just over half (52 per cent) the residents in relation to flooding in the next 50 years. But our respondents were uncertain about the long term: the proportion of residents unable to give a rating increased from 8 per cent for the five years question to 20 per cent for 50 years. It is clear that whilst a large proportion of residents here

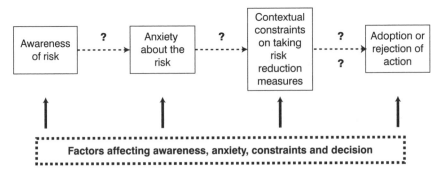

Figure 8.2 The simple model of decision making underlying the survey

Question: Using this scale, from what you know or have heard, how much, if at all, is your home at risk of flooding?

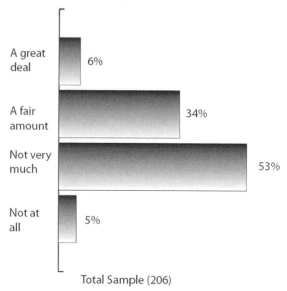

Total Sample (206)

Figure 8.3 Perception of future risk of flooding

Table 8.2 Perceived likelihood of being flooded

Question: How likely or unlikely do you think it is that your home will be flooded?

Response	In the next 5 years (%)	In the next 50 years (%)
Certain to be flooded (+1)	2	16
Very likely (+2)	9	13
Fairly likely (+3)	14	23
Fairly unlikely (+4)	38	19
Very unlikely (+5)	26	7
Certain not to be (+6)	4	2
Don't know	8	20
Mean (+1 to +6)	3.95	2.93
Sample	(206)	(206)

acknowledge some risk, this understanding does not relate to an immediate flood risk to their homes.

This profile of residents' perception of the risk needs to be seen in context, particularly in relation to the small proportion of residents who had experienced flooding here. Some 12 per cent of our respondents claimed to have experienced flooding of their properties above floor level (excluding flooding in outhouses and garages). For this 'flood experienced' group, just over a third (36 per cent) had been flooded in the last three years and 37 per cent claimed to have been flooded more than once. So for the majority of this small sub-group of residents the experience of flooding was not fresh in their minds. This is reflected in residents' appreciation of the risk in terms of anxiety or, as measured in this survey, the degree of 'worry' about the risk, as discussed below.

Figure 8.4 shows that, on average, respondents showed a low level of worry. This might at first be thought to result from residents' perception profile of the risk and the level of their flood experience. However, it is not as clear cut as that: only one-third of residents (38 per cent) who rated the risk to their home as 'great' or 'fair' also expressed anxiety about the risk. The proportion of residents who were both aware and worried accounted for only 16 per cent of the total sample. Interestingly, these were not exclusively those residents who had experienced flooding of their homes, who accounted for only a third of this sub-group. Sources of awareness and knowledge other than direct flood experience of residents' homes are clearly informing the responses of the majority of the sub-group. Overall in terms of the location of the residents' homes in relation to the river – on or away from the river bank – we found no statistical differences in levels of perception, worry and claimed experience of flooding. This may be due to the low level of flood experience in the sample.

Question: Using this card, how worried, if at all, are you about the possibility of your home being flooded during the next 12 months?

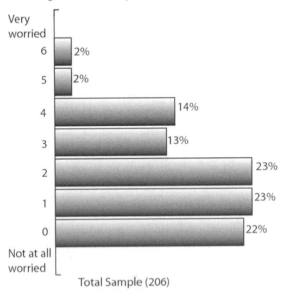

Figure 8.4 Respondents' rating of the degree of worry about future risk

Awareness of the flood risk was not new. Just under half the sample (49 per cent) who perceived a risk of flooding to their homes (as a 'great' or 'fair' amount) claimed to have been aware of that risk before they moved there, a figure that is not untypical of results from other surveys we have undertaken (Tunstall 2006). In addition only 10 per cent of those risk-aware residents claimed they would not have moved to the area if they had known about this risk and 15 per cent were undecided about if they should or should not have moved. This has to be viewed in the light of the low level of experience of flooding across the sample. The majority of all the residents (75 per cent) claimed they, knowing about the risk, still would have moved there. This illustrates the strong 'pull' of amenity for living in the area where the river is an important landscape feature, to be balanced against the 'push' in terms of the risk of flooding.

For the risk-aware residents their claimed initial source of knowledge of the risk (Table 8.3) either originated from their wider experience of flooding (39 per cent) or simply living in the area, rather than any specific formal or institutional source. The higher claimed rate of experience of flooding derived from this question might be explained by the additional residents who only experienced flooding of gardens, outhouses and neighbouring areas such as roads and fields.

Table 8.3 Initial source of knowledge about the risk

Question: How did you first find out about the risk of flooding to your home? (Multiple answers allowed)

Response	%
I have been flooded	39
Its just obvious	14
Told by local residents	11
Environment Agency letter	8
Neighbours been flooded	6
Media	6
Solicitors' search	6
Environment Agency website	4
Local paper	3
Council	2
Water company	2
Other	36
Live in area/near river	13
Garden/road flooded	9
Insurance application	3
Flood warning	2
Sample: those who believe themselves to be at some risk (112)	

In summary, we judge that this population is moderately well aware of the risk of flooding in this locality, despite their lack of personal experience of a flood, and is not unduly worried by the flood threat. Many have accepted that flooding is a risk to be balanced against living in an area with high amenity value, a result in parallel with Fordham's earlier research in similar Lower Thames locations (Fordham 1992, 2000).

'Engagement' with the flood risk: flood warnings and the public's pre-flood preparations

Even with the limited experience we found of flooding in their homes, our respondents were surprisingly flood-aware: 53 per cent claimed they had received some form of flood warning in the past, endorsing multiple forms of warning communication, on average mentioning 1.6 sources (Table 8.4). The main channels of communication cited were the Environment Agency but some residents (13 per cent) claimed that their flood warning was taken from their own observations of the river water rising. The results overall (Table 8.5) show that only just under a third of residents (30 per cent)

Table 8.4 Cited source of flood warnings

Question: How did you receive this warning? (Multiple answers allowed)

Response	%
Any Environment Agency	23
Personal telephone call	17
Telephoned Floodline	15
Recorded telephone message from	14
Warning via fax	4
Local authority	21
Neighbour/friend/relative	18
Television announcement/broadcast	13
Personal observation/water rising	13
Radio announcement/broadcast	6
Warning siren or loudspeaker	5
From flood warden	4
From the police	4
From fire brigade	1
From BBC Ceefax	1
From ITV Teletext	1
Other	27
Postal leaflet/pack	19
Sample: those who received a flood warning (110)	

Table 8.5 Respondents' rating of their reliance on authorities for warnings

Question: How much, if at all, do you currently rely on the authorities or your own judgement of when the River Thames is going to flood?

Response	%
Completely rely on the authorities (+2)	10
Mainly rely on the authorities (+1)	20
About half and half (0)	35
Mainly rely on own judgement (–1)	26
Completely rely on own judgement (–2)	7
Mean position (+2 to –2)	0.0
Total sample (206)	

claim to rely 'mainly' or 'completely' on the authorities, with no significant difference if residents lived on or away from the river bank.

Hazard-response research (Tunstall 2006) tends to show that awareness of flooding – or other hazards – generally leads to some level of preparation (Figure 8.2) although the process of decision-making is often complex and far from clear. Residents can make physical changes to their homes and modify their behaviour. This can stop water entering properties or reduce the resulting flood damages and disruption. We enquired about both 'physical' and 'behavioural' measures, as possible actions, and found that, even with low levels of risk awareness and worry about the risk here, two-thirds of the sample endorsed and had adopted at least one of the actions (Table 8.6), with insurance dominating the list.

Two-thirds of residents claimed that they had made sure that their home contents insurance covered flooding, although 17 per cent did not know. With a local population substantially biased towards the more affluent AB/ C1 socio-economic groups, it was perhaps unsurprising to find such high levels of insurance cover. But there was an additional 17 per cent of residents who said that they had not ensured that their contents insurance covered flooding, which is perhaps more worrying. Some 17 per cent of residents also claimed that they had experienced problems renewing or obtaining their contents insurance because of the risk of flooding to their home. Only a minority of the residents experiencing problems were those who said that their insurance did not cover flooding. Of the total sample, only 2 per cent of respondents both claimed to not have contents flood cover and had experienced difficulties obtaining insurance due to their flood risk.

Fieldwork showed that some residents' houses have been raised above ground level as a historic precautionary measure against flooding rather than, as suggested in the list of actions, an active measure on the part of the

Table 8.6 Claimed physical preparations undertaken

Question: Have you undertaken any of these flood prevention measures at this address? (Multiple answers allowed)

Response	%
Any mention	65
Made sure I am covered by insurance against flooding	49
Raised the property off the ground	12
Keep sandbags at the property	9
Have flood boards/floodgates/airbrick covers	3
Made permanent changes to the house interior	1
Have permanently raised furnishings	1
Purchased water pumps	1
Put up walls or changed grounds around property	1
Flood-proofed exterior walls	–
Total sample (206)	

current residents. How living in a raised property contributes to residents' perception and worry about the risk might be of interest but is unfortunately beyond robust analysis on this survey's sample size (see Penning-Rowsell and Smith 1987).

Behavioural preparatory measures – at a 71 per cent adoption rate – were more common (Table 8.7), focusing on heightened personal awareness of weather indicators at the time of a possible flood. Residents relied on their own interpretation of the risk with few claiming to have signed up to the Environment Agency's telephone-based automated warning messaging system. As noted for the physical actions, where insurance cover dominated, few residents had adopted measures to limit disruption or loss in advanced preparation for water actually entering their home. However, only 18 per cent of residents did not endorse any of the statements across both the physical and behavioural question lists.

Encouragingly, therefore, these results indicate that the majority of residents have sought to take some level of preparatory action. This could form a base upon which to widen the proportion of residents involved in self-help flood risk reduction and broaden the activities they undertake. However, this will be limited by their evaluation of the hazard and their concern for the risk they face: the decisions leading to the adoption of these measures are far from automatic.

The acceptance of community-based flood risk reduction approaches

Having gauged our respondents' level of engagement with the flood risk that they face, we used a combination of photographic and descriptive prompts

Table 8.7 Claimed behavioural preparation undertaken

Question: Have you undertaken any of these actions because of the risk of flooding at this address? (Multiple answers allowed)

Response	%
Any mention	71
I keep an eye on the level of the river	52
Keep alert for flood warnings during high-risk months	40
Make sure I am aware of bad weather forecasts	32
I listen for reports of other areas flooding	31
Keep ditches and drains around the property clean	12
Have a plan of what to do when my home might flood	7
I know where I would evacuate to	6
Signed onto the flood warning messaging service	6
Have a supply of canned and bottled provisions	3
Avoid keeping irreplaceable items or goods of sentimental value on ground floor of my home at all or at certain times	3
Total sample (206)	

in our survey to elicit residents' attitudes towards a range of community-based flood risk reduction approaches (CBFRR).

In terms of the demountable/temporary defence concept, we showed photographs of three possible designs and asked respondents to rate how strongly they agreed or disagreed with the concept (Figure 8.5). The result was support from over three-quarters of residents and only minimal disagreement (Figure 8.6). The steel slat design was preferred by 44 per cent of the residents, followed by the wooden pallet design (32 per cent), with the tube design the least liked (11 per cent).

However, the proportion of residents indicating disagreement increased substantially when the concept description added that there might be a permanent change to the 'natural' appearance of the local surroundings so as to form a solid foundation for the works. Disagreement increased from 4 per cent initially to at least a quarter of residents when this change was mooted (Table 8.8). This heightened level of opposition was not significantly different between those who lived on the river bank and residents who did not, and there were no specific differences between the different interview localities in the Reach.

In addition to the demountable flood barriers, residents were also shown a list of 'communal measures' (Table 8.9). Just over half the sample (52 per cent) raised an objection to at least one of these measures. The measure attracting the highest level of objection was road humps, probably reflecting residents' general dislike of road humps as traffic speed control measures

(a) Taken down   In use

(b) Taken down   In use

(c) Taken down   In use

Figure 8.5 Three possible designs of demountable/temporary defence

Question: How much do you agree or disagree with the principle of using temporary flood defences in this area?

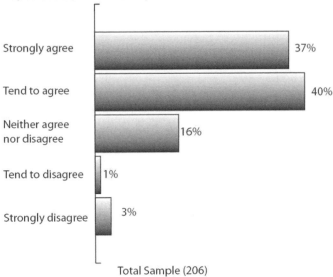

Total Sample (206)

Figure 8.6 Respondents' rating of the principle of using demountable/temporary measures

Table 8.8 Respondents' ratings of the principle of demountable/ temporary designs.

Question: How strongly would you now agree or disagree with the use of each of these types of flood defences?

Response	Steel slat (%)	Wooden pallet (%)	Tube (%)
Strongly agree	10	11	17
Tend to agree	37	31	27
Neither agree nor disagree	21	21	21
Tend to disagree	11	13	12
Strongly agree	14	16	15
Total sample (206)			

locally. However, this highlights how an understanding of local context is important: sensitivities sometimes unrelated to flooding affect attitudes to adopting possible CBFRR measures, a finding that has echoes in other chapters in this volume.

In terms of individual household measures, a photograph of a selection of household flood-protection measures that could be employed to 'dry proof' homes was shown to the residents (Figure 8.7). In terms of their adoption, we found low levels of possession of individual household measures but almost three-quarters (73 per cent) of the residents said that they would

Table 8.9 Objections to communal CBFRR measures

Question: Which, if any, of these approaches to reducing the risk of flooding do you think you might have an objection to? (Multiple answers allowed)

Response	%
Any mention	52
Road humps to divert possible water	33
Permanent earth bank that would not restrict current access	18
A raised curb to the paving	10
Flood gates at pedestrian access points	8
The strengthening and extending of walls	6
Don't know	3
Total sample (206)	

consider using at least one of these measures (Table 8.10). However, the cost to the householder of purchase and installation of these measures was not included in our questions, due to the likely differences for each property. Also the contribution of some measures – such as one-way sewer valves – may have required more detailed explanation of their role in flooding than was possible during the interview, and with clarification might have attracted higher endorsement. This indicates how a more involved engagement with residents might be required regarding implementing some of the more complex CBFRR measures, again a result that mirrors findings in other chapters in this volume.

Adoption of self-help actions is likely to be related to the financial context, not least because more traditional flood-risk measures are usually state funded in the UK. Our sample of residents was asked who should pay for these risk-reduction measures (Figure 8.8) and the majority of residents (67 per cent) favoured 'the tax payer' paying all or most of the costs of undertaking communal-scale works. However, interestingly, 66 per cent of residents suggested the person at risk should pay all or most of the cost of the individual household measures. The motivation behind this might be that our respondents wanted greater control and not want the state to be involved in modifications to the house that they owned or modifications and potential benefits to a home are viewed as 'normally' the responsibility of the home owner. Analysis of different socio-economic groups does not show significant differences here, so relative affluence is not the driving force. No example of the costs associated with both the works or the individual household measures were given with the question, so to a certain extent the question remains hypothetical in that it gauges only attitudinal directions rather than an indication of the eventual level of acceptance.

Flood boards

Flood skirts

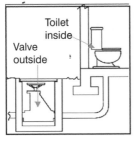

Waterproof exterior walls

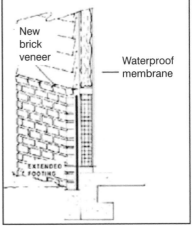

New brick veneer

Waterproof membrane

One-way valves

Toilet inside

Valve outside

Air brick covers

Figure 8.7 Selection of household flood protection measures that could be employed to 'dry proof' homes

Table 8.10 Use and consideration of individual household measures

Question: These are some products that can be bought to protect individual homes. Which, if any, of these products do/would you consider?

Response	Currently have (%)	Consider in future (%)
Any mention	9	73
Flood boards	3	54
Air brick protectors	3	39
Flood skirts	2	32
One-way valves for sewer pipes	1	24
Waterproof veneers for exterior walls	–	18
Don't know	–	10
Total sample (206)		

Question: Which of these statements best reflects your view of who should pay the costs of ...

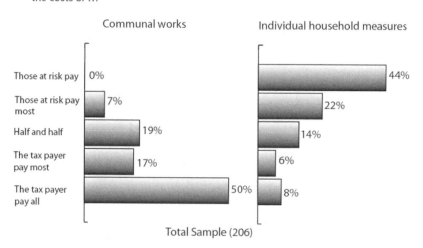

Figure 8.8 Respondents' ratings of who should pay for measures

Attitudes to communal measures and self-help are likely to be affected by attitudes to state-sourced assistance. In this regard opinion was divided amongst residents as to whether the authorities were 'doing enough' to reduce the risk of flooding in the area. Equal proportions of residents (32 per cent) suggested that 'the authorities' were either not doing enough or were doing enough to reduce flood risk, although 36 per cent of residents were unable to offer an opinion. Such a three-way split should not be taken to indicate indifference. Our results indicate that residents show a strong involvement with local organisations and causes (Table 8.11) – perhaps reflecting their

Table 8.11 Residents' involvement in local issues

Question: Which, if any, of these things on the list have you done in the last two or three years for a local charity or campaigning organisation? (Multiple answers allowed)

Response	%
Any mention	69
Attended meetings related to specific local issues	39
Visited/ written a letter to an MP/ councillor	28
Been a member of a local community group, even if you joined 2–3 years ago	25
Given up my time on a voluntary basis, to raise funds or work in some other way	21
Been involved in a campaign on a local issue	20
Been involved (with a charity or other non-profit making organisation) in my local area	20
Been a member of an action group about a specific issue, even if you joined more than 2–3 years ago	12
Displayed a poster in your window, car or place of work	8
Written (to the charity or other non-profit making organisation) asking for information	7
Written a letter for publication to a local newspaper	7
Taken part in a demonstration or protest about a local issue	6
Total sample (206)	

socio-economic status – and therefore potentially in the issues surrounding CBFRR approaches, and 29 per cent of our sample agreed that they would actively object to a CBFRR scheme if for technical or financial reasons they were excluded from it but their neighbours were included.

The island residents

Many decades ago the island houses in the Thames were simple holiday cottages or huts for daytime use, whereas now many are permanently occupied. Our focus group discussion with a selection of Thames Ditton Island residents gives a flavour of the possible other contextual issues, and highlights similarities and differences with the 'mainland' residents' findings.

In contrast to the 'mainland' residents' survey, the islander group appeared more aware and worried about the risk of flooding. This may have been a product of a self-selected group, although the flood experiences recounted were not just given in the context of the residents taking part in the discussion but related to all the island residents.

The group claimed that the island had experienced 'serious' floods in 1947 and 1968 but none of the group respondents had lived on the island during those periods. They had experienced flooding in 2000 and 2003 but there was much discussion about what actually defines flooding for the island. Is it when they have to wear waders, or when the water enters the house, or when the whole island is covered, or when the Electricity Board (*sic*) turns off the power to the island? In this respect the focus group members did not consider flooding to be a 'normal' consequence of living on the island. Similar to the mainland residents there was limited past experience of flooding but, in comparison, most of the island community had experienced some form of flooding (e.g. in gardens or on roadways).

From a flooding perspective, the islands do not appear to be a rational place in which to live. However, the attractions here for this group of residents were both the community spirit on their island and the amenity value of environmental aspects of living close to the river. People like to live near rivers (and are prepared to pay for the benefit in many cases), for the aesthetic and recreational benefits that rivers can offer. For example, upstream, along the non-tidal stretch of the Thames, some 12,000 houses are within 500 metres of the riverbank, and their riverside location adds £580m to the value of these properties (McGlade 2002). Also mentioned a number of times throughout the discussion was the unique type of person who would live in their community for which tenacity was a valued characteristic. There were indications, however, that the balance could tip against amenity considerations if the risk was judged to be greater than the advantages of living there.

There was an appreciation among this group of their special situation as an island community compared with the mainland residents. This was reflected in the group's attitudes to the different flood preparation approaches (Figures 8.5 and 8.7). Residents were open to considering any ideas to increase their resilience. Temporary and demountable measures were felt to be impractical for reasons of cost and implementation in a fast-rising and fast-flowing river. Certain communal measures such as bunds were also thought to be unsatisfactory, as likely to be ugly landscape features harming the amenity value of living on the island by obstructing key river views. The group felt, however, that they should not be excluded from decisions made about such structures on the mainland as they might be affected by the resulting heightened flood levels.

The group was encouraging with respect to property-based risk reduction measures. Some measures had already been investigated by residents who had relevant professional qualifications. Such ideas included a range of suggestions from tanking houses, houses that would float up from their secured foundations, to increasing the stilt height of their current properties. Regarding stilt heights, the residents felt that they were unfairly restricted by planning regulations which limited the roof height of their homes. The group also bemoaned the fact that ideas were raised among the residents but never fully investigated by them. Simple examples for coping with floods were given, such as hand rails or ropes for the paths to help and guide people through the floodwater.

While such measures were enthusiastically considered, some residents felt they would need to be flooded frequently to justify the cost if they had to pay for them themselves. Others judged that if that situation arose – or threatened to arise – they would simply move away, not least to protect the investment they had already made in buying the property in which they now lived if the value of that property were threatened by flood risk or a reputation of such a risk. However, on a positive note, final comments showed an openness reflected by most of the group for a dialogue and engagement with the authorities to discuss flood management and the risk of flooding to their island and homes. But the *status quo* appears to be one where flood risk reduction measures are continually being considered but not actively pursued. This appears to be partly because the level of flood experience does not warrant the necessary investment, despite the very obvious threat posed by the river, and partly because their idyllic island location could thereby be threatened.

Conclusions

Table 8.12 gives a summary of residents' views as to the merits and demerits of community-based flood risk reduction approaches in this location.

Our social survey reveals a high level of acceptance amongst respondents of temporary/demountable flood defences as a concept which might be adopted in this area. There is also support from these floodplain residents for further advice on flood resilience and resistance measures for the houses that they occupy, although the take-up of these measures is currently low. This low take-up appears to reflect low levels of anxiety about flooding, despite (or because) the population being relatively well aware of the risks that they face.

Clearly local context is also important and with it the socio-economic status of the population at risk and the range of choices for them that this determines. The strength of support for temporary/demountable flood defences depends upon their detailed design, location and implementation. But our results show that the public is willing to be persuaded: support for such defences can be garnered by the manner in which proposals are presented to and negotiated with residents in particular locations (i.e. both the residents and the authorities may well be prepared to compromise on specific details when they recognise that they might be left out of a scheme).

Other contextual factors are clearly significant. For many respondents the amenity value of their immediate environment could be threatened by the installation of small-scale flood risk reduction measures. For others the costs are not justified by the frequency of the flooding that they face. For others

Table 8.12 The merits and demerits of community-scale flood risk reduction measures from the public's perspective in the Thames case study area

Merits	Demerits
1. The public unacceptability on environmental grounds of large-scale engineered defences (e.g. flood diversion or flood=relief channels)	1. Reduced levels of resident support for physical changes when it was suggested that these might affect the amenity value of localities (meaning that detailed design and implementation will require significant detailed consultation which has its costs and resource issues)
2. Relatively intense floodplain development in pockets presenting high flood damage potential; rising damage potential in existing buildings (with high socio-economic status of residents)	2. Current relatively low flood frequency, with flood experience limited to a minority of those interviewed, limiting engagement with issues
3. Apparent high degree of acceptance of temporary/demountable options as a concept which could be applicable, with an aversion to being excluded from a scheme when implemented	3. An apparently restricted level of concern for future flooding, again limiting engagement
4. Residents do not believe that sufficient is being done by authorities to address the flood risk	4. Lack of a severe flood event in the area implies that greater work will be required to garner public and political support for schemes
5. If other coping strategies by residents such as insurance was withdrawn then the CBFRR approach becomes all the more valuable	5. Some localities will be left out of schemes or have reduced schemes, with accompanying resident concerns
6. Such measures, while reducing damage and disruption, may also provide increased time for preparation and evacuation	
7. CBFRR could make a valuable contribution to sustainable floodplain management in this area where few other options appear to be feasible	

again, it is 'the authorities' who should be providing flood protection, not the individual householder.

Notwithstanding all this, there appears to be a measure of public support for a variety of local community-scale flood risk reduction measures. However, the way that a strategy for these measures is implemented will be crucial to its success. In terms of our model (Figure 8.2) it will be important in any situation – as other authors in this volume have found – to understand the levels of awareness of risk, and of anxiety or worry, and the local factors affecting the public's acceptance of particular measures and designs. Indeed a multiplicity of local issues will always need to be engaged with, using a variety of methods and a 'hand-stitched' approach related to the range of risk that individuals face. Without such a full understanding of the context for the acceptance of such measures in a particular community, obtained through using meaningful participatory methods, a strategy for the engagement with the public is likely to be undirected and possibly ineffective.

References

Defra (2004) *Making space for water: Developing a new government strategy for flood and coastal erosion risk management in England: A consultation exercise*, London: Department for the Environment, Food and Rural Affairs.

Defra (2005a) *Making space for water: Taking forward a new Government strategy for flood and coastal erosion risk management in England. First government response to the autumn 2004 Making space for water consultation exercise*, London: Department for the Environment, Food and Rural Affairs.

Defra (2005b) *Making space for water: Taking forward a new Government strategy for flood and coastal erosion risk management in England. Delivery Plan*, London: Department for Environment, Food and Rural Affairs.

Environment Agency (2001) *Bewdley flood alleviation scheme: your questions answered*, Bristol: Environment Agency.

Environment Agency (2007) *Managing flood risk: Lower Thames Strategy Study, Phase 3 Final report*, Bristol: Environment Agency.

Fordham, M. (1992) *Choice and constraint in flood hazard mitigation: the environmental attitudes of flood plain residents and engineers*, Unpublished PhD, London: Middlesex University.

Fordham, M. (2000) 'Participatory planning for flood mitigation: models and approaches', in Parker, D.J. (ed.) *Floods*, London: Routledge.

Harries, T. (2008). Feeling secure or being secure? Why it can seem better not to protect yourself against a natural hazard. *Health, Risk and Society*, 10(4) special issue.

McCarthy, S., Parker, D.J. and Penning-Rowsell, E.C. (2006) *Preconsultation social survey – community based flood risk reduction options, Reach 4: Walton Bridge to Teddington*, London: Middlesex University, Flood Hazard Research Centre.

McGlade, J. (2002) 'Thames navigation and its role in the development of London', *Proceedings of the 2002 London's Environment and Future (LEAF) Conference*, London: University College London.

Maxwell, J.A. (1996) *Qualitative research design: an interactive approach*. London: Sage.

Mitchell, J.K., Devine, N. and Jagger, K. (1989) 'A contextual model of natural hazard', *Geographical Review*, 79: 391–409.

Penning-Rowsell, E.C. and Smith, D.I. (1987) 'Self-help flood hazard mitigation: the economics of house-raising in Lismore, NSW, Australia', *Tijdschrif voor Economische en Sociale Geographie*, 78(3): 176–89.

Tunstall, S. (2006) *Vulnerability and flooding: A re-analysis of FHRC data.Country report England and Wales*, Floodsite Project, Report Number T11-2006-11, Project Contract No: GOCE-CT-2004-505420, Project website: www.FLOODsite.net.

Chapter 9

Facing the future by designing in resilience

An architectural perspective

Jacqueline Glass

Introduction

This chapter considers the relevance of an architect's role within disaster risk management, but not in the context of designing emergency shelters or housing reconstruction as is often the case. Instead it focuses on the consideration of resilience during the early decision-making stages in the conventional building design process, as is experienced in most countries. The importance of designing built assets that are resilient to natural and human-induced hazards has really come to the fore in recent years and so it is appropriate to consider the role of the architectural profession in providing the resilient built environment that governments and others seek.

It should be noted at this point that while the chapter is grounded in the experience of the construction industry in many typical Western cultures, the general discussion and conclusions relating to the everyday work of an architect (i.e. design activities and processes) will be broadly relevant and applicable throughout the world. So, to place this chapter within the disaster risk management and resilience context is important, but fairly straightforward.

Disaster risk management (DRM) can be considered in terms of the definition offered in the United Nations International Strategy for Disaster Reduction, i.e. 'a systematic process of using administrative decisions, organization, operational skills and capacities to implement policies, strategies and coping capacities of the society and communities to lessen the impacts of natural hazards and related environmental and technological disasters' (UN/ISDR 2004: 17). This document suggests that there are four basic phases of pre-emergency, preventive and mitigating actions; emergency plans and preparedness activities; emergency response; short-term rehabilitation and longer-term reconstruction. Within this scenario, it is perhaps preventive, structural mitigation actions that are of most relevance at the early stages of a building project. The Hyogo Framework for Action 2005–2015 (UN/ISDR 2005) urges that DRM should be addressed in urban planning and housing provision and calls on governments to mainstream this into procedures for major infrastructure works, but little research exists on other types of projects or activities such as urban planning and housing (Wamsler 2004, 2006; Burby 1998; Burby *et al.* 2000).

Recent research has indicated that awareness of hazards tends to be much more prominent in stakeholders who govern or advise on the built environment, rather than those who actually design, build and operate it (Bosher *et al.* 2006). It is quite credible therefore that Hamelin and Hauke (2005) have identified a clear need for the construction industry at large to possess a much better understanding of how to avoid and mitigate the effects of disasters. Thus, Bosher *et al.* (2007b) open a fresh debate about how resilience can be systematically 'built-in' to planning and design processes, stating that proactive risk assessment could be the solution to two key shortcomings: a failure of foresight (Toft and Reynolds 1994) and a failure to learn (Weir 2002). In attempting to refine the language of DRM as applied to building construction, the concept of resilience has proved pertinent. Burby *et al.* (2000) and Mileti (1999) state that resilience applies to the minimisation of losses and damages when a disaster occurs, but the term is being used increasingly as shorthand for a holistic approach to the four main components of DRM, as described above.

According to Bosher *et al.* (2007a), 'resilient built assets' can be defined as those that have been designed, located, built, operated and maintained in a way that maximises the use of sustainable materials and processes while at the same time maximising the ability of the built asset to withstand the impacts of extreme natural and human-induced hazards. The scope of such a definition is very broad and so needs to be researched and discussed in a manner appropriate to each of the various professional groups within the construction industry.

For this reason, this chapter considers the position of the architect in relation to resilience; it focuses on the early design stages of a building project (where the architect is typically most influential), but stops short of offering specific recommendations on actual building design and specification for any specific hazard or category of hazard. It is the aim of the chapter to consider integration of resilience into mainstream architectural design practice in principle, rather than in practice. With this in mind, the nature of architectural design, the typical building design process and the relevance of skills development are all discussed and the chapter concludes with some outline findings on the best ways in which architects could improve their input to the design of resilient built assets.

The nature of architectural design

This section begins with an explanation of the role of an architect and thereafter considers that contemporary architectural practice is typically characterised by two core activities: creative design and information management. Only by understanding this delineation can an appropriate strategy for the integration of resilience issues be developed.

The Royal Institute of British Architects (RIBA) describes the role of an architect as a highly influential, professional expert who works closely with related built environment professions to 'define new ways of living and working' and 'develop innovative ways of using existing buildings and creating new ones' (RIBA 2007a). The architect is typically characterised as someone with creative talent and problem-solving skills, someone who can advise on the most effective way to develop new buildings, refurbish older properties and perhaps also make the most of spaces which surround buildings. As a result of this breadth, the period of professional training to become an architect is understandably lengthy (up to seven years, depending on the country in which he/she is based).

Wikipedia, the on-line encyclopaedia, offers a pragmatic definition of the role – that the architect acts as a translator or go-between, listening and interpreting the owner/client's needs into information that is sufficient for a builder (contractor) to proceed with construction (Wikipedia 2007). However, it also states that part of the architect's role is to reach a compromise, such that the client's needs can still be met whilst being cognisant of 'cost and time boundaries'. This suggests that an architect must work within a decision space that is specific to a given client and his/her building project and this explanation begins to capture the essence of daily life for many practitioners, but importantly fails to represent the criticality of its inherent variety (e.g. different costs, programmes, building types, needs, sites, climates, styles etc). Indeed, the RIBA has recently recommended that its members change the way they calculate their fees, because of 'the more varied and complex world that architects now have to operate in' (RIBA 2007b). Lawson (1997) describes design problems as multi-dimensional and uses the design of a window to illustrate how physical, physiological and psychological factors come into play. In their book on the design of building elements, Rich and Dean (1999) discuss how the list of principles applied to buildings (or functional requirements) has increased steadily and they express concern that, in trying to satisfy all such demands, the ultimate purpose and thus the beauty of architecture may be lost.

With such a typically vast set of variables to deal with in any building project, it is understandable that the solution for some architects and/or their practices is to specialise in particular building types (such as offices or hospitals) or base their portfolio in specific, limited geographical or climatic regions, thereby developing a niche market for themselves. In most cases, however the more obvious outcome to all of this is a clear need for tools and techniques to manage all the various factors involved and thereby facilitate effective design decision-making. Hence, one reaches a point where an architect can be seen as a manager of design information, liaising between client and contractor – the lynchpin in the design process. Indeed, Lawson

(1997: 12) cites De Bono's idea that design has more to do with having the skills to be able to think one's way around a problem, rather than being in possession of all the facts necessary to solve it.

That Kendra and Wachtendorf (2003) emphasise the value of creative actions of organisations in the aftermath of a disaster suggests there may also be some relevance in exploring the creativity or ideas development process inherent in an architect's role. There is of course extensive research and related literature on building designs, materials, layouts for emergency response, aid and recovery, but somewhat less consideration has been given to the role of creativity at the early design stages. In exploring this idea a little further, recent research by Tubaila (2006) identified an interesting list of attributes which architects associated with creativity in themselves and their colleagues; among these were:

- ability to synthesise
- divergent thinking
- hard-working
- imaginative
- highly-motivated
- confident.

Like many others who have researched the creative personality, Tubaila (2006) also links creativity with innovation, ingenuity and the generation of new ideas, but these traits do not score as highly in his study as one might expect, perhaps acknowledging that Thomas Edison's famous quote: 'Genius is one per cent inspiration and ninety-nine per cent perspiration' is particularly appropriate.

Although the various stages of the design process are discussed broadly in the next section, it is worth focusing at this point on one of the critical, early design phases in a project during which time the architect plays an important and perhaps unique role. This is the concept design development stage, when the architect begins to hone basic ideas for the form, layout or character of a building design. Laseau (2001) describes this discovery stage as part invention and part concept formation, explaining that many architects seek inspiration (or indeed validation) from analogies, which may be physical, organic or cultural in nature. He believes that four important functions of an architectural concept (or parti, as it is sometimes called) are as follows (Laseau 2001: 149):

- The first synthesis of the designer's response to the determinants of form (programme, objectives, context, site etc.)
- A boundary around the set of decisions that will be the focus of the designer's responsibility

- A map for future design activities in the form of a hierarchy of values and responding forms
- An image that arouses expectations and provides motivation for all persons involved in the design process.

This list gives a useful insight into the attitude of the architect during the early design development stages. Importantly, it recognises the inter-relationship between invention per se and the role of the decision space in determining what is actually possible for a particular project. Laseau's list also captures the notion that the architect has a personal response to the project site, client and brief, that this is a valid part of his/her role and a recognisable, specific deliverable from an architect to a project.

The architect therefore operates concurrently in two distinct, complementary ways to produce a building design: first, as a manager of information, responding systematically to a client's requirements, and second, as a creative individual, developing design ideas that add a personal input to the project. This duality is extremely advantageous on the whole, enabling architects to both debate creatively and identify critically, but of course a perfect balance is unlikely in practice. Hence, the nature of architectural design tends towards an extraordinary blend of subjectivity and objectivity which makes the building design process both fascinating and frustrating at times. With this in mind, the next section examines how the concept of resilience can be considered in relation to the process of building design and how best it might therefore be integrated in architectural practice.

Integrating resilience in the design process

Having described the role of the architect and the nature of his/her approach to design, this section examines some of the challenges of considering resilience within the building design process, which leads to a proposal that both management-led and design-led approaches are necessary.

Numerous researchers and authors have attempted to describe, depict and criticise the design process, whether or not buildings are involved. Lawson (1997) explains that many conceptual models try to represent the various activities in the design process, but often fail because it is a creative process carried out in an iterative fashion, thus making it highly complex and very difficult to capture. To produce a model relevant to building design, Lawson develops a model with three basic steps (analysis, synthesis and evaluation), with each step having a feedback loop to the previous ones to show the iterative nature of design (Lawson 1997: 35–6).

- Analysis – an exploration of relationships, looking for patterns in the information and classification of objectives.

- Synthesis – an attempt to move forward and create a response to the problem (i.e. the generation of solutions).
- Evaluation – a critical appraisal of suggested solutions against the objectives identified in the analysis stage.

For the UK architectural profession, the formalisation of this process culminated in the 1960s with the RIBA Plan of Work, which has been revised regularly to reflect changes in building procurement practices, but remains the benchmark against which architects manage their projects and their practices (Philips 2000; RIBA 2007c). The RIBA Plan of Work identifies various stages in the life of a building project, as shown below.

A Appraisal
B Design brief
C Concept
D Design development
E Technical design
F Production information
G Tender documentation
H Tender action
J Mobilisation
K Construction to practical completion
L Post-practical completion

Overall, the RIBA Plan of Work is divided into five main phases; Preparation (A, B), Pre-construction Design (C–E), Pre-construction (J–K) and Use (L). The longevity of the Plan of Work is explained partly by its emphasis on specifying deliverables that a client can expect at the end of each stage, thus clarifying the nature of the contractual relationship between an architect and his/her client. However, it did not necessarily apply to all forms of contract (such as design build, partnering, Private Finance Initiative etc) and until recently was considered somewhat limited in scope.

Kagioglou et al. (2000) and Cooper et al. (2005) present an alternative model, called the Generic Design and Construction Process Protocol (GDCPP), which relates well to the Plan of Work, but is much more generic in nature and contains much less emphasis on the role of the architect. The GDCPP takes a much more strategic, process-driven view of the management of the delivery of a building and so has been found particularly appropriate for use on complex, major building projects, such as hospitals. It is divided into a series of stages, just like the Plan of Work, but importantly recognises that a series of activities take place concurrently on a building project (e.g. project management, resource management, production management and, of particular relevance here, design management). Further to this, Cooper et al. (2005: 142) explain that the 'design management' activity zone in the

GDCPP is 'responsible for the design process that translates the business case and project brief into an appropriate product definition'; that it involves not only design professionals, but also materials suppliers and representatives from the main contractor and sub-contractors. Hence, it focuses on all design activity in a project. This then offers a useful distinction between the GDCPP as a management-focused approach to architectural input to building design and the RIBA Plan of Work as an architectural approach to building design management. It is arguably this distinction which holds the key to integrating resilience in building design.

In the previous section, it became clear that building design is a multi-dimensional problem (Lawson 1997) and that the architect needs to develop skills to be able to deal with a significant variety of issues that might arise during the course of a particular building project. That the RIBA recognises the existence of such levels of complexity is important and welcome, but the fact that this complexity is increasing remains cause for concern and it may now be appropriate to ask the question 'how much is too much?' The real danger of trying to genuinely adapt the design process to take any new issues into account in today's complex building projects is that the pressure on certain individuals simply becomes too great. In this scenario the subject might be ignored, treated cursorily or become part of everyday practice only through legislation. Perhaps there is a way in which any pressure can be prevented or at least alleviated, such that important issues like resilience are given due consideration and integrated appropriately in the design process without undue recourse to legislation.

Interestingly, Bosher et al. (2006) suggest that the construction industry should 'embrace and pre-empt regulatory changes regarding resilient and sustainable construction and use them as an opportunity to compete in the sector nationally and globally'. While this might seem rather ambitious, there is evidence in Bosher et al. (2007b) that some leading companies have already recognised the potential, but interestingly are also calling for legislation:

> I believe that involvement needs to be regulatory driven; we need to make people consider these issues, such as tick off these issues at each RIBA process stage to make sure that people are compliant. It will add to the complexity of the building process but it is a very complex process at the moment anyway. This issue needs to be incorporated into the early design stages. At the moment without the required regulations, if someone sticks their head over the parapet and offers this consultancy service it will be a risk but it could be very useful to be a leader in the game. At the moment organisations can offer bits and pieces of such a service but no one I am aware of can offer the full holistic package.

Certainly, within the more limited scope of building design, Gray and Hughes (2001) acknowledge a growing complexity in procurement, fee agreements and contractual relationships and state that traditional approaches to the management of design are therefore inadequate, lacking sufficient integration and coordination to cope. They propose that the role of a 'design manager' is to act as a single point of contact, organising the design process for the task at hand, drawing in design information from the architect, liaising with the client and ensuring that the right quality and level of design information is provided to other parties such as the contractor. This is clearly a management role with an appreciation of architectural design, rather than the RIBA Plan of Work approach (devised for an architect with an appreciation of management). The definition of design management as 'the application of a process of management to the process of innovation and design' is particularly apposite (Cooper and Press 1995).

The logical conclusion therefore is that a two-pronged approach is needed to embed resilience within the design stages, at least, of the building project procurement process. Evidence from Bosher *et al.* (2007a) supports the development of a new framework within which all the relevant professions related to DRM are able to work together, understanding one another's roles and responsibilities. This type of systematic approach would seem most useful and attractive to someone in a design manager's role, particularly if the said framework were to align with conventional models of the project process, such as the GDCPP (i.e. a management-led approach) because this would make it easier to understand and integrate with existing practice.

On the other hand, there needs to be some recognition of the role of resilient design ideas within the development of the architectural concept/ parti, for which no one as yet seems to have devised a solution. However, on the basis of Laseau's (2001) observations on the use of analogy in architectural design, there is some sense at least in the possible development of case study information. That said, there are a number of practical challenges in actually delivering practical design and specification data to architects via case studies, such as a lack of availability of up-to-date design guidance (for non-government/military facilities) and the potential compromise to existing critical infrastructure or buildings should information on government and military facilities be transferred to other building types. Anecdotal evidence suggests that architects and other consultants engaged on sensitive projects are not able to divulge information to anyone outside the immediate project team, so published case studies of benefit to other designers are hardly likely to become readily available.

The use of charrettes to embed the principles of resilient design within the architectural practice (i.e. a design-led approach) may offer an interesting alternative model. The design charrette is a collaborative workshop in which a group of designers (sometimes with other groups, or laypeople) work together on a specific design problem. It is often used in community planning to encourage participation and involvement from local stakeholders. For example, the National Charrette Institute (USA) describes a charrette as 'a collaborative planning process that harnesses the talents and energies of all interested parties to create and support a feasible plan that represents transformative community change' (NCI 2007). The technique has also become popular within the sustainable building design field: the US Department of Energy website contains some very useful information on this in a charrette handbook produced by the National Renewable Energy Laboratory (NREL 2003: 11). This document suggests that conducting a charrette early in the design/decision-making process will:

- Establish a multidisciplinary team that can set and agree on common project goals.
- Develop early consensus on project design priorities.
- Generate early expectations or quantifiable metrics for final energy and environmental outcomes.
- Provide early understanding of the potential impact of various design strategies.
- Initiate an integrated design process to reduce project costs and schedules, and obtain the best energy and environmental performance.
- Identify project strategies to explore with their associated costs, time considerations, and needed expertise to eliminate costly 'surprises' later in the design and construction processes.
- Identify partners, available grants, and potential collaborations that can provide expertise, funding, credibility, and support to the project.
- Set a project schedule and budget that all team members feel comfortable following.

Examples of design charrettes for emergency work can be found quite readily. Burns (2002) describes a charrette on sustainable settlement design undertaken by the Rocky Mountain Institute (USA), which tackled the hypothetical situation of a 10,000 strong refugee population arriving at the Afghan–Pakistan border; the presence of a range of disciplines within the charrette was undoubtedly critical to its success. This provokes the idea that architects could invite resilient buildings experts into charrette-type meetings within the practice or through the client body, which would enhance the quality of the design being produced and give time to their learning about designing in resilience in general (and need not necessarily require access to sensitive data from other projects).

The next section considers further how it may be a case of developing new skills or working in new ways, rather than having access to new information, which may offer the most effective way forward in embedding resilience in architectural design.

Appropriate skills and decision-making for the long term

This section considers the inter-related issues of legislation, tools and education in the mainstreaming of resilient building design. It becomes apparent that the complexities of inter-disciplinary relationships and professional boundaries could hamper future progress on this matter.

Taking a broad view of resilience, Bosher et al. (2006) explain that there is an urgent need for the construction industry to change its ad hoc response to disaster risk management (DRM), but like Lorch (2005) they acknowledge the complexity of trying to do so in an environment with typically fragmented relationships between the various actors. This highlights the differences between the disparate professions involved with DRM, as reported by Trim (2004), but does not really shed any light on the architectural paradigm.

To examine the problem in detail, Bosher et al. (2007b) undertook a multi-disciplinary survey of attitudes towards the integration of DRM in construction. There was a lack of common understanding of who is involved at present, but more than half of the respondents stated that urban planners, designers, engineers (civil and structural), developers, clients and architects should be more involved with DRM than they are currently. In fact, in a series of interviews following the survey, it was urban planners who were seen as having the critical role in appropriate decision-making and enforcement to ensure a more integrated approach to resilience in local community planning (Bosher et al. 2007b). One analysis of the existing UK planning system suggests that it may be already overburdened in its 'impossible' role: 'to resolve contradictions and deep-seated conflicts between, on the one hand, competing private interests over the use and development of land and, on the other, private interests and community interests over property rights and development priorities' (Gillingwater and Ison 2003: 561). Likewise, Monbiot (2000) questions the integrity of the UK's arrangements for granting planning permission, but while none of these issues affect architects directly, neither do they absolve them of their own responsibility towards development.

A helpful parallel on the subject of integrating resilience within design can be found in the sustainable buildings paradigm. Following the publication of the World Commission on Environment and Development report (Brundtland 1987) which included what is now considered to be the seminal definition of sustainable development, there was a time lag of about 15–20 years in making the concept of sustainable buildings mainstream in principle, let alone in practice. In fact, the lack of a rigorous mechanism to legislate for either sustainability or resilience within buildings in the UK remained cause for concern until relatively recently, when the 'Sustainable and Secure Buildings Act 2004' (the Stunnell Bill) was passed with the aim of making buildings 'greener' and 'safer' by strengthening regulations on new, extended and altered buildings, requiring sustainability and crime reduction measures to be applied as a matter of course (The Stationery Office 2004). In a study of the effectiveness of regulation as a means to reduce the impact of natural disasters, Spence (2004) explains that appropriate regulation can have a positive effect, but it can be difficult to formulate and may require incentives to encourage change. The introduction of new mechanisms or incentives to deliver legislation is undoubtedly important, but will typically put pressure on architectural practitioners to become informed about resilience so that they can advise a client appropriately. Hence, the popularity of tools such as BREEAM (Building Research Establishment Environmental Assessment Method), which make concepts in policy and legislation more easily operable in practice (BRE 2007), but even with such tools available, only in the past few years has the attainment of a specific level on the BREEAM scale begun to be cited in planning requirements as a simple criterion for approval.

Returning to an earlier theme of design being a multi-dimensional problem-solving exercise (Lawson 1997), one tool that has proved remarkably successful in dealing with design decision-making is 'Design Quality Indicators' (DQI). This assessment tool to evaluate design and completed buildings was developed in response to ongoing calls from construction clients for greater involvement in the design decision-making process and to facilitate due consideration of end users' views (Gann *et al*. 2003). The DQI is exactly the sort of framework that appeals to the design manager as a facilitator and aligns with his/her aim to provide direction to the design team, but arguably not at the expense of their creativity (although Rich and Dean (1999) express concerns about such checklists). By automating client and user feedback on 97 indicators, the DQI provides an extraordinarily thorough, equitable and auditable commentary on the progress of a building design. One can easily envisage that such a system could also be useful in determining the 'resilience' of a building design, provided the appropriate quality indicators were to be used.

These examples provide interesting evidence of the potential for architects/design managers to act as facilitators (and for local planners to act as arbiters), all with the caveat that credible methods or tools become available. The case of sustainability also shows the possible extent of the time-lag involved in mainstreaming resilient design in policy and practice. Another part of the change necessary in practice of course, is the development of appropriate professional and subject-specific skills in resilient design. A recent review of skills undertaken by Sir John Egan, in this case in relation to sustainable communities, identified the importance of several built environment professions as 'core professions' (Egan 2004). However, it was the exploration of 'skills, ways of thinking and ways of acting' that could underpin good practice in the planning, design and occupation of sustainable communities which provides an insight into the potential type of personal and professional development that might be required. For example,

Egan suggests that inclusive visioning, breakthrough thinking and leadership are among the most important skills and that these should be coupled with behaviours such as creativity, humility and cooperation (Egan 2004: 56–7). This has interesting parallels with some of the core skills of a design manager (e.g. creativity and problem solving) (Cooper and Press 1995). Again, we can only come to the conclusion that developing new skills or working in new ways, rather than having access to new information, is an effective way forward. Hence, the question arises of involving professional institutions (such as the RIBA or equivalent) in the enhancement of architects' skills. One of the key findings from research by Bosher *et al.* (2007b) was to make available trans-disciplinary training for construction industry professionals and emergency managers; Lorch (2005) argues that change within higher education will be correspondingly vital. Sterling (2001: 38) acknowledges the difficulties in trying to introduce sustainability into the existing educational system and advocates a move from transmissive (instructive/ imposed) to transformative (constructive/participative) models of learning. There is some resonance between both Egan's and Sterling's ideas and the further integration of the resilience agenda in the educational system. A three-part model for a transformative educational model proposed by Sterling (2001: 84–5) is shown below.

1   Extended: appreciative, ethical, innovative, holistic, epistemic, future-oriented and purposeful.
2   Connective: contextual, re-focused, systemic, relational, pluralistic and multi- and trans-disciplinary.
3   Integrative: process-oriented, balancing, inclusive, synergetic, open and enquiring, diverse, a learning community and self-organizing.

One way of visualising the way in which such an approach could be applied to integrate resilience into design education and architectural practice is shown in Figure 9.1. UN/ISDR (2004: 17) puts forward four basic phases for DRM: pre-emergency, preventive and mitigating actions; emergency plans and preparedness activities; emergency response; short-term rehabilitation and longer-term reconstruction. Leaving aside 'response', these aspects can be combined with a model from Laseau (2001: 86) that views design problems as being solved by the interrogation of three areas: need, context and form. Figure 9.1 therefore suggests that key phases in DRM lie at the interstices of these well-understood areas of architectural design. Hence, this might offer a viable way of at least introducing the subject into teaching and practice in a holistic, systemic and balanced way.

Nevertheless, it is clear that there is no quick-fix solution to the mainstreaming of resilience within building design or indeed in the education of building professionals, but there are definite parallels with sustainability and lessons to be learned. The complexity inherent in design decision-

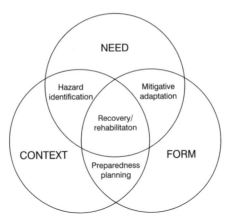

Figure 9.1 A construct for designing in resilience: integrating the four phases of DRM into a model of architectural design priorities (after Laseau 2001 and UN/ISDR 2004)

making, professional relationships and higher order issues such as planning legislation, provide some specific problem areas for future research and policy-making to address.

Conclusions

A clear case for all built environment professionals to have a better understanding of natural and man-made hazards and their effect on built assets has already been made, within the context of the UK and similar countries such as the USA. However, there remains a genuine schism in knowledge; those who govern and advise on the built environment are better informed and influential and there is a disjointed relationship between emergency managers and the construction industry. The presence of new legislation should enable government to enact change in the way planning and building level decisions are made; correspondingly, the role of urban planners, land-use planners and housing specialists has been identified as of particular importance. That said, in their role as innovative designers and managers of design information, architects can also be said to have a key role to play, certainly in the UK context. This is because their role is to interpret the client's physical, physiological and psychological needs within, of course, the overall legislative context. Hence, design has been described as a multi-dimensional problem-solving exercise and the professional body, the RIBA, has warned about the difficulties associated with design given the growing complexity of today's buildings.

Given the need for a framework within which legislation can be made operable and practitioners can respond appropriately thus meeting clients' needs, there is scope within this to develop a specific methodology to address the needs of architects and design managers. The aim would be to adapt their skills and working practices to the resilient building agenda. 'Resilient built assets' have been defined as those that have been designed, located, built, operated and maintained in a way that maximises the use of sustainable materials and processes while at the same time maximising the ability of the built asset to withstand the impacts of extreme natural and human-induced hazards. So, it would be credible to conclude that there may well be a trend for architectural practices to decide to specialise in resilient building design, just like some currently specialise in particular building types. This is plausible – perhaps there is simply too much to learn to treat resilience as just another issue and integrate it within regular building design. That said, one can envisage the architectural studio tutor devising appropriate exercises or conceptual design projects based on the interstices model shown in Figure 9.1.

In any case, there is a need for training, cross-fertilisation and skills development within industry and education; some suggest this could be sufficient, that it is more a matter of skills than of creating new technical information. More specifically, an example from the sustainable communities paradigm supports the notion of 'soft' skills development, such as creativity, humility and cooperation, rather than technical training as a means to improve participatory decision-making.

To improve and integrate the consideration of resilience within building design specifically, an appreciation of the duality of the roles of the design-focused architect and the management-focused architect/design manager is useful and leads to some more specific conclusions.

First, the existence of relevant legislation and a framework through which it is operationalised will be vital; without this the resilience agenda is simply too broad for an inexperienced individual or practice to tackle. If a framework were to be developed, the design manager or management-focused architect would be well placed to use it in his/her role as the single point of contact, organising the design process, drawing in design information, liaising with the client and ensuring that the right quality and level of design information is provided to other parties. This offers a good degree of congruence between the management of information required relating to resilience and the roles/ skills available. It is possible that a company with effective strategy and staff in this area could become a leader in the field quickly.

Second, the creative architect who is responsible for concept design development, deals with the client at an early stage and is known for their innovation, ideas generation and ability to manipulate multiple criteria, is a lynchpin in the design process. It is important that resilience does feature in the thinking behind his/her early response to a project, site and client's brief. Perhaps the best way to address this would be align with the principle of ideas generation through analogy, but using design charrettes, with input from emergency managers for instance, or using a system of resilience indicators. The idea of a transformative model of learning provides the basis for a new agenda (learning at the interstices of conventional architectural design practice) and the charrette example offers at least one potential solution. Thus, a fully extended, connective and integrative model of learning could offer a good chance of embedding the right skills to create resilient built assets.

References

Bosher, L.S., Dainty, A.R.J., Carrillo, P.M., Glass, J. and Price, A.D.F. (2006) 'The construction industry and emergency management: towards an integrated strategic framework', Paper presented at the Information and Research for Reconstruction (i-Rec) Third International Conference on 'Post-disaster Reconstruction: Meeting Stakeholder Interests', University of Florence, Florence, Italy, 17–19 May.

Bosher, L.S., Carrillo, P.M., Dainty, A.R.J., Glass, J. and Price, A.D.F. (2007a) 'Realising a resilient and sustainable built environment: towards a strategic agenda for the United Kingdom', *Disasters: The Journal of Disaster Studies, Policy & Management*, 31(3): 236–55.

Bosher, L.S., Dainty, A.R.J., Carrillo, P.M., Glass, J. and Price, A.D.F. (2007b) 'Integrating disaster risk management into construction: a UK perspective', *Building Research and Information*, 35(2): 163–77.

Brundtland, G. (ed.) (1987) *Our Common Future: The World Commission on Environment and Development*, Oxford: Oxford University Press.

BRE (Building Research Establishment) (2007) *Building Research Establishment Environmental Assessment Method (BREEAM)*, Watford: BRE. Online. Available: HTTP: <http://www.breeam.org/> (accessed 25 January 2007).

Burby, R. (ed.) (1998) *Policies for Sustainable Land Use, Cooperating with Nature*, Washington, DC: Joseph Henry Press.

Burby, R., Deyle, R.E., Godschalk, D.R. and Olshansky, R.B. (2000) 'Creating hazards resilient communities through land use planning', *Natural Hazards Review*, 1(2): 99–106.

Burns, C. (2002) *Sustainable Settlements Charrette: Rethinking Encampments for Refugees and Displaced Populations*, Snowmass, CO: Rocky Mountain Institute. Online. Available: HTTP: <http://www.rmi.org/images/other/Con-SusSettleCharRpt.pdf> (accessed 25 January 2007).

Cooper, R. and Press, M. (1995) *The Design Agenda: a Guide to Successful Design Management*, Chichester: John Wiley & Sons.

Cooper, R., Aouad, G., Lee, A., Wu, S., Fleming, A. and Kagioglou, M. (2005) *Process Management in Design and Construction*, Oxford: Blackwell Publishing.

Egan, J. (2004) *Skills for Sustainable Communities*, London: Office of the Deputy Prime Minister.

Gann, D.M., Salter, A.J. and Whyte, J.K. (2003) 'Design quality indicator as a tool for thinking', *Building Research and Information*, 31(5): 318–33.

Gillingwater, D. and Ison, S. (2003) 'Planning for sustainable environmental futures', in D.A. Hensher and K.J. Button (eds) *Handbook of Transport and the Environment*, Oxford: Elsevier.

Gray, C. and Hughes, W. (2001) *Building Design Management*, Oxford: Butterworth-Heinemann.

Hamelin, J.-P. and Hauke, B. (2005) *Focus Areas: Quality of Life – Towards a Sustainable Built Environment*, Paris: European Construction Technology Platform.

Kagioglou, M., Cooper, R., Aouad, G. and Sexton, M. (2000) 'A generic guide to the design and construction process protocol', *Engineering Construction and Architectural Management*, 7(2): 141–53.

Kendra, J. and Wachtendorf, T. (2003) *Creativity in Emergency Response to the World Trade Center Disaster on and beyond September 11th. An Account of Post-disaster Research*, Special Publication No. 39, Boulder, CO: Natural Hazards Research and Information Center, University of Colorado.

Laseau, P. (2001) *Graphic Thinking for Architects and Designers*, 2nd edn, New York: John Wiley & Sons.

Lawson, B. (1997) *How Designers Think: The Design Process Demystified*, 3rd edn, Oxford: Architectural Press.

Lorch, R. (2005) 'What lessons must be learned from the tsunami?', *Building Research and Information*, 33(3): 209–11.

Mileti, D.M. (1999) *Disasters by Design: A Reassessment of Natural Hazards in the United States*, Washington, DC: Joseph Henry Press.

Monbiot, G. (2000) *Captive State: The Corporate Takeover of Britain*, London: Macmillan.

NCI (National Charrette Institute) (2007) *The National Charrette Institute home page, What is a Charrette?*, Online. Available: HTTP: <http://www.charretteinstitute.org/charrette.html> (accessed 25 January 2007).

NREL (National Renewable Energy Laboratory) (2003) *A Handbook for Planning and Conducting Charrettes for High-Performance Projects*, Online. Available: HTTP: <http://www.eere.energy.gov/buildings/highperformance/pdfs/charrette_handbook/33425rep.pdf> (accessed 25 January 2007).

Philips, R. (2000) *The Architect's Plan of Work*, 2000 edn, London: RIBA Publications.

RIBA (Royal Institute of British Architects) (2007a) *Shaping the Future: Careers in Architecture*, London: RIBA. Online. Available: HTTP: <http://www.careersinarchitecture.net/> (accessed 22 January 2007).

RIBA (Royal Institute of British Architects) (2007b) *A Client's Guide to Engaging an Architect*, London: RIBA. Online. Available: HTTP: <http://www.riba.org/go/RIBA/Member/Practice_304.html> (accessed 22 January 2007).

RIBA (Royal Institute of British Architects) (2007c) *Outline Plan of Work 2007*, London: RIBA. Online. Available: HTTP: <http://www.ribabookshops.com> (accessed 18 January 2008)

Rich, P. and Dean, Y. (1999) *Principles of Element Design*, 3rd edn, Oxford: Architectural Press.

Spence, R. (2004) 'Risk and regulation: can improved government action reduce the impacts of natural disasters?', *Building Research and Information*, 32(5): 391–402.

The Stationery Office (2004) *Sustainable and Secure Buildings Act*, London: HMSO.

Sterling, S. (2001) *Sustainable Education: Re-visioning Learning and Change*, *Schumacher Briefing No. 6*, Devon: Green Books (for the Schumacher Society).

Toft, B. and Reynolds, S. (1994) *Learning from Disasters*, Oxford: Butterworth-Heinemann.

Trim, P. (2004) 'An integrated approach to disaster management and planning', *Disaster Prevention and Management*, 13(3): 218–25.

Tubaila, R. (2006) 'Measuring creativity in building design', Unpublished Masters thesis, Loughborough University.

UN/ISDR (United Nations International Strategy for Disaster Reduction) (2004) *Living with Risk: A Global Review of Disaster Reduction Initiatives*, Geneva: UNISDR.

UN/ISDR (United Nations International Strategy for Disaster Reduction) (2005) *Hyogo Framework for Action 2005–2015: Building the Resilience of Nations and Communities to Disasters*, Geneva: UN/ISDR.

Wamsler, C. (2004) 'Managing urban risk: perceptions of housing and planning as a tool for reducing disaster risk', *Global Built Environmental Review*, 4(2): 11–28.

Wamsler, C. (2006) 'Mainstreaming risk reduction in urban planning and housing: a challenge for international aid organisations', *Disasters*, 30(2): 151–77.

Weir, D. (2002) 'When will they ever learn? The conditions for failure in publicly funded high technology projects: the R101 and Challenger disasters compared', *Disaster Prevention and Management*, 11(4): 299–307.

Wikipedia (2007) *Architect – from Wikipedia*, Online. Available: HTTP: <http://en.wikipedia.org/wiki/Architect> (accessed 22 January 2007).

Part III
Non-structural adaptation

Chapter 10

Community-based construction for disaster risk reduction

Marla Petal, Rebekah Green, Ilan Kelman, Rajib Shaw and Amod Dixit

Speedy reconstruction and building codes are not panaceas

When disaster mitigation advocates reported in 2006 that a well-known international non-governmental organisation (NGO) had to tear down several hundred newly-built houses because they were found to be unsafe, some viewed this as the 'good news'. The bad news was that many more unsafe reconstruction projects would remain undetected until the next disaster. Similarly, in 2004, a consultant to the United States Agency for International Development (USAID) reported 'good news'. In Kabul, he had convinced USAID to retrofit hospitals and schools built since the end of the 2001 war which overthrew the Taleban government, in order to make them more seismically resistant. The bad news was that other international donors said that there were not enough funds for such earthquake safety.

Large-scale, top-down, technologically-driven reconstruction projects that typify post-disaster reconstruction engage outside engineers and builders, introduce new and expensive construction technologies, supplant both local knowledge and local labour, and do not necessarily reduce vulnerability overall. Homes are often built without regard for their effects on the lives and livelihoods of the people who will live in them. The nature and extent of participation is frequently limited (Twigg 2004). In these, and many similar but smaller projects, the focus is on physical structures (houses) rather than on living and livelihoods (homes).

The adverse consequences can be observed almost immediately. Confidence and trust in viable traditional construction approaches are undermined. Because inhabitants are dissatisfied with the housing that results, and because communities do not have an opportunity to be involved in making safer construction and retrofitting a priority, the resulting structures can have unnecessary vulnerabilities (Twigg 2004). Neither safety nor sustainability is attained (Schilderman 2004). The 'myth of speed' that affects most aid agencies must be resisted (Anderson and Woodrow 1998: 49).

Poverty is often suggested as breeding fatalism with regards to disasters. In reality, when informed choices are permitted with regards to building, most people tend to incorporate safety concerns (Maskrey 1989). People who have homes built for them – without consultation, without information and without choice – will naturally adopt a fatalistic view of the product, including with regards to safety.

School construction faces similar challenges. While pursuing the goal of 'Education for All' (WCEA 1990; WEF 2000; UNESCO 2004), rapid new school construction projects could focus on the immediate product of new schools without considering longer-term and wider contexts of education and safety. Children stand to be put at risk from disasters just by going to school (Wisner *et al.* 2007).

To overcome these challenges, shelter and other buildings which serve communities must be viewed as processes rather than as products only (after Davis 1978; UNDRO 1982). Architects, engineers, and planners are frequently tempted to look at the vulnerabilities of people and the environment as simply the result of poorly designed and poorly built structures. They focus on the technical task of fixing the offending buildings and community design to avert future disaster, working assiduously on designs, materials and construction details. Meanwhile, physical scientists have improved the accuracy and precision of estimates of, for example, ground-shaking for micro-zonation and storm surge heights to the nearest centimetre. Increasingly detailed and now performance-based building codes have been produced while pioneering advocates from these professions have pursued their adoption as public policy in many countries.

Most of these efforts have focused on changing the structures while little attention has been paid to effecting needed change within specific social, political, cultural and economic environments. The consequence is that the people who are the intended beneficiaries of these advances in both technical knowledge and policies often become steadily more vulnerable. This tragic irony suggests the necessity for a community-based approach to construction for disaster risk reduction.

This chapter provides a foundation for a community-based approach to construction for disaster risk reduction. The motivation is detailed, principles are outlined, and case studies of the approach in practice are provided. Common threads are drawn from the discussion and future challenges are articulated. This chapter begins the process of filling in the identified gaps by using past experience, successes and failures, to suggest a community-based future for construction for disaster risk reduction.

Motivation for community-based disaster risk reduction in construction

Embracing a community-based approach to construction emerges from three principal observations: (a) the failure of past approaches, (b) examples of adopting other safety measures suggest that bottom-up educational approaches are most successful in bringing about behavioural changes and powerful social policies, and (c) to ignore community-based approaches is to perpetuate oppression and vulnerability.

In such work, it is important to continually ask 'what or who is the community?' Communities are rarely homogeneous with every individual being treated equally because divisions occur by gender, age, experience, leadership capability, leadership style, culture, and religion amongst other factors. An important step is identifying how a community is defined by itself and by others (e.g. geographically or culturally); who purports to represent the community and how they achieved that position (e.g. elected, appointed, or hereditary); who is assumed to be in the community by themselves and by others; and who is marginalised from a community. The answers might not be clear, but asking the questions assists in understanding the different views and characteristics of communities, communities within communities, and sectors of communities.

Failure of past approaches

Most evidence suggests that the 'structures focus' on buildings, building codes, and building code enforcement has succeeded moderately well in wealthier economies. Building codes and building practices have been steadily improved in wealthier nations with professional engineers and builders, well-educated users, well-developed code-enforcement systems, and well-publicised consequences of failure. Notwithstanding the upward revision of minimum standards following each new disaster along with the legacy of older building stock built to a lower standard, positive results have been achieved.

Disaster-induced structural damage and human casualties in countries such as Japan and the USA have decreased over the long-term. A single event, such as the 1995 Kobe earthquake in Japan and Hurricane Katrina in 2005 in the USA can easily reverse this apparent success. Other unsettling evidence from wealthier countries exposes inadequacies in the purely structural approach. For example, in 1992, Dade County, Florida had one of the toughest building codes in the USA, but much of the approximately US$ 30 billion of damage occurred that year when Hurricane Andrew struck because buildings were not designed according to the code and poor enforcement practices failed to uncover the problems (Coch 1995). Similarly, examples from several tornadoes in eastern Canada indicated that buildings in which people were killed or seriously injured did not satisfy requirements of Canada's Building Code (Allen 1992).

Building code enforcement as a strategy is most likely to succeed where there is social demand for safe construction; the educational resources for builders to know about and implement the standards; the financial resources for meeting these standards; and a large, well-trained, and adequately paid cadre of licensed professionals, technical and enforcement, with a manageable caseload who are able to respond rapidly to problems.

These conditions are absent in the majority of less affluent countries. The needed professionals are rarely tested or licensed and most construction is done without professional training, input or monitoring. Shifting away from the structures focus suggests a community-based imperative that emphasises knowledgeable users and educated builders with the goal of complying with building codes.

The starkest failure of the structures focus can be traced in countries with well-educated engineering elites who have promulgated these same universal standards, yet where most of the construction is semi-engineered or not engineered. Considering earthquakes, India, Mexico, Iran and Turkey have suffered heavy loss of life from disasters, and large vulnerabilities remain, despite their adoption of building codes suited to modern construction techniques.

For instance, the 17 August 1999 Turkey earthquake disaster resulted in more than 17,000 deaths, mostly in urban construction built after the 1980 adoption of modern building codes which considered seismic vulnerabilities. The disaster revealed systematic inadequacies in site planning, design, construction detailing, and monitoring and enforcement of regulations (Sengezer and Koç 2005) and almost total lack of awareness of the risk exposure and ignorance regarding the purpose of various construction measures (see also Lewis 2003).

With increasing global cultural homogeneity and rural-to-urban migration, traditional disaster-resistant materials and design and construction techniques have been increasingly undervalued. As skylines become dominated by reinforced concrete structures, diminishing numbers of artisans apply their skills in working with adobe, stone and rammed earth. Modern forms of construction, perceived as 'development' and 'progress', have undercut the value of traditional apprenticeships, degraded traditional construction and demanded technical knowledge and skills that builders have not yet acquired (Langenbach 2007). The lack of formal educational opportunities combined with high illiteracy make it challenging to communicate knowledge and techniques. The impact of poverty and rapid urbanisation and further discussion of the social complexities of building enforcement are detailed by Rebekah Green in the following chapter.

The simplest response to this abject failure despite the best intentions, is to shrug at the intractability of rapid urbanisation and poverty, to sigh at the tragedy of corruption in construction industries with no transparency, to refer obliquely to 'inadequate governance' and 'failure of political will', and to repeat that the solution lies in better technology, better codes, and better enforcement. Socio-economic, political, and educational contexts which hinder more effective code development, monitoring, and enforcement are presented as unfathomable domains in which mitigation advocates are helpless and for which design professionals bear no responsibility. A community-based approach helps to overcome these challenges.

Appropriateness of the community-based approach

A 'community-based imperative' is needed in which construction and design professionals learn to share their knowledge with, and at the same time learn from, the users of the structures. These users include owners, renters, teachers, school children, activists, construction tradespeople and government workers. This knowledge exchange would yield a bottom-up demand for safe construction, voluntary compliance with standards, and public, government, and private sector expectation and support for enforcement.

The concept of 'building-code compliance' may usefully replace 'building-code enforcement'. This approach shifts away from punishment as a primary motivator and instead points toward a community-based imperative that emphasises users and builders who are educated sufficiently to take the lead in voluntary compliance and in developing a leadership from the grass roots up. Building codes are not a panacea, they are but one important tool among many in the task of developing broad-based user awareness and demand for safety.

The histories of several successful public adoptions of safety measures do not include legislation as the main step. Examples are washing hands and brushing teeth. Despite the abstract concept of microbes that transmit disease, wealthier countries in particular generally succeed in teaching hygiene to young children for illness and disease prevention. Laws have been useful for reinforcing and maintaining this behaviour, mostly for those in the food production, health and public service industries (Glanz *et al.* 2002).

In many cases, public demand through zealous advocacy led to safety improvements. Disasters are often necessary catalysts. The sinking of the Titanic in 1912 inspired universal demand and expectation for life jackets and led to the formation of the international Safety of Life at Sea Committee (Transport Canada 2003). The 11 September 2001 terrorist attacks in northeastern USA forced changes to air travel security which had before been considered too expensive and unnecessary. At other times, individual advocates are the major influence. For instance, in the USA, Ralph Nader and Mothers Against Drunk Drivers applied pressure to manufacture cars with seat belts, and influenced the public consciousness to wear them, long before legislative mandates (Fell and Voas 2006).

Public awareness and voluntary adoption of safety measures have typically preceded strong and well-enforced social policies. Public pressure then generates sufficient 'political will' to codify the approaches and to introduce the ideas into school curricula. By this time, children remind parents how to act, such as always wearing seat belts, and formal support programmes emerge, such as providing child car seats. Policy enforcement serves chiefly to capture outlying behaviour. Some aspects of the safety measure may remain overlooked, as exemplified by the absence of seat belts from most public buses and trains. These lessons can be applied to building construction for disaster risk reduction.

Top-down approaches perpetuate oppression and vulnerability

How do well-intentioned advocates unwittingly participate in oppression? Laws (1994) argues for an important addition to Young's (1988, 1990) five classic faces of oppression (exploitation, marginalisation, powerlessness, cultural imperialism and violence): adding knowledge denial (Table 10.1). Knowledge denial is a phenomenon rooted in the hierarchical organisation of knowledge in society. Social injustice is perpetuated in part because of the weight and credibility we give to the voices of scientific and technical 'experts', to the exclusion and denigration of knowledge acquired through experience. Written, logically-presented material is too-often valued above knowledge transferred orally and the unwritten experiences of community elders or the visual evidence left by vernacular construction.

Indigenous peoples, those with grass-roots knowledge, and marginalised populations tend to be outside the powerful knowledge-production environments and appear only as objects and not as subjects, agents or authors. Weichselgartner and Obersteiner (2002: 76) mourn that 'disaster schemes and programs still treat people as "clients" in disaster management processes where science and technology do things to them and for them, rather than together with them'. The most significant reason that scientific 'expert' knowledge and indigenous knowledge of design and construction for disaster risk reduction are sometimes not fully used is that neither are yet owned by the people who must use them.

For disaster risk reduction, Shah (2003) issued an appeal, using the classic telecommunications metaphor, exhorting those with valuable knowledge to 'go the last mile' to reach those most in need of vulnerability reduction techniques and strategies. He argues that despite the availability of a wealth of material for making communities safer from disasters and the extensive research supporting that knowledge, a gap exists in reaching the right people with the right strategies for disaster, risk and vulnerability reduction.

However, 'the last mile' has been critiqued for placing last those who should be considered first. Instead, the phrase 'going the first kilometre' is suggested here to emphasise that connecting with those who will directly experience disaster should be the primary goal, not the last endeavour. Indigenous peoples, those with grass-roots knowledge, and marginalised populations should be the first point of contact in discussing or developing disaster risk reduction strategies – not the last point of contact on whom established notions are bestowed. Shah (2003) rightly addresses the need

Table 10.1 The faces of oppression in construction for disaster risk reduction

Face of oppression	General manifestation	Examples and counterexamples
Exploitation	Poor people are forced to live in unsafe areas with limited livelihood opportunities and where they are exposed to multiple hazards.	In Dar es Salaam, Tanzania, workers living near the Kunduchi quarry site are exposed to the effects of unregulated mining activities. The pollution results in disease, injuries, and deaths.
Marginalisation	Urban land use forces rural migrants and poor people into areas known to be unsafe. Building safety regulations may not exist and disaster-related insurance is unavailable or unaffordable.	In New Orleans, USA, poorer people were settled in known flood zones, where insurance was either not recommended, unavailable, or unaffordable. After Hurricane Katrina in 2005, they had no resources to reconstruct or to move.
Powerlessness	Failure to enforce existing building regulations and unresponsiveness to civil initiatives for improving the regulations make it impossible for many to improve infrastructure safety, site selection, or construction standards. Laws are unevenly applied.	In El Salvador, a court case by one community challenged a plan for deforestation to make way for luxury homes. The case failed. On 13 January 2001, an earthquake-induced landslide along the deforested area killed 700 people in the same community.
Cultural imperialism	External and internal forces promote the superiority of western construction materials such as steel-reinforced concrete, ignoring the indigenous heritage of disaster-resistant designs and materials that often have economic, climatic, environmental, and aesthetic advantages.	In Yunnan Province, China, concerted efforts are being made to re-introduce and update vernacular bamboo construction. Meanwhile, knowledge regarding disaster-resistant properties of traditional timber-frame houses in Turkey, reinforced adobe structures in Iran, Dhajii in Kashmir and other such wisdom is almost lost.
Violence	Violence forces displaced people into inadequate shelter where long-term housing needs cannot be met. Squatters are forcibly evicted.	In 2005, hundreds of thousands of homes in Zimbabwe's shanty-towns were burnt and bulldozed.
Knowledge denial	In mega-cities, hazardous self-built and semi-engineered structures continue to be constructed and tolerated, although many low- and no-cost measures could make them substantially safer. Design and construction processes exclude occupants from both decision-making and acquiring the skills for safe construction.	In post-disaster reconstruction worldwide, transitional and permanent settlements are often designed and constructed without consulting the people who will live in them. In Turkey, India, Nepal, Aceh, and Central Asia, efforts are under way to overcome knowledge denial with community-based construction initiatives.

to overcome knowledge denial, but the 'the last mile' metaphor can only reproduce the existing hierarchy while 'the first kilometre' integrates outside 'experts' with the communities seeking to implement construction for disaster risk reduction.

The community-based imperative suggests that advocates go not only 'the first kilometre' but the entire distance side-by-side with the community. People vulnerable to disasters must be accorded the respect and equal footing with those interested in helping to reduce vulnerability. Scientific and technical 'experts' are often surprised and encouraged by the nature and robustness of local values and knowledge. As Laws (1994: 11) argues, 'Oppressive situations can be changed by listening to those on the bottom of the knowledge hierarchy. Their knowledge ... can indeed lead to new and different ways of seeing.'

For community-based construction, it is important to combine external and internal knowledges through some form of participatory action-oriented community risk assessment. The result will encompass the risks which the community is aware of, and has usually experienced, along with those which have not been identified or of which people may have marginal awareness but no direct experience. Some solutions will be provided locally, some externally, and many will be developed through collaboration and interaction. Cronin et al. (2004a,b) provide case studies of this approach for small islands which are particularly vulnerable to volcanic hazards.

While this approach might seem to be unnecessarily laborious, as mutual trust is gained and as knowledge flows back and forth, innovative and unexpected solutions are found, tested, and owned by communities. The most successful outcomes have broad support and action from local residents, rather than emerging wholly from external specialists, professionals, or interventions (Petal 2004; Twigg 2004; Green 2005). 'Bridging' individuals and organisations are able to take the ideas and work to 'scale-up' to reach larger populations. Knowledge exchange rather than knowledge transfer is enacted.

Successful community-based disaster risk reduction

Experience from successful community-based disaster risk reduction construction efforts suggests these principles are at work:

- *Support local ownership and participation:* Community-based efforts are locally-developed and locally-owned, even if the impetus comes from external advocates. Broad participation is encouraged. Participants come to understand that they can control their own fates and witness the impact of their actions on their own safety and security. They frequently reach out to share this experience with others. Disaster risk reduction takes on meaning and relevance because it is understood to have positive and tangible impacts on daily life. Cost-effectiveness is clear (Brown *et al.* 1997; FEMA 1997, 1998; PAHO 1998; ASFPM 2002; BTRE 2002; Twigg 2003) and often leads directly to improved choices, education, livelihoods and community. Resources gained can be quantified and re-invested.

* *Share information:* The discussion of housing and construction is widened to include building users, so that informed choices can be made.
* *Build relationships:* Dialogue is required between building users and builders and artisans, as well as between local communities and outside facilitators and 'experts'. Knowledge exchange and education are reciprocal. Trust and respect, the foundation for sound decision-making, are forged over time.
* *Build capacity:* Long-term capacity-building is seen as a primary goal of external interventions. This refers to developing human skills and societal infrastructures to reduce ongoing risks (UNISDR 2006). Documenting and sharing lessons learned plus impacting formal education are key indicators. Create long-lasting educational resources.
* *Repeat and evaluate, continuously:* Risk reduction is an ongoing and iterative process, rather than being about single or one-off actions. At the individual level, the process is part of usual, day-to-day life rather than being a special activity to be completed and then forgotten. At the community level, it seeks to be 'mainstreamed', that is, adequately incorporated into policies and practices across many areas (LaTrobe and Davis 2005).

Many tools exist for putting these principles into practice:

* Generic, such as ProVention's Community Risk Assessment Toolkit (ProVention 2006).
* Regionally specific, such as for Asia (Abarquez and Murshed 2004), and the Comprehensive Hazard and Risk Management (CHARM) programme for Pacific small islands (SOPAC 2005).
* Sector specific, such as transitional settlement and shelter (Corsellis and Vitale 2005), vernacular architecture conservation (Langenbach 2007), confined masonry adobe, and other construction handbooks for masons (Brzev 2007; Stephenson and Schacher 2006; Blondet 2005; Blondet *et al.* 2003; Dixit 2003), and shorter public awareness materials to introduce principles of earthquake-safe adobe construction in Central Asia (GHI 2006).
* Multi-country building codes and standards (IAEE 2004) and retrofitting standards (USAID-OAS 1997).
* The *World Housing Encyclopedia* (EERI 2006) and the World Adobe Forum (EERI 2007).

- Compilations of successful case studies from around the world (UNISDR 2004, 2005).

Caution is suggested because transferability amongst locations and sectors might be limited. These examples nevertheless provide insight into past achievements in community-based construction education and action for disaster risk reduction.

A specific example of success occurred in the late 1990s when the Intermediate Technology Development Group (ITDG) worked with local NGOs on housing projects for disaster risk reduction in Peru, the Philippines and Bangladesh, amongst other locations. The focus was placed on maintaining and improving local and traditional techniques. In Peru, initial expectations of local participants were that new houses would be built with modern materials and external aid. As the participatory process developed, including showing pictures to share vernacular construction types throughout the region, the people realised that affordability and sustainability meant retaining local approaches. This process took six months, a length of time usually not tolerated by aid agencies (Clayton and Davis 1994; Aysan et al. 1995; Coburn et al. 1995).

Community-based structural awareness, construction education and practice

Five specific examples of community-based construction programmes for disaster risk reduction are described, with a focus on increasing earthquake resistance. Table 10.2 lists past and future earthquake disasters to which these programmes apply. While these examples are heartening, most post-disaster reconstruction and most development construction continues to disregard long-term disaster risk reduction. Much remains to be achieved, but there are good practice case studies to emulate.

Public awareness, shake tables and mason training in Nepal

In Nepal in the 1990s, the National Society for Earthquake Technology (NSET; http://www.nset.org.np) – a small, local NGO that now has international reach – was launched with support from a small US-based NGO, GeoHazards International, and several international donors. One of their objectives has been to dispel the myth that a poor country can do little to reduce disaster risk before being hit. NSET has demonstrated that disaster risk reduction can be effectively interwoven into development work while simultaneously meeting the most pressing needs for education, health and other basic services.

Table 10.2 Case studies of community-based construction education programmes for disaster risk reduction

Date and event	Location	Consequences
17 August 1999 Earthquake	Western Turkey	Over 17,000 people killed, over 3,000 buildings collapsed, and 23,400 buildings condemned (Erdik 2003)
26 January 2001 Earthquake	Gujarat, India	Over 20,000 people killed, 300,000 buildings collapsed, and 600,000 buildings severely damaged
26 December 2004 Earthquake followed by tsunamis	Indian Ocean coastal areas	In Aceh, over 160,000 dead with dozens of villages flattened and 100,000 buildings damaged
8 October 2005 Earthquake	Kashmir, Pakistan and India	Over 70,000 dead. In Pakistan, 203,579 housing units destroyed and another 196,574 damaged (Asian Development Bank and World Bank 2005)
Earthquake yet to come (as of 2007)	Kathmandu Valley, Nepal	More than 1.5 million people could be affected by the next earthquake. Current community-based risk-reduction efforts are improving construction practices
Earthquake yet to come (as of 2007)	Central Asia	Major cities of Central Asia are expected to be affected by a major earthquake. Tashkent (population 2,000,000+) or Almaty (population 1,000,000+). Current community-based risk-reduction efforts are improving construction practices

The first forays into community-based disaster-resistant construction in Nepal eschewed formal training programmes and lengthy technical manuals. A shake table was constructed with a half-scale model to compare aesthetically attractive but seismically unsound local construction with seismic-resistant construction designs using the same layout and materials (Figure 10.1). Shaking accomplished in early versions was through a tractor-induced jolt. The shaking lasted less than a minute, but brought down the unsafe house and left the robust construction standing. Masons and villagers who witnessed the demonstrations were instantly convinced of the necessity for seismic-resistant construction.

The programme has been developed further to address school construction. In Nepal, schools are built and financed locally and informal construction will remain prevalent for the foreseeable future. NSET's School Earthquake Safety Initiative involves communities in school construction and retrofitting, focusing on informal, site-based, mason training. The school construction process remains under local control with engineering guidance and mason training as the only add-ons to an otherwise intact and familiar process. Financial transparency allowed residents to experience the cost-effectiveness

Figure 10.1 A shake table demonstration for Earthquake Safety Day 2007 in Nepal
(Photograph courtesy of NSET)

and affordability of the efforts. Teachers, students, and parents gain awareness
and knowledge about disaster risk reduction.

At each site, a minimum of six masons receive training in seismic-
resistant construction and retrofitting. A visiting engineer and an on-site
mason supervisor are the only NSET-funded staff. NSET reports that the
kin of masons often attend mason-training sessions held after the end of
the working day. The interest and pride of family members has a positive
influence on changing the culture of construction far beyond the bounds of
the school yard. Each mason brings their improved knowledge and skills to
between 10 and 15 new construction sites every year. NSET has published
a Mason Training Manual and a colourful poster and calendar illustrated
for a general audience to raise awareness of disaster-resistant construction
techniques (Dixit 2003).

To spread these lessons to urban areas, NSET seeks invitations to the
meetings of social clubs and other community-based organisations and offers
free orientation programmes on earthquake risk. These invitations inevitably
lead to questions about do's and don'ts and invitations to return. NSET
engineers make regular presentations to government officials, planners,
municipal workers and other authorities to communicate earthquake risks
and to convince them of the need to mitigate these risks. Meanwhile, NSET
also provides a free weekly consultation to house owners who attend with

their engineers/architects and masons. Detailed discussions are led by NSET engineers on aspects of design and construction.

Evaluative research has yet to be undertaken to learn how to maximise and monitor the ripple effects of local mason training. Another challenge is adapting the training for wider-scale implementation, beyond just the local level, without losing useful vernacular approaches. Unknowns include the relative success of pre- vs. post-disaster timing and how to determine the critical mass of trained masons or of villages per district which would allow the practices to influence a wider region.

To streamline the process, using videos of shake table demonstrations is being considered to replace live scale-models. Volcanologists have had reasonable successes using videos of volcanic hazards to convince rural populations with low literacy levels of imminent dangers that had never before been witnessed in the area. Could that success translate into videos of shake table results or of post-earthquake damage assessments? How could rural approaches be applied in urban environments? The successes in pre-disaster Kathmandu Valley demonstrate the tremendous potential of educating for community-based construction for disaster risk reduction.

Co-learning in India

India has an enormous problem with buildings vulnerable to natural hazards, especially since more than 80 per cent of buildings are from the informal construction sector. Most private residences still follow practices based on available knowledge and skills of the local construction workers. The technical–legal regime has largely remained ineffective in most medium- and small-sized towns, as well as in rural areas. The Sustainable Environment and Ecological Development Society (SEEDS, http://www.seedsindia.org), an Indian NGO, is addressing these challenges, working with some of the most vulnerable communities to provide appropriate technologies and techniques in building construction to reduce disaster risk.

SEEDS refers to its approach as 'co-learning'. Building safety is promoted through multi-pronged participatory strategies. The organisation's field practitioners and the communities where they work are each aware of the other's contributions to learning, applying and promoting know-how for safer construction. Community participation and local capacity building are key. Disasters, too, are used as opportunities for catalysing desired change in prevalent construction practices. The principal goal is that reconstruction efforts not only yield safer buildings but also ensure that the safer building practices become normal for the local building industry.

SEEDS was involved in reconstruction following the 2001 Gujarat earthquake and the 2004 Indian Ocean tsunami. Following the 2001 Gujarat earthquake, SEEDS joined with the United Nations Centre for Regional Development (UNCRD) and the Earthquake Disaster Mitigation Research

Center (EDM) in Kobe, Japan to implement *Patanka Navjivan Yojana*, a rehabilitation programme in Patanka village. The programme allowed house owners to lead reconstruction of their own homes with assistance provided by architects and engineers. The goal was to build local knowledge capacity for earthquake-resistant construction with minimum modifications to existing construction practices and material use.

Each house's design was different, reflecting the taste, culture, size, and economic capacity of each household. Moreover, each family retained the freedom to choose the pace and design. Steel and wood workshops were established within the village to provide training to the village's construction workers as well as to willing house owners. The training urged each family to adhere to the minimum building standards which led to a transparent self-monitoring mechanism. Architects and social workers worked closely with the families to support them in making their living spaces safer and more efficient. This reconstruction process, not just the reconstruction product of a house, promoted each family's understanding and appreciation of the usefulness of 'safer and more sustainable' building design. Community ownership and hands-on learning were key factors for the initiative's success. The process further enabled the community to share their learning with other neighbouring villages as local builders circulated to work on construction projects in the district.

This Gujarat story led to related initiatives. In India's Himalayan region, at risk from regular earthquakes and landslides, communities shared experiences with people from Patanka, engaged in group discussions, and considered scenarios related to their own contexts. Following the 2004 Indian Ocean tsunami and the 2005 Kashmir earthquake, SEEDS team members included masons from Patanka who shared both their empathy and knowledge with fellow disaster-affected citizens. This peer-level exchange led to instant trust and friendship that became the starting point for co-learning in order to share knowledge and build safer shelters.

To consolidate learning, to create a support network, and to institutionalise the promotion of building safety, the SEEDS Masons Association was founded. By late 2006, the Association had more than 800 members in four regions of India. Members are provided with training and exposure to the best construction standards. A database enables house owners to find Association members while the Association's newsletter, *Buniyaad* (Foundation), keeps members informed and connected.

Housing facilitators in Aceh

The cataclysmic 2004 Indian Ocean earthquake and tsunami disaster struck hardest at Aceh, Indonesia. The World Bank accepted a community-driven reconstruction policy in order to support the Indonesian Reconstruction Agency's in-situ settlement rehabilitation and reconstruction in 400 villages.

Together with UN-Habitat (the United Nations Human Settlements Program) they adopted a 'housing facilitators' approach.

Four major players are involved in this community-based reconstruction process:

- Housing facilitators: young professionals, mainly engineers and architects, working with local communities to facilitate the housing construction process, especially ensuring construction quality.
- Community contractors: the representatives of local residents. In Aceh, each neighbourhood was divided into a cluster of 10–15 houses. Each cluster selected one person as their representative. That representative is the community contractor whose responsibility it is to ensure the proper construction of houses.
- Masons and workers: employed by the community contractors for constructing houses. In many cases, the masons and workers were from outside Aceh.
- House owners: need to be heavily involved in the construction process, although frequently that does not occur.

As a first step to implementing housing facilitators, surveys were undertaken in order to understand training needs and to develop training programmes. These surveys identified technical, social, and management concerns.

Technical issues related mainly to quality control of the buildings. Although all the buildings are reinforced, the quality of the reinforcement displayed major problems. Construction materials – particularly the size specification on the iron bars, cement sand mixing, and specification of the concrete – were highlighted as concerns along with workmanship which is always a key issue for masonry construction. Responsibility for oversight of these issues lies with the housing facilitators, but informal interviews and interactions with them indicated that they were not properly aware of such details nor of the justification for being concerned about these details. Local masons, carpenters, and labourers also did not necessarily know what they were aiming to achieve and why.

Social issues highlighted the involvement of local house owners in reconstruction. In ideal community-based reconstruction, the homeowner's role is crucial for establishing ownership of, and confidence in, the design and for ensuring quality control of the houses. In the Jeulingke area, some homeowners were heavily involved in the construction process, visiting the sites with their families almost every day. In contrast, in the Lambung area, homeowners hardly visited the sites, leaving the construction entirely to the community contractors and housing facilitators. Simple but accurate information needed to be provided to the homeowners; for example, guidelines or checklists highlighting key points, what to look for, and how to become positively involved in the construction of one's own home. This material must be in the local language(s) and be sensitive to local cultural needs.

Management issues revealed concerns about coordination and information flow. The interviews showed that housing facilitators spent approximately 60 per cent of their time on disbursement issues, 20 per cent on procurement, and 20 per cent on technical aspects. So while the housing facilitators were busy with procurement, disbursement, and other budget-related items, technical issues were relatively neglected. Moreover, a system did not exist for reporting questions and concerns to higher levels, such as the District Management Committee or the Provincial Management Committee. Coordination could also have been improved with the community contractor, who is often assisted by the village team supervisor and community volunteers.

The reconstruction approach began officially in December 2005, a year after the disaster, and was scheduled to be completed at the end of 2007. These issues were raised within the first six months and efforts were initiated to address them. Construction quality and the programme's success will depend on a balanced approach of the interaction amongst the four above-mentioned players. Properly monitoring the effectiveness of these partnerships will assist in learning from Aceh's reconstruction in order to be ready for reconstruction following disasters elsewhere and to use this approach for retrofitting and new construction before a disaster strikes (Schilderman 2004 also discusses such partnerships for other case studies).

Non-structural earthquake mitigation in Turkey

In the context of post-disaster humanitarian assistance, following the 1999 Kocaeli and Düzce earthquakes in Turkey, the American Friends Service Committee (AFSC) funded a participatory needs assessment with ten local neighbourhood organisations in Istanbul. Citizens expressed pressing concerns about the need to raise public awareness and to engage citizens in disaster risk reduction and preparedness for a large earthquake expected in the Marmara region, near Istanbul. Their priorities were basic disaster awareness and training citizens to be disaster responders.

To address these priorities, Boğaziçi University's Kandilli Observatory and Earthquake Research Institute launched the Istanbul Community Impact Project (later to become the nationally-focused Disaster Preparedness Education Unit) with support from USAID's Office of Foreign Disaster Assistance. Project staff delivered blunt messages about the known dangers of semi-engineered and un-engineered houses: earthquakes don't kill people, buildings do, but the specific reasons for earthquake-related building collapse were not initially explained. In the programme's first two years, community educators, engineers, and a community-based advisory committee found structural safety to be too daunting a subject to translate into a public education campaign. Instead, the team chose to focus first on a more manageable topic, non-structural risk reduction, hoping to use that material as a basis for tackling structural safety.

The team started community engagement with an epidemiological field study of the causes of deaths and injuries in the Kocaeli earthquake. The survey's results illuminated the high incidence of falling, cutting, piercing and crushing injuries caused by household furnishings and objects during the main shock and aftershocks. These injuries occurred in buildings at all damage levels (Petal 2004).

The team used these results in workshops and seminars. Adult and youth participants in basic disaster awareness seminars (45–180 minutes) were introduced to the need for individual and household measures for physical protection. The 'Earthquake Hazard Hunt' became a cornerstone activity to identify, prioritise and then move or secure building contents to protect the inhabitants. A youth leadership project engaged teenagers from four neighbourhoods in learning practical steps, such as securing furnishings, and in demonstrating these activities. A table-top model was created and reproduced to demonstrate these measures in four main rooms of a house (Figure 10.2).

Some of the attachment devices needed to secure furniture to the house's structure were not being marketed, so the team showed samples of American

Figure 10.2 School children try a hands-on model to see the effects of shaking on building contents (Photograph courtesy of Bodrum IMMOD)

devices to seminar participants. A demonstration project was carried out to secure furnishings and equipment in a large public hospital. This activity began to spur the development of locally-produced devices.

As the concept of non-structural mitigation became more widespread, instructors and participants of basic disaster awareness seminars and pilot projects began to ask detailed questions about applying non-structural mitigation in Turkey. Turkish residents were uncertain how to use imported products in their concrete frame and masonry infill homes and were unaware of where they could find comparable Turkish devices. Those who did know about the fledgling entrepreneurs making these products locally were wary of their quality. Could Turkish products really be as effective as the vastly more expensive imported items?

With both trainers and participants needing further information on non-structural mitigation, a project to develop practices for Turkish non-structural mitigation and to create a public training seminar on non-structural mitigation was funded with support from the Turkish Red Crescent Society and the American Red Cross. A local Turkish trade school teacher who was to become an instructor trainer and an American engineer led the project. They formed a cross-disciplinary, cross-sectoral, and cross-cultural partnership with inputs from a similarly diverse and concerned group of community-based basic disaster awareness instructors. The project leaders used engineering design calculations, local construction techniques, and cultural sensitivity as the bases for recommendations.

Initial research on the availability of non-structural mitigation devices in neighbourhood hardware stores helped to locate potential devices and to catalogue adaptable products available. A search into non-structural mitigation guidelines both locally and internationally unearthed a plethora of often-contradictory advice. The team felt that a systematic method was necessary to test products and guidelines. While a full-size shake table was not yet available in Turkey, Kandilli Observatory and Earthquake Research Institute arranged for use of the small educational shake table provided by a local chapter of the Rotary Club to be used for testing.

Demonstrations with this equipment showed non-structural mitigation in action and provided a qualitative assessment of the performance of various devices and guidelines (results are documented in Green 2003). Moreover, the demonstrations presented an opportunity for developers of adapted products to observe their products in action and to discuss the development of appropriate guidelines for their products. These products could then be compared with each other and with other local items. This dynamic, community-centred testing environment proved to be an excellent setting for allowing the manufacturers to assess and modify their products, despite the qualitative nature of the shake table tests.

The team followed up the shake table tests with conversations with volunteer instructors and target audience members. People renting properties told project leaders that they were reluctant to perform non-structural mitigation when it required drilling holes in walls because they did not want to incur liability for damaging the owner's property. Women, never before having used power tools, felt uncomfortable doing so on their own. Some women also explained that they preferred to move tall and heavy furniture out of daily living space and place it spare rooms rather than securing it to walls. They wanted solutions that didn't mar the visible exterior of their furniture. These conversations showed that to be effective, the programme would need to address diverse concerns about aesthetics and efficacy as well as gender differences in sensibilities and skills.

Based on this work, the team developed simple guidelines for local non-structural mitigation. These guidelines took the form of a 36-page brochure that could be understood by non-engineers and formed the basis of a community education training programme. The team also designed a large portable toolkit for instructors to share locally-available and home-made devices for securing furnishings. Continued evaluation of the training shaped the programme, expanding the content and offering it to 11th or 12th grade trade school students throughout Istanbul. Feedback helped to simplify and to shorten the guidelines for non-technical audiences attending disaster preparedness training programmes. Moreover, these materials formed the basis for rapid localisation and adaptation leading to the introduction of the subject in settlements in Central Asia which have not recently experienced an earthquake.

User awareness of adobe construction in Central Asia

The challenges of encouraging citizens to invest in disaster risk reduction before a disaster are notorious (Tierney 2006). The Central Asia Earthquake Safety Initiative (Figure 10.3), supported by USAID and launched in 2003 by GeoHazards International with local implementing partners in Tashkent, Almaty, Dushanbe and Bishkek, placed extraordinary demands on a workgroup of regional experts in disaster-resistant construction. Young disaster mitigation advocates identified from public, non-profit, academic, and community organisations needed to spread the word and engage new participants, conduct needs assessments, and become 'basic disaster awareness' trainers.

In Dushanbe, Tajikistan's capital, representatives of several community-based advocacy organisations gathered to learn what they could do about the earthquake threat facing them. Some local experts were discouraging disaster education efforts with comments including 'don't make any more problems for these people', 'they have their hands full', and 'they don't know where their next day's meal is coming from, let alone have energy to think about what to do about an earthquake'. Likewise, international experts demurred that not much could be accomplished until after a disaster strikes. Despite this lack of enthusiasm from experts, a handful of pioneers felt otherwise.

СОХТМОНИ ТАҲКУРСИИ (ФУНДАМЕНТИ) МУСТАҲКАМУ БЕХАТАР

Майдони мувофиқ барои сохтмони хона

Биноро БОЯД дар майдони ҳамвор сохт. Масофаи девори берунии хона аз лаби чарй на кам аз 1м (ва дар ҳолати аз ҳад зиёд будани моилии чарй – боз ҳам зиёдтар), ва на кам аз 3 м дар ҳолате, ки хона дар таги чарй чойгир шудааст.

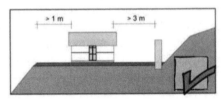

  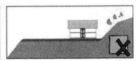

Хонаҳои гилиро НАБОЯД ба чарй часпонда сохт, чунки фишори хоки чарй дар вақти фуруд рафтан деворҳоро чаппа мекунад.

Хонаҳои гилиро НАБОЯД дар лаби чарй сохт, чунки дар ҳолати канда шудани лаби чар деворҳо ҳам канда мешаванд.

Хонаҳои гилиро НАБОЯД дар зери кӯҳҳо сохт, чунки имконияти канда шудан ва фаромадани сангпораҳо вучуд дорад.

Таҳкурсии (фундаменти) хуб

Ҳама намуд хонаҳои гилии ба заминларза тобовар бояд таҳкурсии мустаҳками бо деворҳо пайвастшуда дошта бошад. Беҳтарин таҳкурсӣ - таҳкурсии бетонии яклухт (монолитный) мебошад. Агар дар ин намуд таҳкурсӣ аз санг истифода барем бояд қисми болоии онро бо торҳои пулодӣ (арматура) мустаҳкам кард. Таҳкурсиро аз сатҳи замин дар чуқурии на кам аз 40 см гузошт. Барои ҳимоя намудани деворҳо аз наммй дар вақти боридани борон ва ё об шудани барф таҳкурсиро бояд 50 см аз сатҳи замин боло сохт.

Бино бо таҳкурсӣ (фундамент) бояд чунон пайваст бошад, ки дар вақти заминларза деворҳо аз болои он лағжида наафтанд. Пеш аз ҳама ба дохили бетони таҳкурсӣ сангҳоро чунон чойгир кардан лозим аст, ки ягон пораи онҳо аз бетон берун бошанд ва андозаи он ба баландии як хишт баробар бошад. Хиштҳои девор бояд ба тавре чинда шаванд, ки ин сангҳо дохили онҳо ба монанди дандон монад. Барои нагузаштани наммй аз таҳкурсӣ ба деворҳо дар бисёр мавридҳо мардум аз тол ва руброид истифода мебаранд, ки қобилияти лағжидан доранд. Беҳтараш аз хамираи (раствор) сементу-рег бо таносуби (1:2) истифода бурдан лозим аст. Ин хамира (раствор) худ намиро намегузаронад ва девор аз болои он намелағжад.

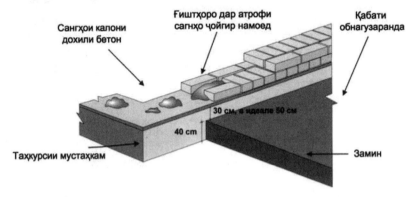

Figure 10.3 Central Asia Earthquake Safety Initiative, Adobe Residential Construction brochure, page 1. Note the simple illustrations clearly marking 'correct' and 'incorrect' with locally-tested, widely-understood symbols (Image courtesy of GeoHazards International)

An inaugural meeting attracted representatives of several voluntary community-based groups. Examples included an organisation working with new homeowners in properties that had recently been converted from state-owned housing to condominiums; a health advocacy organisation working with women and children; an environmental organisation called 'Women and Earth'; and a group of young ecology advocates working with school children. Once introduced to a basic disaster awareness curriculum available on flipcharts and with short handouts, each of these community activists enthusiastically identified ways to integrate this curriculum into their own outreach programmes. The groups immediately expressed interest and urgency in communicating structural safety awareness, especially related to the most vulnerable type of construction, adobe.

Since literacy levels are greater than 90 per cent throughout Central Asia, written material could be emphasised. Leading engineers started with a 50-page manual which was then shortened to 30 pages. The community activists were actually seeking a document which they could understand and convey, not just to builders, but also to neighbours, relatives and friends. An eight-country collaboration and more than a dozen drafts eventually produced a four-page brochure – including several caveats.

Among the most challenging issues were the compromises between 'better' and 'best' and between conflicting rights. Schilderman (2004: 417) notes that '… rules should not aim for the best but for what is optimal in a given context. If rules could improve on local building traditions, rather than imposing alien solutions, they would also be easier to accept'. In Central Asia, as a matter of policy, adobe is ruled out as a construction material in the highest seismic risk areas, so no codes exist for adobe construction. The 'best' approach to saving the maximum number of lives is never using adobe.

On the other hand, the reality is that hundreds of thousands of people cannot afford any other kind of housing. To complicate matters, materials that would provide the needed vertical reinforcement for adobe structures are not available locally. So either no changes are made and people will continue living in highly vulnerable adobe structures. The alternative is to explore and support feasible changes which will save many, but not all, lives. This is the 'better' rather than 'best' approach. Of course, the choice is not simple because many other factors and trade-offs are involved. Community-based construction for disaster risk reduction is frequently complicated, involving difficult and subjective choices.

Producing appropriate material for community awareness of structural safety engendered much lively discussion; deciding which salient details needed to be conveyed, deciding which new vocabulary should be introduced, figuring out how to synthesise the technical knowledge and express it in plain language, and finding an engineer/illustrator team to collaborate on visuals and the painstaking field testing of the results.

This process illustrates that producing technical and non-technical material for community-based construction for disaster risk reduction is not simple, but with patience and persistence, it can be done. Building on this past experience will facilitate the process for other areas, for non-adobe structures, and for non-earthquake concerns.

Common threads and future challenges

The five case studies provide examples of achievements in community-based disaster risk reduction. The projects demonstrate the power of local ownership and participation, sharing information, building relationships, and building capacity. A culture of safety and disaster risk reduction is embedded and integrated into ideas, policies, practices, and daily life. While process orientation, dialogue, exchange, and the development of relationships require time, they ensure a greater potential for sustained success. They represent an investment, not a cost, because the future savings will far outweigh the present investment.

Directly involving building occupants and other users is essential for the long-term success of community-based disaster risk reduction through construction and construction-related activities. Informed owners tend to be the strongest players in building safety. Renters have more limited decision-making authority, but can also be a conduit to involving owners. While squatters have more fundamental needs, such as tenure and resources, safety can be addressed simultaneously, respecting their desire and abilities to work collectively to improve their living conditions and livelihoods – through a culture of safety and disaster risk reduction as an important aspect.

The experience presented here and in the references highlights new insights and a cogent set of priorities:

- **All humanitarian assistance, reconstruction and development projects should include disaster risk reduction.** The alarming increase in risk and vulnerability can and must be halted. Donors, advocates, banks, insurance companies, taxpayers and users amongst others can and should demand that every school, hospital and home built with public funds, insurance proceeds, loans or charitable contributions must be built to meet or exceed performance-based standards for construction for disaster risk reduction.
- **Building-code compliance rather than building-code enforcement should be the goal.** The Nepali experience illustrates that attempts to spread awareness and to exchange knowledge on construction for disaster risk reduction will find a receptive audience, even before a catastrophe, even where a disaster has not occurred in living memory, and even amongst poorer households. Schools provide a universal focal point from which entire communities can discover and practise safer construction.

- **Local skills, traditions and concerns are a starting point upon which to build confidence and capacity.** The SEEDS Masons Association, formed following the Patanka Navjivan Yojana project, holds great promise for enabling ideas and skills to be transferred elsewhere, for creating a sustainable demand for and supply of trained masons and for scaling-up and institutionalising this work. The training of school teachers in Turkey to convey non-structural mitigation principles and techniques has the potential to institutionalise this new capacity for future generations.
- **Individual and collective efforts are both important.** Aceh's reconstruction demonstrates the effectiveness of collective processes through community participation. Turkey's education programme demonstrates how individuals with different perspectives and skills need to form a team. Individuals can and do contribute, especially at the level of house owners and technical expertise, but collective efforts add benefit to quality control and creativity in the approaches adopted.
- **Educational materials can be developed and sensitively localised by local and international collaborators. These materials can be shared, tested for effectiveness and improved.** Community activists and people with technical expertise in Nepal, India, Aceh, Turkey and Central Asia collaborated to create tools for sharing knowledge that previously had been denied to vulnerable people. Efforts to synthesise, codify, and attractively market the feasibility of household and community-based disaster risk reduction activities require time, thought and multiple contributions. They need to be collaborative and iterative, based on existing strengths, developed with and tested by users.
- **Models and live demonstrations have an important role to play in community-based construction education for disaster risk reduction.** Shake-table models have been essential and effective in piquing imagination and engaging commitment. They have also been important for building confidence in non-engineered construction using local materials and further developing traditional methods. Videos hold similar promise. Involving Turkish manufacturers in the testing of their products ensured that the manufacturers accepted their products' deficiencies, having witnessed it directly, and were willing to tap into the available expertise to improve. Table-top models and toolkits enabled school-children and local craftspeople and volunteers to relate to their new understanding of non-structural risk reduction.
- **The process of introducing innovative technologies to local construction practices remains to be explored.** There are old and new technologies with tremendous potential to increase the safety of houses and other buildings. Bamboo, which can be cultivated in many climates, could be introduced to provide affordable and plentiful vertical reinforcement materials, subject to ecological considerations. Experiments in Japan with inexpensive polypropylene bands for reinforcement of adobe

construction provide another alternative, again subject to ecological considerations, this time of polypropylene's life cycle. However, independent of commercial interests, little is known about the social processes by which new construction practices might be tested, introduced and accepted without creating new dangers through misuse of the technology or technique.

The challenges ahead involve developing, testing, evaluating and championing old and new community-based construction approaches for disaster risk reduction. The process will require acculturating a generation of technical specialists and field practitioners to higher education techniques and on-the-job training. Non-governmental organisations, the private sector, and government departments involved in shelter, housing, planning, community design, and development amongst other areas must be involved.

Promising pilot efforts deserve to be mined to determine factors in success and transferability of those factors. Full-scale theoretical, evaluative and experimental research will be needed alongside the development of frameworks and guidelines that make community-based construction for disaster risk reduction a standard operating procedure.

From vast and dispersed rural landscapes to dense and increasingly perilous urban slums, community-based construction can save lives through disaster risk reduction. We must be proactive in bringing about the partnerships to make this possible.

References

Abarquez, I. and Murshed, Z. (2004) *Community-Based Disaster Risk Management: Field Practitioners' Handbook*, Bangkok: ADPC (Asia Disaster Preparedness Centre).

Allen, D.E. (1992) 'A design basis tornado: discussion', *Canadian Journal of Civil Engineering*, 19(2): 361.

Anderson, M. and Woodrow, P. (1998) *Rising from the Ashes: Development Strategies in Times of Disaster*, London: Intermediate Technology Publications.

ASFPM (2002) *Mitigation Success Stories in the United States (Edition 4)*, Madison, WI: Association of State Floodplain Managers.

Asian Development Bank and World Bank (2005) *Preliminary Damage and Needs Assessment*, Bangkok and Washington, DC: Asian Development Bank and World Bank.

Aysan, Y., Clayton, A., Cory, A., David, I. and Sanderson, D. (1995) *Developing Building for Safety Programs: Guidelines for Organizing Safe Building Improvement Programmes in Disaster-prone Areas*, London: Intermediate Technology Publications.

Blondet, M., Garcia, G.V. and Brzev S. (2003) *Earthquake-resistant Construction of Adobe Buildings: A Tutorial*. Contribution to the *World Housing Encyclopedia*, Available HTTP: <http://www.world-housing.net.asp>.

Blondet, M. (2005) *Construction and Maintenace of Masonry Houses: For Masons and Craftsmen*, Lima: Pontica Universidad Catolica del Peru, Available HTTP: <http:world-housing.net>.

Brown, D.W., Moin, S.M.A. and Nicolson, M.L. (1997) 'A comparison of flooding in Michigan and Ontario: 'soft' data to support 'soft' water management approaches', *Canadian Water Resources Journal*, 22(2): 125–39.

Brzev, S. (2007) *Earthquake Resistant Confined Masonry Construction*, Kanpur: National Information Center of Earthquake Engineering, Available HTTP: <http://www.nicee.org>.

BTRE (2002) *Benefits of Flood Mitigation in Australia (Report 106)*, Canberra: Bureau of Transport and Regional Economics, Department of Transport and Remedial Services.

Clayton, A. and Davis, I. (1994) *Building for Safety Compendium: An Annotated Bibliography and Information Directory for Safe Building*, London: Intermediate Technology Publications.

Coburn, A., Hughes, R., Pomonis, A. and Spence, R. (1995) *Technical Principles of Building for Safety*, London: Intermediate Technology Publications.

Coch, N.K. (1995) *Geohazards: Natural and Human*, Englewood Cliffs, NJ: Prentice-Hall.

Corsellis, T. and Vitale, A. (2005) *Transitional Settlement: Displaced Populations*, Oxford: Oxfam.

Cronin, S.J., Gaylord, D.R., Charley, D., Alloway, B.V., Wallez, S. and Esau, J.W. (2004a) 'Participatory methods of incorporating scientific with traditional knowledge for volcanic hazard management on Ambae Island, Vanuatu', *Bulletin of Volcanology*, 66: 652–68.

Cronin, S.J., Petterson, M.J., Taylor, M.W., and Biliki, R. (2004b) 'Maximising multi-stakeholder participation in government and community volcanic hazard management programs: a case study from Savo, Solomon Islands', *Natural Hazards*, 33: 105–36.

Davis, I. (1978) *Shelter After Disaster*, Oxford: Oxford Polytechnic Press.

Dixit, A. (2003) 'The community based program of NSET for earthquake disaster management', Paper presented at the International Conference on Total Disaster Risk Management, Kobe.

EERI (2006) *World Housing Encyclopedia*, Earthquake Engineering Research Institute, Oakland, Available HTTP: <http://www.world-housing.net>.

EERI (2007) *World Adobe Forum*, Earthquake Engineering Research Institute, Oakland, Available HTTP: <http://www.worldadobeforum.eeri.org>.

Erdik, M. (2003) 'Report on 1999 Kocaeli and Düzce (Turkey) earthquakes', in F. Casciati and G. Magonette (eds) *Structural Control for Civil and Infrastructure Engineering*, Istanbul: World Scientific.

Fell, J.C. and Voas, R.B. (2006) 'Mothers Against Drunk Driving (MADD): the first 25 years', *Traffic Injury Prevention*, 7(3): 195–212.

FEMA (1997) *Report on Costs and Benefits of Natural Hazard Mitigation*, Washington, DC: Federal Emergency Management Agency.

FEMA (1998) *Protecting Business Operations: Second Report on Costs and Benefits of Natural Hazard Mitigation*, Washington, DC: Federal Emergency Management Agency.

GHI (2006) *Seismic Safety for Adobe Homes: What Everyone Should Know*, Palo Alto, CA: GeoHazards International.

Glanz, K., Rimer, B.K. and Lewis, F.M. (2002) *Health Behavior and Health Education: Theory, Research and Practice*, San Francisco: Jossey-Bass.

Green, R. (2003), 'Home and office non-structural mitigation: lessons learned from shake table testing', Paper presented at Fifth National Conference on Earthquake Engineering, Istanbul.

Green, R. (2005) 'Negotiating risk: earthquakes, structural vulnerability and clientelism in Istanbul', Unpublished doctoral thesis, Cornell University.

IAEE (2004) *Guidelines for Earthquake Resistant Non-engineered Constructions*, revised edn reprinted in English by National Information Center of Earthquake Engineering of India, Tokyo: International Association for Earthquake Engineering.

Langenbach, R. (2007) *Don't Tear It Down: Preserving the Earthquake-Resistant Vernacular Construction of Kashmir*, Delhi: UNESCO.

Latrobe, S. and Davis, I. (2005) *Mainstreaming Disaster Risk Reduction: A Tool for Development Organisations*. London: Tearfund.

Laws, G. (1994) 'Oppression, knowledge and the built environment', *Political Geography*, 13(1): 7–32.

Lewis, J. (2003) 'Housing construction earthquake-prone places: perspectives, priorities, and projections for development', *Australian Journal of Emergency and Management*, 18(2): 35–44.

Maskrey, A. (1989) *Disaster Mitigation: A Community-based Approach*, Oxford: Oxfam.

PAHO (1998) *Natural Disaster Mitigation in Drinking Water and Sewerage Systems: Guidelines for Vulnerability Analysis*, Washington, DC: Pan American Health Organization.

Petal, M. (2004) 'Urban disaster mitigation and preparedness: the 1999 Kocaeli earthquake', Unpublished doctoral thesis, University of California, Los Angeles.

ProVention Consortium (2006) *Provention Community Risk Assessment Toolkit*, ProVention Consortium, Geneva: Available HTTP: <http://www.proventionconsortium.org/?pageid=39>.

Schilderman, T. (2004) 'Adapting traditional shelter for disaster mitigation and reconstruction: experiences with community-based approaches', *Building Research and Information*, 32(5): 414–26.

Sengezer, B. and Koç, E. (2005) 'A critical analysis of earthquakes and urban planning in Turkey', *Disasters*, 29(2): 171–94.

Shah, H. (2003) 'The last mile: earthquake risk mitigation assistance in developing countries', Available HTTP: <http://www.radixonline.org/resources/haresh-shah-lastmile.doc>.

SOPAC (South Pacific Applied Geoscience Commission) (2005) 'Pacific CHARM,' in UNISDR (ed.) *Know Risk*, Leicester and Geneva: Tudor Rose and United Nations Secretariat for the International Strategy for Disaster Reduction.

Stephenson, M. and Schacher, T. (2006) *Basic Training on Dhaji Construction*, Islamabad: ERRA, SDC, UN Habitat.

Tierney, K. (2006) 'Foreshadowing Katrina: recent sociological contributions to vulnerability science', *Contemporary Sociology*, 35(3): 207–12.

Transport Canada (2003) *Survival in Cold Waters: Staying Alive*, Ottawa: Transport Canada Marine.

Twigg, J. (2003) 'Lessons from disaster preparedness', Notes for presentation to 'Workshop 3: It Pays to Prepare' at International Conference on Climate Change and Disaster Preparedness, 26–28 June 2002, The Hague, Netherlands.

Twigg, J. (2004) *Disaster Risk Reduction: Mitigation and Preparedness in Development and Emergency Programming*, London: Overseas Development Institute, Humanitarian Practice Network.

UNDRO (1982) *Shelter After Disaster: Guidelines for Assistance*, Geneva: United Nations Office for the Coordination of Humanitarian Affairs.

UNESCO (2004) *Education for All (EFA): Global Monitoring Report 2003/2004*, Paris: United Nations Educational, Scientific and Cultural Organization.

UNISDR (2004) *Living With Risk*, Geneva: United Nations International Strategy for Disaster Reduction.

UNISDR (2005) *Know Risk*, Leicester and Geneva: Tudor Rose and United Nations Secretariat for the International Strategy for Disaster Reduction.

UNISDR (2006) 'Terminology: Basic terms of disaster risk reduction', Geneva: United Nations International Strategy for Disaster Reduction, Available HTTP: <http://www.unisdr.org/eng/library/libterminology-eng%20home.htm>.

USAID-OAS (1997) *Basic Minimum Standards for Retrofitting: USAID-OAS Caribbean Disaster Mitigation Project*, Washington, DC: Organization of American States, General Secretariat, Unit for Sustainable Development and Environment.

WCEA (1990) 'World Declaration on Education For All', Declaration adopted at World Conference on Education for All, Jomtien, Thailand, 5–9 March.

WEF (2000) 'Education for all: Meeting our collective commitments', Dakar Framework for Action, text adopted at World Education Forum, Dakar, Senegal, 26–28 April.

Weichselgartner, J. and Obersteiner, M. (2002) 'Knowing sufficient and applying more: challenges in hazard management', *Environmental Hazards*, 4(2–3): 73–7.

Wisner, B., Kelman, I., Monk, T., Bothara, J.K., Alexander, D., Dixit, A.M., Benouar, D., Cardona, O.D., Kandel, R.C. and Petal, M. (2007) 'School seismic safety: Falling between the cracks?', in C. Rodrigué and E. Rovai (eds) *Earthquakes*, New York: Routledge.

Young, I. (1988) 'The five faces of oppression', *Philosophical Forum*, 19: 270–90.

Young, I. (1990) *Justice and the Politics of Difference*, Princeton, NJ: Princeton University Press.

Chapter 11

Informal settlements and natural hazard vulnerability in rapid growth cities

Rebekah Green

Introduction

Every day, humans face immediate health risk and experience low-probability, high-loss hazards. Over the last century, there have been close to 8,000 naturally triggered disasters (droughts, floods, earthquakes, windstorms and landslides) that have affected over 100 people, caused 10 or more deaths or resulted in a state of emergency. Earthquakes in the past century, which have predominantly occurred along seismic hotbeds that run from the Himalayas to south-eastern Europe and along the Pacific Rim, have caused the deaths of approximately 2 million people (CRED 2005). At the same time, diarrhoeal diseases cause close to 2 million deaths in a single year, with over 80 per cent of these deaths linked to poor sanitation and access to safe drinking water. Nearly 17 per cent of the global population lacks access to an improved water source and over twice that lack improved sanitation (WHO 2006).

Much of the mortality in the last century has been concentrated in less economically developed countries. In rapidly industrialising urban centres of Asia and Latin America, poor quality housing and hap-hazardous land-use has placed a large percentage of the population at risk of structural collapse and death at the onset of earthquakes, tsunamis, landslides, and meteorological events. At the same time, this unsafe housing can result in immediate hazards from communicable diseases, fire, and security. Yet, the underlying roots of this vulnerability are not found in the poorly built slums themselves or in simple lapses in building and health code enforcement. Rather, social, political and economic processes often make code violation an acceptable alternative or even the only option for many residents at risk. In studies of risk perception, Asgary and Willis have found that 'safety measures enforced without considering people's preferences fail to be adequately adopted in practice' (Asgary and Willis 1997: 613). In the same way, a close examination of economic and social realities in less economically developed countries is critical to understanding the continued construction of highly vulnerable housing in the face of natural hazards.

This chapter will first explore the links between hazard, vulnerability and national economy by examining global records of past disasters. This will be followed by a discussion of population growth, urbanisation and industrialisation, processes that have led to an explosion of poor-quality informal housing in many cities of less economically developed nations. The diverse outcomes of these processes, and their implications for human vulnerability, will be examined through case studies of three cities: Nairobi, Lima, and Istanbul. Each highlights different risk – both immediate and long term – to which residents of informal housing are exposed. The chapter concludes with a short discussion of vulnerability reduction. Natural hazard vulnerability reduction must address both long-term infrastructure fragility and immediate security needs to reduce vulnerability in the lives of the world's urban poor.

Natural hazard, vulnerability, and development

Many cities in Latin America, the Middle East, and Central and Southern Asia are sites where high exposure to rapid-onset hazards intersects with a loosely controlled construction industry. Construction in these cities is noted for rampant illegal building, violation of urban planning guidelines and poor design, construction and inspection practices – factors that heighten vulnerability to the geophysical and meteorological hazards that many of these cities face. Death tolls that exceed 10,000 for a single urban earthquake or landslide are commonplace in these regions and contrast with low mortality from similar hazard events in the United States, Japan, and New Zealand (Guha-Sapir et al. 2004; UNDP 2004a). Thus, recent studies of natural hazard risk suggest that global hotspots for natural hazards are increasingly concentrated within less economically developed nations (Dilley et al. 2005).

A review of geophysical hazards over the last 26 years confirms a concentration of mortality in countries with low GDP per capita. The EM-DATA global database (OFDA/CRED 2006) catalogues mortalities from landslides, earthquakes, volcanoes and tsunamis from self-reported disasters globally. When mortalities from geophysical disasters over the period 1980 to 2005 are tabulated and ranked according to 2005 Gross Domestic Product per capita, purchasing power parity,[1] it is clear that loss of life is concentrated in nations of the third and fourth quintile. Table 11.1 clearly shows that countries in the third and fourth quintiles, often referred to as middle and lower-middle income countries, account for over 90 per cent of global deaths from geophysical hazards over this time period. Moreover, because countries registering geophysical-induced disasters over the period 1980 to 2005 are more numerous in low-middle and middle income countries, there is a possibility that geophysical hazard may be retarding economic development for countries exposed to these hazards.

The concentration of mortality in low-income regions is not a new phenomenon, but it may be a growing one. In a study of disasters between

Table 11.1 Loss of life from geophysical hazards by GDP (PPP) per capita, 1980–2005

Global economic quintile	Average GDP per capita (PPP), 2004	No. countries reporting geophysical disaster	Total loss of life	Average loss of life per million inhabitants	Countries with highest loss of life per quintile
1st quintile	US$ 30,900	15	14,193	19	Japan, Italy, Taiwan, US, Greece
2nd quintile	US$ 13,900	15	13,451	29	Mexico, Russia, Chile, Malaysia, Puerto Rico
3rd quintile	US$ 6,700	20	149,890	74	**Iran**, Columbia, Turkey, Thailand, Philippines
4th quintile	US$ 3,200	27	374,054	180	**Indonesia, Pakistan, India, Sri Lanka**, Armenia
5th quintile	US$ 1,900	21	14,706	23	Afganistan, Nepal, Yemen, Tajikistan, Kyrgzstan
Total		98	566,294		**Bold** countries indicate top countries with loss of life globally

1947 and 1989, Degg (1992) found that the average loss of life was lower in North America and Western Europe than in the rest of the world. More importantly, in a comparison of data from two sources, one tracking disaster outcomes from 1947 to 1967 and another from 1969 to 1989, he found that average mortality per disaster had gone down in developed countries; whereas the reverse was true outside of North America and Western Europe. On average, fatality rates rose outside these two geographic regions (Degg 1992).

Vulnerable construction in low economic developed countries

The last 50 years has seen an increased depth in technical understandings of the behaviour of modern structures under dynamic loadings. This better understanding has supported the development of new technologies that reduce structural vulnerability to these hazards. Through consensus building, model building codes have incorporated newer empirical and analytical understandings of the performance and failure of structures. In areas where political will and public perception have been favourable, these model codes have been adopted as mandatory building codes. Deaths due to natural hazards such as earthquakes and hurricanes have been steadily declining in the United States, Japan and New Zealand with the adoption of progressively more detailed building codes and hazard assessments over the last half century. Comparison of hurricane damage in Florida before and after the adoption and rigorous enforcement of strict residential building codes following Hurricane Andrew in 1992 attest to the effectiveness of this approach in the United States (Coch 1995).

Despite the success in nations with high economic development, the promotion of building code reform has not been uniformly effective. In countries with low and moderate economic development, natural hazards in the last 15 years have resulted in heavy loss of life despite the adoption of modern building codes. For example, in Turkey, where most urban construction occurred after the adoption of modern seismic codes in 1980, a 1999 earthquake revealed inadequate design, construction detailing, and site planning, resulting in over 17,000 deaths (Gülkan 2000). The success in reducing disaster casualties in economically advanced nations is a result of a particular social, political, and economic context, not simply a result of building code reform.

The persistent vulnerability of enormous populations to the impacts of natural hazards, in the face of a tremendous body of accumulated expert and technical knowledge, requires some examination. The confluence of physical hazard and social vulnerability is especially significant in areas where the building stock contains a high percentage of un-engineered, often unauthorised, construction. This building stock is concentrated in low and middle income nations, much of it in their urban centres. The vulnerability of unauthorised, un-engineered construction, while varying in detail geographically, has risen in accordance to social and economic realities of development and urbanisation.

Informal construction in rural settings

Recent earthquakes in rural and small-town settings the world over have revealed the high seismic vulnerability of adobe, stone, and rammed earth dwellings (Turkey, Iran, Peru, Pakistan) (Jain *et al.* 2002; Naeim *et al.* 2004; Rai and Murty 2005). Yet, many traditional forms of construction do have effective disaster-resistant construction techniques (e.g. seismic belts, lintel bands, through stones, cross beams, and timber bracing). The un-engineered structures of the Americas and Asia that rigorously incorporate aseismic components have consistently fared well in seismic events, often even better than more recent construction of masonry, steel and concrete (Kaushik *et al*. 2006; Langenbach 2005a, 2005b; Tolles *et al*. 2000). Rural vernacular construction also addresses issues of social interaction and the regulation of the external environment. Thick-walled masonry construction conserves heat in winter and remains cool in scorching summers (Chalfoun and Michal 2002). In the humid coastal regions of Asia, bamboo and palm frond offer critical ventilation (Vries 2002). While rural vernacular construction remains less resistant to rapid-onset hazards than well-built steel and concrete structures, it makes use of readily available material and addresses cultural and location-specific needs.

Despite the cultural, economic, and environmental appropriateness of much vernacular construction, it is in decline within many rural communities worldwide. In some communities, economic collapse, such as the collapse that resulted from Angola's prolonged civil war, has left residents with little memory of construction techniques. New construction has simply not occurred for decades (Traub 2006). In other communities, economic growth has led to a similar decline. For instance, in Northern India, a rapid rise of the tourist industry and a ban on forest cutting precipitated the use of masonry and reinforced concrete construction in replacement of seismic-resistant *ikra* and timber vernacular construction (Gupta *et al*. 2006; Kaushik *et al*. 2006). In central Turkey, an influx of factory-produced building materials has reduced the cost of modern construction methods and made them more widely used in an expanding radius from urban centres (Green 2002). Moreover, for several decades now, young men have routinely worked in the urban centres of Turkey to support rural family members, bringing back knowledge of modern construction materials and sensibilities to their rural relatives (Stirling 1965). Throughout many rural regions of low and middle income nations, untrained labourers often lead the design and construction of modern masonry and concrete construction without the education, training and tools necessary to ensure that these concrete structures meet basic hazard-resistant minimum standards.

Modern forms of construction, perceived as development and progress, undercut the value of traditional apprenticeships with master builders, and degrade traditional construction methods that contribute to disaster risk reduction. Each year, fewer master builders are available to ensure the incorporation of these techniques within new construction in the traditional vernacular style or to maintain the ageing stock of traditionally constructed buildings. Modern rural construction outside the vernacular tradition also remains vulnerable. Those who can afford more urban forms of construction, often do so with master builders and labourers that are untrained in the techniques that can make these forms of construction as or more durable than the traditional vernacular. However, despite the significant risks associated with rural construction – both degraded traditional forms and approximations of modern forms – the relative proportion of risk represented by this sector diminishes as rural residents migrate to urban centres in increasing numbers.

Informal construction in urban settings

Much of the loss of life due to rapid-onset natural hazards is concentrated in rapidly urbanising centres of low and middle income countries. Degg found that 88 per cent of the world's fastest growing cities are exposed to natural hazards and all of these cities are located in developing countries (Degg 1992). While most of these nations have modern building codes, regulatory systems, and expanding economic sectors, their urban centres have experienced intensive population expansion. During the second half of the twentieth century, the largest growth of urban populations in the world's history, city dwellers increased from 29.1 to 47.1 per cent of the world population. (Angel 2000; Gulati 1985; Sachs 2005; UNDP 2001, 2004a, 2004b). Similarly, the Housing Indicators Program funded by the United Nations found that the average growth rate for eleven cities located in countries with low and moderate economic growth was 4.4 and 3.9 per cent respectively in 1990 (Angel 2000).

Like its rural counterpart, urban construction in low and middle income countries has also seen rapid shifts from traditional construction materials to modern materials of steel and concrete (Merkezi 1998). Prestige and changes in material availability have partly driven this change. And, likewise, construction has occurred at the hands of poorly-trained and ill-equipped workers. Yet, there are also problems of urban construction and hazard vulnerability that are unique to urban settings. Unlike their rural counterparts, urban construction is subject to environmental regulations, land-use planning, and building code restrictions that, while meant to ensure health and safety, also drive up the cost of construction. Where these regulations have met with rapid population growth, construction has often developed in ways that often directly impact the vulnerability of urban populations to natural hazards.

Urban population growth in the late twentieth century was predominantly among the urban poor whose primary source of income was low-paying jobs. Many migrants could not be integrated into the formal economies of these cities because there was insufficient investment in industrial development. Thus, migrants were unable to find jobs in the formal industrial sector and were forced to search for work in the lower-paying informal sector, often with much lower levels of safety and economic security (Pugh 1995). The wages received were close to basic food security, thus necessitating cost savings in the area of housing (Neuwirth 2005; Pugh 1995; Schusterman and Hardoy 1997; Turner 1977).

Post-colonial industrialisation itself also contributed to the demands for inexpensive housing. Sassen (1998) argues that post-colonial industrialisation led to specific patterns of land use that exacerbated housing insecurity for the poor. Industries in low and middle income countries tended to establish along the outskirts of major urban centres to maximise access to natural resources, linkages to global markets, labour supply and inexpensive land. Industrial workers, needing to be near their source of income, constructed housing in the urban periphery where municipal oversight and services did not exist. Likewise, export-oriented economies necessitated an abundant supply of inexpensive labour and placed downward pressure on wages, increasing the imbalance between wages and housing costs (Shatkin 2004; Wisner 2003).

Land-use patterns have further exacerbated housing scarcity in urban centres. When transportation systems are ineffective or expensive, there is often a need for cheap, centrally located housing for low-wage workers servicing the urban core. As Shatkin notes, 'the poor lack access to one basic necessity – centrally located urban space' (Shatkin 2004: 2471). This scarcity has fuelled the creation of backyard tenements, pavement dwellings, and illegal upward construction in legal districts of the urban core. For example, stall cleaners at a South African race track began squatting on undeveloped government land adjacent to the track. Squatting allowed them to live close to their employment even though housing in the wealthy suburb around the track was well beyond their salaries (Saff 1996). Similarly, several thousand families who work in Manila's International Container Port have erected illegal shacks along the port's sea wall (Shatkin 2004). Over a million urban poor pay for squatting rights upon the roof-top terraces of modern apartment buildings in Cairo (Soliman 1996); those who cannot, squat in the graveyards (El-Messiri 1985). Even in the economically booming island of Hong Kong, housing shortages have forced the urban poor to illegally convert the air-wells of buildings into living space (Smart 1995). Centrally located informal sector housing, like that in peripheral settlements, is often affordable and close to income sources, making it an ideal choice for many urban poor. The spectacular growth of sprawling peripheral squatter settlements, pavement dwellings, and inner city illegal housing attests to this expanding problem (Angel 2000; Pelling 2003).

Case studies of informal settlement patterns and hazard vulnerability

The UN Millennium Project estimates that in 2001 more than 900 million people were living in urban slums (Garau *et al.* 2005). While urban informal settlement and slum growth has been fuelled by population growth, migration, land scarcity – and inadequate policies to address these issues – each is also the unique outcome of historical, economic, and cultural processes at the local, regional and national level. The following case studies will highlight the unique and often divergent outcomes of these processes.

Land invasion and self-constructed settlements: Lima, Peru

Like many developing nations, Lima's unauthorised housing construction rose sharply in the period between 1950 and 1990, during the country's industrialisation and economic liberalisation process. Rural migrants flooded the capital city, first residing in dilapidated inner-city tenements. However, from 1950 to 1960, the first major land invasions of governmental and private land in the hills surrounding the capital began. These land invasions were not individually carried out, but planned in advance and executed in large communal groups. The cooperative nature of such invasions made eviction and demolitions difficult (Oliver-Smith 1999).

Peru and other Latin American countries were early adopters of housing policy that specifically addressed informal housing. Starting in the 1950s Peru initiated the programme for Popular Urbanizations of Social Interest, setting aside land, subdividing, and providing infrastructure for the urban poor. Other programmes relocated squatters from invaded land to governmental lots. These programmes were followed in the 1960s with laws that legalised squatter settlements and gave basis for residents to demand municipal services. Through this process, Lima grew over tenfold in 60 years and by the year 2000 approximately half of the residents lived in 'young towns', the squatted sections of the city (Williams 2005).

Lima's informal housing followed a trajectory similar to many informal districts world-wide. At the time of land invasions, squatters set up temporary shacks from straw mats and other temporary materials. As the land invasion gained formal recognition, or at least informal acceptance, squatters began an incremental process of construction – building as time, family size and income permitted. Typically permanent walls of brick and cement were constructed first, followed by a concrete roof. As the family grew, lots were subdivided. New storeys were also added to existing buildings, leading to the consolidation of the neighbourhood building stock (Turner 1977). In Lima, this process of invasion, incremental construction, and consolidation was carried out in a political environment that both recognised the rights of squatters and, on occasion, even facilitated the process by which they obtained land, services, and permanent housing. Today much of the city's middle class lives in these vast, informally constructed, but legally recognised, districts of one to three storey masonry and reinforced concrete frame construction, at many different levels of completion (Williams 2005).

Despite the success of informal housing in Lima, issues of hazard vulnerability remain. Peru has witnessed death tolls related to flash floods and landslides that are the highest of all countries in the Western Hemisphere (Ferradas 2006). Much of this vulnerability is located in the Andean highlands where the rural poor have self-constructed homes of adobe. Yet, risks in the capital also exist. Lima is closer to a major earthquake subduction zone than any other city of comparable size in the Hemisphere (Martin and David 2005). Tsunamis also threaten the city's Port of Callao (Oliver-Smith 1999). The vulnerability of the city's informal settlements is of primary concern. A study of one informal settlement found that only 13 per cent of residents had received professional assistance in designing their homes. Rather, residents and local master builders had sketched out structural plans prior to each phase of construction. Because of the flexible and incremental nature of the construction, plans often changed in ways that could impact future vulnerability to hazards (Williams 2005). Many structures lack basic hazard-resistant technology, such as continuous framing and adequate ties between floors.

With a 90 per cent probability of Lima neighbourhoods experiencing seismic intensities of up to VIII on the Modified Mercalli Intensity Scale in the next five decades, concerns regarding the vulnerability of Lima's informal housing are legitimate (Martin and David 2005). Vulnerability to seismic and landslide hazard will increase significantly if future consolidation leads to further informal vertical expansion beyond the relatively earthquake-resistant single and double storey masonry homes of today. So too will exposure increase as new informal settlements are allowed to expand into the steep slopes of the Andean foothills surrounding Lima.

Tenement settlements: Nairobi, Kenya

Not all informal settlements are vulnerable to geo-physical natural hazards. Some are not exposed to such hazards; others are in such early stages of consolidation that the structures do not, in and of themselves, pose a threat in such events. Nairobi, Kenya is one such example. Understanding the arrested development of Nairobi's informal settlements, helps to clarify how resistance to hazard-prone informal self-help housing may reduce vulnerability to natural hazards while still leaving squatters in a vulnerable state.

Nairobi, Kenya is located in the world's fastest urbanising region, and is another example of the outcome of population pressure, poverty, and rapid urbanisation. In Kenya, poverty rates within urban areas have been growing faster than in rural areas. Much of this urban poverty is located in the slums of Nairobi, the capital city. Of the city's approximately 2 million inhabitants, between a third and a half live in these slums (Gulyani et al. 2006). Over a quarter of the city's population live in one vibrant, but extremely congested slum, Kibera. It is the largest and poorest slum in Africa (Harding 2002). The density within this district is close to 300,000 people per square kilometre.

Kibera, like many unauthorised settlements, is the outcome of a specific historical process that left the poor little option but to create their own housing. During colonial rule, the government overwhelmingly funded housing construction for Europeans living in Nairobi. What little housing funds were made available for native Kenyans working as domestics, menial labourers, and rail yard workers, resulted in bachelor houses where single men shared overcrowded rooms (Neuwirth 2005). Only native Kenyans with housing passes were allowed to live within the capital; rural migrants to the city were viewed as a 'drone and parasitic class of native' to be rounded up and repatriated in tribal lands (Hake 1977, quoted in Macharia 1992). Despite these obstacles, native Kenyans found ways to bring their families to the capital, though slum clearance and repatriation continued to keep their numbers low. A relaxation of the pass system – without corresponding rural development efforts in the soil-poor, over-cultivated hinterlands – led to an extraordinary urban population boom in the mid-century. Given the trickle of funds for native-Kenyan housing, illegal shanty towns boomed (Macharia 1992).

With Kenyan independence in 1963, untenured housing, including that of the Kibera squatters, was declared illegal. Despite unauthorised status and continued demolitions, Kibera and other Nairobi slums continued to grow rapidly. While in other countries, democratic elections have allowed slum dwellers to demand infrastructure improvement and legalisation, voters in Kenya are not tied to place of residence. There was little political power squatters could wield, though some squatters gained a semblance of land security through the entrenchment of their tribe in powerful administrative positions (Gulyani et al. 2006).

A lack of infrastructure investment, continued land insecurity, and low political power shifted Nairobi's slums from squatter sites to tenement slums. Squatters had little incentive to upgrade their housing and continue living in Kibera. Rather, they became 'landlords', renting out rooms and plot segments to both incoming rural migrants and to other urban residents from other slums. Despite the established nature of the unauthorised settlements, most housing in Kibera and other Nairobi slums continues to be of the poorest quality (Macharia 1992). Over 90 per cent of Nairobi residents of unauthorised housing settlements are rent-paying tenants. So-called landlords often own many plots of land, employing local intermediaries to collect rent while they live elsewhere in the city (Gulyani et al. 2006).

Recent studies of Nairobi's unauthorised neighbourhoods, including the slums of Kibera, offer an alarming picture of human poverty and vulnerability. Settlements remain largely un-integrated into the urban infrastructure networks, lacking governmentally funded systems to distribute potable water, pick up trash, set up public latrines and collect sewage (Otiso 2003). Only 19 per cent of Nairobi slum residents have access to piped water, either through in-house lines or neighbourhood taps, much lower than the more than 70 per cent of residents who have piped water city-wide (Gulyani et al. 2006). Policing occurs through vigilante groups that exhort payments. Crime is rampant. The slum is famous for 'flying toilets', plastic bags which residents have defecated in at night and thrown from their doors and windows because it is not safe to use public latrines, sewage ditches or trash piles at night (Harding 2002). While housing costs are high in proportion to income, they remain below that of other wards of the city (Gulyani *et al.* 2006), making Kibera and other slums an attractive housing option for those in low-wage and informal jobs (Macharia 1992).

These extremely dense areas of substandard housing disproportionately expose Nairobi's poor to fire, flood, hazardous materials, and massive health epidemics (Gulis *et al.* 2004). These threats are immediate, contrasting with the low-probability, high-loss risk Lima's young town residents' face from rapid-onset geophysical and meteorological hazards. Because of the lack of municipal refuse collection, residents periodically burn trash heaps against the sides of squatter houses. Risk from fire outbreak is ever present (Neuwirth 2005). Raw sewage, open trash heaps, and a lack of adequate refrigeration contribute to periodic breakouts of gastric and respiratory diseases (Gulis *et al.* 2004). During the rainy season, this sewage and trash is swept down muddy streets into the houses of low-lying residents (Yap 2004). Moreover, only slightly over 10 per cent of the unauthorised housing in Kibera is made of permanent materials like brick, stone or concrete block (Gulyani *et al.* 2006). The rain also undermines the stability of impermanent housing materials such as wattle and daub, threatening structural failures (Gulyani, personal communication).

A lack of land security and the consolidation of control in the hands of a few absentee landlords have suppressed housing quality in Kibera well below that of Lima's young towns. Moreover, laws that bar municipal development in these areas have resulted in exposure to a host of health, security, and environmental hazards. As Kibera illustrates, ignoring informal settlements because they do not conform to governmental regulations does not necessarily reduced human vulnerability to natural hazards; rather, it leaves residents facing a host of daily threats and living in appalling conditions.

'Formalised' informal construction: Istanbul, Turkey

Residents in integrated illegal urban settlements have the fastest growing long-term vulnerability and the greatest illusion of safety. Huge numbers of urban dwellers live in un-engineered and semi-engineered construction located in the world's largest cities (Angel 2000; Coburn and Spence 1992). As residents of un-integrated urban settlements become economically embedded in the markets of the urban centre, housing development often shifts from the self-help housing documented in Peru, to the heavily commercialised illegal housing of mega-cities like Tunisia, Istanbul, Cairo, Tehran, and New Delhi. While these residents have expanded access to municipal services and a reduced vulnerability to the immediate health and security concerns of Nairobi's slum dwellers, they are increasingly vulnerable to rapid-onset natural hazards such as earthquakes, landslides, tsunamis and cyclones.

Istanbul is an example of a city with a long history of self-building and commercialised illegal construction, both in informal and formal neighbourhoods. These types of construction have resulted in high vulnerability to the mid- and large-scale earthquakes that periodically occur along the North Anatolian Fault just south of the city. Between the years 1930 and 2000, Istanbul was transformed from the capital of the ruined Ottoman Empire, a city of 800,000 residents, to a teaming modern metropolis of 10 million, the economic and cultural capital of Turkey (Merkezi 1997). This rapid growth brought new forms of construction and a form of political clientelism that helped transform the physical landscape of the city and greatly shape the vulnerability of its housing. Starting in the 1940s, agricultural workers began migrating to Istanbul in search of work, only to scramble to find housing in a city whose population had grown faster than its formal housing market (Heper 1982; Şenyapılı 1982; Stirling 1963, 1965). During the 1950s and 1960s, the inner core of middle and high income districts along the Marmara Sea and the Bosphorus coast were surrounded by a proliferation of unauthorised housing. Structures were made out of tin, plastic, discarded construction material and adobe, much like current housing in Kibara, Kenya. They were built by new migrants and other low-income residents themselves similar to Lima's informal housing settlements (Keles 1990; Yasa 1973).

Early attempts to demolish these shanty towns did little to curb the growth of Istanbul's unauthorised housing market (Gulati 1985; İMO 1976). By the 1970s, the rising land value in unauthorised shanty towns facilitated a transformation of the urban landscape. Shanties were sold or demolished by the squatters themselves and replaced by mid-rise reinforced concrete apartment buildings (Merkezi 1998) built primarily by small, one-man construction companies and self-builders on the contested land of the unauthorised settlements (Baharoglu 1996; Dener 1994; Gülöksüz 2002; Öncü 1988). The building boom was financed by a wide segment of the Turkish population as land speculation and the financing of cheap construction became the primary method of protecting savings against skyrocketing inflation (Baharoglu 1996; Öncü 1988; Yildirim 2000).

While some families continued to practice self-help construction in peripheral settlements, much of Istanbul's construction became commercialised. No longer could individual families squat as a means of obtaining land. Undeveloped land in Istanbul was cordoned off, measured, and illegally sold by pirate developers. For-profit contractors began purchasing illegal land and constructing apartments to rent or sell to third parties. Similarly, early squatter housing was illegally rented out, sold or redeveloped as higher density, and higher profit illegal housing. In other cases, residents formed cooperatives that hired shady contractors promising to construct apartment flats for them at below market rates. In each case, residents of informally constructed housing were no-longer intimately involved in the construction or supervision of their housing (Green 2005).

The consolidation process brought greater density, modern construction materials, and increased building height to centrally located informal settlements in Istanbul. Yet, these changes did not improve the seismic resistance of the building stock. If anything, the consolidated, illegal settlements of Istanbul became more prone to seismic hazard. Unplanned growth within unauthorised squatter settlements overwhelmed the city's capacity to monitor and regulate the quality of construction, both in authorised and unauthorised settlements.[2] Under-trained, under-funded, and over-extended, building officials became targets for bribery and corruption. Absence of clearly delineated lines of responsibility added to the poor conditions for assuring that housing met minimum standards of hazard resistance (Balamir 2001; Özerdem and Barakat 2000). By the late 1990s, the majority of the housing in Istanbul did not meet even minimum building standards specified in the earthquake design codes introduced in 1944 and later updated in 1953, 1968, 1975, and 1998 (Gülkan 2000). The building stock that resulted provided inexpensive housing for an expanding population, but one with an uncertain capacity to withstand seismic hazards.

With little or no ability to personally oversee the construction of their homes and with little governmental oversight of the construction, Istanbul residents in illegal districts have been exposed to untold scams. Low-income families have funded the construction of what they believed were legal apartment flats, only to find that their land deeds had been forged. Residents have been left with incomplete (illegal) construction, their contractor having left in the middle of the night. Like in Lima's 'young towns', unauthorised construction in Istanbul often lacks professional design and municipal oversight. Construction has occurred according to construction site 'rules of thumb' that often are based upon older code provisions or devised with an eye to profits gained by cutting corners (Green 2008).

The result of these processes is housing that defies published land-use planning guidelines, is often several stories higher than district height restrictions, and has material quality that is only a fraction of design code minimums (Aschhiem *et al.* 2000; Bruneau 1999; Gülkan 2000; Safak *et al.* 2000; Sezen *et al.* 2000). Erdik and Aydınoğlu (2002) estimate that the building stock in Turkey is 10 times more vulnerable to earthquakes than similar multi-storey reinforced concrete buildings in California exposed to the same level of hazard. Much of Istanbul's housing – built during rapid development and through the commercialised informal housing market – poses a great threat to human life during earthquakes.

Development and vulnerability

Vulnerability to geophysical hazards such as earthquakes, landslides and tsunamis is not uniform globally. Rather, economically less developed countries are more likely to have experienced disasters in the preceding decades and experienced far greater mortality rates per capita. Much of these countries' vulnerability is related to rapid industrialisation and urban migration. Of particular concern are urban centres of lower-middle and middle income countries where the urban poor have constructed informal housing using incremental construction techniques or purchased housing built through commercialised informal housing markets. These structures have been built largely without professional design or critical construction detailing that can greatly reduce vulnerability to natural hazards.

Economically less developed countries have not been successful in keeping urban populations from building in hazardous locations and outside formalised inspection channels. Despite a variety of approaches – demolishing unwanted slums, tacitly accepting their development, engaging in clientelistic politics, and actively managing site and service programmes – vulnerable informal settlements continue to expand uncontrollably. The processes of industrialisation, population growth, and enforced economic liberalisation have overwhelmed meagre building department funding and technical capacity. Moreover, continued economic inequality has compelled many residents to seek out informal housing on the one hand and, on the other, to engage in clientelistic politics and administrative corruption that further reduce hazard-wise choices.

While countries like Peru, Turkey, Egypt, and Iran have modern design codes that prescribe robust design and construction detailing, the formal construction process is beyond the economic means of many urban residents. With low wages, poor residents of Nairobi can often only afford un-serviced rental housing that not only fails minimal standards of structural integrity, but also fails minimal standards of health safety (Gulyani, personal communication). Residents of Lima's 'young towns' save over years and decades to complete construction that building codes require to be completed within a much shorter time span (Williams 2005). Even in the advanced informal housing sector of Istanbul, residents seek out housing made affordable by bending zonation laws, bribing building department officials, and constructing buildings with under-specified construction materials (Green 2005).

Informal housing sectors – upwards of 50 per cent of many urban centres in low and middle income countries – are outcomes of historical processes beyond individual, local, and at times even national control. When considering the immediate needs for low-income, centrally-located housing and the multiple pressing needs of health, education, and security, it is important to be cognisant of future hazards. Vulnerability is built into informally constructed homes through their location on exposed sites, their incorporation of vulnerability-heightening construction techniques, and their lack of adherence to codes intended to reduce vulnerability to health and natural hazards.

In countries where slum demolition and denial of land security is common, informal settlements continue to grow, but housing quality is suppressed. The urban poor, housed in temporary shelters without infrastructure and municipal services, are exposed to a host of health and security threats, in addition to environmental hazards such as flash floods, fires, and landslides. While their houses may not pose a direct threat, they remain an extremely vulnerable population. In countries where land tenure is more secure, housing quality often improves. Structures are made of permanent materials, though often over an extended process of incremental construction. While material quality improves, informal construction often also expands both horizontally into more precarious locations and vertically into un-engineered, multi-storey housing. These homes can pose a graver threat to residents when exposed to geo-physical hazards.

Clearance of informal housing may be necessary in extremely hazardous locations – for example the hurricane-exposed wharfs and drainage canals of Manila, the most landslide-prone slopes of South American capitals, and Turkey's north-western flood planes prone to liquefaction during earthquakes. Yet, in areas where housing is less exposed to extreme hazard, engagement with unauthorised settlements needs to finds ways to support the adoption of hazard-resistant construction. More is needed than simply a call for stronger codes, firmer enforcement and more knowledgeable professionals (Comartin et al. 2004). To address vulnerability in these areas, appropriate knowledge needs to reach those engaged in the informal housing sector: contractors, self-builders, renters or buyers.

Vulnerability reduction strategies in less economically developed countries begin in partnership with vulnerable urban communities. They must address issues of land security, immediate health and security issues, and the long-term vulnerabilities resulting from the informal construction process. Examples of such community-based programmes are described in Chapter 10 and follow what Degg and Chester (2005) describe as an 'alternative vulnerability agenda'. This agenda acknowledges that people are often in hazard's way, not for lack of knowledge, but for lack of viable alternatives. Effective vulnerability reduction needs to directly engage with urban communities as partners in multi-faceted vulnerability reduction. This means working together to find no-cost and cost-effective hazard-resistant technologies and increasing residents' ability to make informed choices. Successful vulnerability reduction in low and middle income countries must address not only the physical vulnerability of the housing stock, but also the economic, social, and historical realities of the people who reside in them.

Notes

1 This economic ranking was devised by dividing the 232 countries and territories officially recognised by the United States as of 2005 into quintiles according to

GDP per capita (PPP). While country self-reporting of disasters, redrawing of national boundaries and change in economic ranking over time make precise comparison difficult, the results are suggestive.

2   Changes in the legal construction process since the devastating 1999 earthquake and recent municipal consolidation have strengthened independent oversight of construction. The earthquakes have also heightened public awareness of construction quality. This has helped to curb bribery, corruption and public appetite for low-cost, low-quality housing.

References

Angel, S. (2000) *Housing Policy Matters: A Global Analysis*, Oxford: Oxford University Press.

Aschhiem, M., Gülkan, P., Sezen, H., Bruneau, M., Elnashai, A., Halling, M., Love, J. and Rahnama, M. (2000) 'Performance of buildings', *Earthquake Spectra*, Supplement A to Volume 16: 237–79.

Asgary, A. and Willis, K.G. (1997) 'Estimating the Benefits of construction measures to mitigate earthquake risks in Iran', *Environment and Planning B – Planning & Design*, 24(4): 613–24.

Baharoglu, D. (1996) 'Housing supply under different economic development strategies and the forms of state intervention: the experience of Turkey', *Habitat International*, 20(1): 43–60.

Balamir, M. (2001) 'Disaster policies and social organization', 5th Conference of European Sociological Association, 28 August–1 September, Helsinki.

Bruneau, M. (1999) *Mceer Response: Preliminary Reports from the Kocaeli (Izmit) Earthquake of August 17, 1999 – Structural Damage*, Buffalo, NY: Multi-Disciplinary Center for Earthquake Engineering Research.

Chalfoun, N.V. and Michal, R.J. (2002) *Thermal Performance Comparison of Alternative Building Envelope Systems: An Analysis of Five Residences in Teh Community of Civano*, Tuscon, AZ: University of Arizona, College of Architecture, Planning and Landscape Architecture.

Coburn, A. and Spence, R. (1992) *Earthquake Protection*, New York: John Wiley & Sons.

Coch, N.K. (1995) *Geohazards: Natural and Human*, Englewood Cliffs, NJ: Prentice Hall.

Comartin, C., Brzev, S., Naeim, F., Greene, M., Blondet, M., Cherry, S., D'Ayala, D., Farsi, M., Jain, S.K., Pantelic, J., Samant, L. and Sassu, M. (2004) 'A challenge to earthquake engineering professionals', *Earthquake Spectra*, 20(4): 1049–56.

CRED (2005) *Disaster Profiles Database*, Louvain: Centre for Research on the Epidemiology of Disasters, Available HTTP: <www.em-dat.net>.

Degg, M. (1992) 'Natural disasters – recent trends and future-prospects', *Geography*, 77(336): 198–209.

Degg, M.R. and Chester, D. (2005) 'Seismic and volcanic hazards in Peru: changing attitudes to disaster mitigation', *The Geographical Journal*, 171(2): 125–45.

Dener, A. (1994) 'The effect of popular culture on urban form in Istanbul', in Neary, S.J., Symes, M.S. and Brown, F.E. (eds) *The Urban Experience: A People-Environment Perspective*, London: E. & F. Spon.

Dilley, M., Chen, R.S., Deichmann, U., Lerner-Lam, A.L., Arnold, M., Agwe, J., Buys, P., Kjekstad, O., Lyon, B. and Yetman, G. (2005) *Natural Disaster Hotspots: A Global Risk Analysis*. Washington, DC: The World Bank Hazard Management Unit.

El-Messiri, S. (1985) 'The squatters' perspective of housing: an Egyptian view', in Van Vliet, W. and Fava, S. (eds) *Housing Needs and Policy Approaches: Trends in Thirteen Countries*, Durham, NC: Duke University Press, 256–70.

Erdik, M. and Aydınoğlu, N. (2002) 'Earthquake risk to buildings in Istanbul and a proposal towards its mitigation', in Pankl, H. (ed.) *Second Annual IIASA-DPRI Meeting Integrated Disaster Risk Management: Megacity Vulnerability and Resilience*, IIASA, A-2361 Laxenburg, Austria.

Ferradas, P. (2006) 'Post-disaster housing reconstruction for sustainable risk reduction in Peru', *Open House International*, 31(1): 39–46.

Garau, P., Sclar, E.D. and Carolini, G.Y. (2005) *A Home in the City: Improving the Lives of Slum Dwellers*, London: Earthscan.

Green, R.A. (2002) *The Afyon Earthquake of February 2, 2002: Traditional Homes and Seismic Culture*. Available HTTP: <www.realrag.com/papers> (accessed 14 September 2004).

Green, R.A. (2005) *Negotiating Risk: Earthquakes, Structural Vulnerability and Clientelism in Istanbul*, Ithaca, NY: Cornell University.

Green, R.A. (2008) 'Unauthorised development and natural hazard vulnerability: a study of squatters and engineers in Istanbul, Turkey', *Disasters: The Journal of Disaster Studies, Policy & Management*, 32(3): forthcoming.

Guha-Sapir, D., Hargitt, D. and Hoyois, P. (2004) *Thirty Years of Natural Disasters 1974–2003: The Numbers*, Louvain-la-Neuve: Centre for Research on the Epidemiology of Disasters.

Gulati, P. (1985) 'The rise of squatter settlements: roots, responses, and current solutions', in Van Vliet, W. and Fava, S. (eds) *Housing Needs and Policy Approaches: Trends in Thirteen Countries*, Durham, NC: Duke University Press, 256–70.

Gulis, G., Mulumba, J.A.A., Juma, O. and Kakosova, B. (2004) 'Health status of people of slums in Nairobi, Kenya', *Environmental Research*, 96(2): 219–27.

Gülkan, P. (2000) 'Building code enforcement prospects: the failure of public policy', *Earthquake Spectra*, Supplement A to Volume 16: 351–74.

Gülöksüz, E. (2002) 'Negotiation of property rights in urban land in Istanbul'. *International Journal of Urban and Regional Research*, 26(2): 462–76.

Gulyani, S., Talukdar, D., Potter, C., Biderman, J., Bruce, C. and Bergen, G. (2006) *Inside Informality: Poverty, Jobs, Housing and Services in Nairobi's Slums*, World Bank Report No. 36347-KE, Washington, DC: World Bank.

Gupta, M., Sharma, A. and Kaushik, R. (2006) 'Saving Shimla, North India, from the next earthquake', *Open House International*, 31(1): 90–7.

Harding, A. (2002) 'Nairobi slum life', BBC News, Available HTTP: <http://news.bbc.co.uk/1/hi/world/africa/2297237.hstm> (accessed 4 October 2005).

Heper, M. (1982) 'The plight of urban migrants: dynamics of service procurement in a squatter area', in Kağıtçıbası, C. (ed.) *Sex Roles, Family, and Community in Turkey*, Bloomington, IN: Indiana University Turkish Studies, 249–67.

İMO (1976) *Türkiye'de Konut Sorunu (the Problem of Housing in Turkey)*, Ankara: İnşaat Mühendisleri Odası.

Jain, S.K., Lettis, W.R., Murty, C.V.R. and Bardet, J.-P. (2002) 'Bhuj, India, earthquake of January 26, 2001 reconnaissance report', *Earthquake Spectra*, 18 (Supplement A).

Kaushik, H.B., Dasgupta, K., Sahoo, D.R. and Kharel, G. (2006) 'Performance of structures during the Sikkim earthquake of 14 February 2006', *Current Science*, 91(4): 449–55.

Keles, R. (1990) 'Housing policy in Turkey', in Shidlo, G. (ed.) *Housing Policies in Developing Countries*, New York: Routledge, 140–72.

Langenbach, R. (2005a) 'Performance of the Earthen Arg-E Bam (Bam Citadel) during the 2003 Bam, Iran, earthquake', *Earthquake Spectra*, 21 (Special Issue on Bam Earthquake I).

Langenbach, R. (2005b) *Survey Report on Northern Kashmir Earthquake of October 8, 2005 from the Indian Kashmir Side of the Line of Control*. Available HTTP: <http://www.conservationtech.com> (accessed 4 October 2005).

Macharia, K. (1992) 'Slum clearance and the informal economy in Nairobi', *Journal of Modern African Studies*, 30(2): 221–36.

Martin, R.D. and David, K.C. (2005) 'Seismic and volcanic hazards in Peru: changing attitudes to disaster mitigation', *Geographical Journal*, 171(2): 125.

Merkezi, İ.A. (1997) *İstanbul Külliyatı Cumhuriyet Dönemi İstanbul İstatistikleri I: Nüfus Ve Demografi 1 (1927–1990) (the Complete Istanbul – Republic Era Statistics of Istanbul I: Population and Demographics 1 (1927–1990))*, Istanbul: İstanbul Büyükşehir Belediyesi Kültür İşleri Daire Başkanlığı and İstanbul Araştırmaları Merkezi.

Merkezi, İ.A. (1998) *İstanbul Külliyatı Cumhuriyet Dönemi İstanbul İstatistikleri 6: İnşaat Ruhsatnamelerine Göre 1 (1963–1996) (the Complete Istanbul – Republic Era Statistics of Istanbul 1: According to Construction Permits 1 (1963–1996))*, Istanbul: İstanbul Büyükşehir Belediyesi Kültür İşleri Daire Başkanlığı and İstanbul Araştırmaları Merkezi.

Naeim, F., Mehrain, M., Rahnama, M., Bozorgnia, Y., Enssani, E., Bastani, A., Ostadan, F., Movahedi, H. and Mansouri, B. (2004) *Preliminary Observations on the Bam, Iran, Earthquake of December 26, 2003*, Buffalo, NY: Earthquake Engineering Research Institute.

Neuwirth, R. (2005) *Shadow Cities: A Billion Squatters, a New Urban World*, New York: Routledge.

OFDA/CRED (2006) *EM-DAT: The International Disaster Database*, Center for Research on the Epidemiology of Disasters, Available HTTP: <http://www.em-dat.net/index.htm> (accessed 12 December 2006).

Oliver-Smith, A. (1999) 'Lima, Peru: underdevelopment and vulnerability to hazards in the City of the Kings', in Mitchell, J.K. (ed.) *Crucibles of Hazard: Mega-Cities and Disasters in Transition*, New York: United Nations Press, 248–94.

Öncü, A. (1988) 'The politics of the urban land market in Turkey: 1950–1980'. *International Journal of Urban and Regional Research*, 12(1): 38–64.

Otiso, K.M. (2003) 'State, voluntary and private sector partnerships for slum upgrading and basic service delivery in Nairobi City, Kenya', *Cities*, 20(4): 221–9.

Özerdem, A. and Barakat, S. (2000) 'After the Marmara earthquake: lessons for avoiding short cuts to disasters', *Third World Quarterly*, 21(3): 425–39.

Pelling, M. (2003) *The Vulnerability of Cities: Natural Disasters and Social Resilience*, London: Earthscan.

Pugh, C. (1995) 'The role of the World Bank in housing', in Aldrich, B.C. and Sandhu, R.S. (eds) *Housing the Urban Poor: Policy and Practice in Developing Countries*, London: Zed Books, 34–92.

Rai, D.C. and Murty, C.V.R. (2005) *Preliminary Report on the 2005 North Kashmir Earthquake of October 8, 2005*, Kanpur: Department of Civil Engineering, Indian Institute of Technology.

Sachs, J.A. (2005) *The End of Poverty: Economic Possibilities for Our Time*, New York: Penguin Press.

Safak, E., Erdik, M., Beyen, K., Carver, D., Cranswick, E., Celebi, M. *et al.* (2000) 'Record main shock and aftershock', *Earthquake Spectra*, Supplement A to Volume 16: 97–112.

Saff, G. (1996) 'Claiming a space in a changing South Africa: the "squatters" of Marconi Beam, Cape Town', *Annals of the Association of American Geographers*, 86(2): 235–55.

Sassen, S. (1998) *Globalization and Its Discontents*, New York: New Press.

Schusterman, R. and Hardoy, A. (1997) 'Reconstructing social capital in a poor urban settlement: the integral improvement programme in Barrio San Jorge', *Environment and Urbanization*, 9(1): 91–119.

Şenyapılı, T. (1982) 'Economic change and the Gecekondu family', in Kağıtçıbası, C. (ed.) *Sex Roles, Family, and Community in Turkey*, Bloomington, IN: Indiana University Turkish Studies, 237–48.

Sezen, H., Elwood, K.J., Whittaker, A.S., Mosalam, K.M., Wallace, J.W. and Stanton, J.F. (2000) *Structural Engineering Reconnaissance of the Kocaeli (Izmit) Turkey Earthquake of August 17, 1999. PEER Report 2000-09*, Berkeley, CA: Pacific Earthquake Engineering Research Center, Available HTTP: <http://peer.berkeley.edu/Products/PEERReports/reports-2000/reports00.html> (accessed 12 December 2006).

Shatkin, G. (2004) 'Planning to Forget: informal settlements as "forgotten places" in globalising metro Manila', *Urban Studies*, 41(12): 2469–84.

Smart, A. (1995) 'Hong Kong's slums and squatter areas: a development perspective', in Aldrich B. and Sandhu, R. (eds) *Housing the Urban Poor: Policy and Practice in Developing Countries*, London: Zed Books, 97–111.

Soliman, A. (1996) 'Legitimizing informal housing: accommodating low-income groups in Alexandria, Egypt', *Environment and Urbanization*, 8(1): 183–94.

Stirling, P. (1963) 'The domestic cycle and the distribution of power in Turkish villages', in Pitt-Rivers, J. (ed.) *Mediterranean Countrymen*, Paris and La Haye: Mounton, 201–13.

Stirling, P. (1965) *Turkish Village*, London: Weidenfeld and Nicolson.

Tolles, E., Kimbro, E., Webster, F.A. and Ginell, W.S. (2000) *Seismic Stabilization of Historic Adobe Structures*. Final Report of the Getty Seismic Adobe Project. Los Angeles, CA: Getty Conservation Institute.

Traub, J. (2006) 'China's African adventure', *New York Times*, New York.

Turner, J.F.C. (1977) *Housing by People: Towards Autonomy in Building Environments*, New York: Pantheon Books.

UNDP (2001) *Human Development Report 2001*, United Nations Development Programme, New York: Oxford Press. Available HTTP: <http://www.worldbank.org/depweb/english/modules/social/life/> (accessed 25 November 2005).

UNDP (2004a) *Reducing Disaster Risk: A Challenge for Development*, New York: United Nations Development Programme.

UNDP (2004b) *World Population Prospects: The 2004 Revision*, New York: United Nations Development Programme, Population Division of the Department of Economic and Social Affairs of the United Nations Secretariat, Available HTTP: <http://esa.un.org/unpp> (accessed 25 November 2005).

Vries, S.K. (2002) *Bamboo Construction Technology for Housing in Bangladesh: Opportunities and Constraints of Applying Latin American Bamboo Construction Technologies for Housing in Selected Rural Villages of the Chittagong Hill Tracts, Bangladesh*, International Network for Bamboo and Rattan and Action Aid Bangledesh. Available HTTP: <http://alexandria.tue.nl/extra2/afstversl/tm/vries2002.pdf> (accessed 17 March 2005).

WHO (2006) *Water, Health and Sanitation*, World Health Organization, Available HTTP: http://www.who.int/water_sanitation_health/publications/facts2004/en/index.html (accessed 12 December 2006).

Williams, S. (2005) '"Young Town" Growing Up: Four Decades Later, Self-Help Housing and Upgrading Lessons from a Squatter Neighborhood in Lima', unpublished Masters thesis, Boston: Massachusetts Institute of Technology.

Wisner, B. (2003) 'Changes in capitalism and global shifts in the distribution of hazards and vulnerability', in Pelling, M. (ed.) *Natural Disasters and Development in a Globalizing World*, London: Routledge.

Yap, G. (2004) 'Improving water and sanitation conditions in Nairobi, Kenya', *Water Drops*, Spring, 3–5.

Yasa, I. (1973) 'The impact of rural exodus on the occupational patterns of cities: the case of Ankara', in Kıray, M.B. (ed.) *Social Stratification and Development in the Mediterranean Basin*, The Hague: Mouton, 138–55.

Yildirim, Z. (2000) 'Capital account liberalisation in Turkey', *Proceedings of the Mediterranean Development Forum, Financing Development Workshop*, 5–7 March, Cairo, Egypt.

Chapter 12

The worm in the bud

Corruption, construction and catastrophe

James Lewis

Introduction

It is collapsing buildings that kill people, not earthquakes themselves.
Good construction can withstand most earthquakes but conversely, poor
construction may not do so. It is regrettable for a world-wide industry
capable of constructing bridges, airports and skyscrapers, that a chapter
comes to be devoted to the extent that corrupt construction practice, and
its 'concealment, like a worm i' the bud',[1] may be the cause of death, injury
and deprivation. Together with how and why corruption happens and what
can be done about it, this chapter is relevant to all countries and all hazards;
Italy and Turkey are two of the most earthquake-prone countries and will be
examined in some detail.

Corruption

Corruption is defined as the abuse of entrusted power for private gain
(Transparency International 2006a, 2006b) and as 'offering, giving, receiving
or soliciting of anything of value to influence the action of an official in
a procurement or selection process or in contract execution' (World Bank
2003). These are the corrupt acts but as the OECD Secretary General has
stated 'The impact of corruption goes far beyond the specific misbehaviour
of the actors involved. Its repercussions sweep across entire populations ...
through derailed development plans and incoherent investment decisions.
Unfinished roads, crumbling schools and crippled health systems are but a few
examples which illustrate the impact of this phenomenon' (OECD 2006a).

In failed integrity and by their own contrivance and concealment, those with
power, authority or influence who exert money or favours, cause depletion
of livelihoods and the happiness[2] of others. Decisions are skewed to the
advantage of those already with money, power, authority or influence; others,
including the powerless and the poor, are deprived of their democratic right
to their share of economic transaction, and thereby are denied equability and
personal economic development. Described by a distinguished international
panel as one of the principal causes of poverty and inequality, and a cause of

injustice, disease and death (World Bank 2006a), corruption is an extreme abuse of democratic values.

Corrupt practices, inclusive of bribery, back-handers, fraud and embezzlement, applied politically and commercially, pervade virtually all countries of the world at national and international levels. An internet search (Guardian Unlimited 2006), reveals world-wide examples relating to national, municipal and local governance; political party funding; law, justice, policing and prisons; immigration and emigration; taxation; share trading and financial management; armaments trading; procurement in all sectors; sport; water and sanitation; health care; development aid and reconstruction – and construction.

The reports of Transparency International, the organisation formed as a worldwide coalition to lead the fight against corruption, annually include a corruption perception index (Lambsdorff 2006a). Now in its eleventh year, the index expresses a 'score of confidence' about the degree of corruption within each country ranging from 10, being 'highly clean' to 0, being 'highly corrupt'. The index for 2005 includes 159 countries, with Iceland, Finland, New Zealand, Denmark and Singapore as the five countries at the top of the index, with scores of 9.7 to 9.3, and Haiti, Myanmar, Turkmenistan, Bangladesh and Chad as the five countries at the bottom of the index with scores of 2.1 to 1.3. The United Kingdom tied with the Netherlands as eleventh on the list; the USA was seventeenth. It has been reported that, in the United Kingdom, one in five companies, mostly in mining, gas and construction sectors, lose business to a competitor paying bribes (Fletcher 2006).[3]

Acknowledging that the achievement of changes in perceived levels of corruption is a long-term undertaking, a review of reports prepared annually for ten years (Lambsdorff 2006b) indicates significant improvements between 1995 and 2004 in twelve countries, with significant deterioration in eleven others. Successful identification of corruption involving large amounts of money at national and international levels, has necessarily bequeathed an integrity and clarity in research and description, in direct contrast to its practice which is covert, underhand, unstated, unrecorded and concealed. Corruption exists not only in high places 'clothed with respectability', where its exposure makes media headlines: it pervades downwards to day-to-day transactions which, if observed at all, could be considered endemic and to be expected. As Leslie Palmier (2000) has written, corruption is so widespread that, factually speaking, it can be regarded as normal or, if not normal, as a disease so pervasive and persistent that it becomes normal. Pervasive normality obscures perception of corrupt practice: a corruption in construction survey has reported difficulty of distinction between gifts towards 'harmonious working relationships' and larger non-cash incentives: 'The point where an innocent gift becomes a bribe is not clear' (CIOB 2006: 23).

Upper level exploitation causes those lower down to act similarly, to a lesser degree, due to payments they have been obliged to make to those above them. Lesser influence with lesser power commands lesser payment. Transactions have their costs loaded, to facilitate payment to the person granting permission or providing the job, by which transactions can be made. As in many systems, behaviour at the bottom of the scale emulates that at the top; if the rich and powerful can do it, then so can others attempt, or be obliged, to do it – the offering of a bribe being as corrupt as its expectation or its receipt.

Corruption is by no means a new phenomenon but, more recently, there has been a realisation of its wider implications and a corresponding increase of exposure of corrupt practices. Allegations have regularly been exposed by international media (e.g. Transparency International's *Transparency Watch Newsletter*), by national media in countries of its exposure (e.g. Cevíc 2003), by published research, and by national institutions and some governments internationally declaring their programmes to expose and prevent it (e.g. DFID 2006a, 2006b) '(T)his government's complicity in alleged international bribery having later been revealed' (Transparency International 2007).

Corruption in construction

Corruption in construction is not, therefore, an isolated phenomenon, nor a new issue, but its consequences are severe. Widespread general corruption may lead to disease and death but corrupted construction in earthquake zones may be the direct cause of injury and death, often on a large scale. Most buildings, for their most part, are site-produced by a variety of subcontracted and often itinerant skills, whereas aircraft and cars, for example, are produced in controlled factory conditions.

In 1977, a study of reconstruction following disaster made no mention of corruption (Haas *et al.* 1977), though its authors did refer to the probable need for improved standards of construction and to the possibility of gain, profit and inequality in recovery after disaster. Five years later, Robert Geipel (1982: 17) noted in the Italian context of the 1976 Fruili earthquake, that nothing could change the effectiveness of disaster management which, to outside observers, had seemed astonishingly high – not even accusations of corruption made in a context of 'jealous criticism'. In their book on earthquake protection, Coburn and Spence (2002: 355) make no mention of corruption as such, though responsibility for 'building control' is recognised as extending beyond 'builders as the guilty party' to planning systems, and project and construction supervision.

A more recent report (Chief Technical Examiner's Organisation 2002) in India has observed that in the past, corruption in construction was at 'a lower level of the hierarchy' (ibid.: 7), having involved the payment of bribes for allowing below-standard material or workmanship during construction, and continuing '… the top officials maintained a high level of integrity in those days' (ibid.: 7), going as far as to order the dismantling of defective construction 'then and there'. The report continues that, as time passed '... corrupt practices have not spared even the Chief Executives of the organisations' (ibid.: 7).

These polite observations provide rare perspective on a situation now, in a global construction industry that is considered to harbour higher levels of corruption than any other economic sector (Stansbury 2005). Transparency International's Bribe Payers' Index indicates that construction is one of three industrial sectors most prone to bribery (DFID 2006b: 41) in which '... stakeholders know how rampant it is but choose to keep silent' (Shakantu 2003: 274).

In addition to the size and scope of the industry, and to the size of individual projects for transportation, fuel and power, and sea-protection, for example, large-scale corruption in construction is facilitated by its organisation, from which Stansbury (2005) itemises a further dozen features lending themselves to corrupt practice. Principal among these features are: unique or one-off projects making it easier to inflate costs and hide bribes; extensive regulation requiring permissions and approvals, inviting the extraction of bribes by unreliable officials; a complexity of contractual arrangements and payments, providing myriad opportunities to demand hidden bribes for faulty certification of work; and commercial confidentiality taking precedence over public interest and transparency, with entrenched tendering procedures suggesting a need for bribes to remain on the list. These seemingly elementary but critical shortcomings continue to resonate, for example, in the reports of inspections, audits and investigations into post-war reconstruction in Iraq, which have concluded that the programme has fallen short 'for the most mundane reasons': poorly written contracts, ineffective or non-existent oversight (supervision), needless project delays and 'egregiously poor construction practices' (Glanz 2006).

The size of built infrastructure projects alone, especially when of 'one-off' content, creates opportunities for corrupt practice, not only as opportunity to increase capital cost. A corrupt budget decision maker might prefer large or grandiose projects to smaller, repetitive projects of schools, clinics or housing. Resulting built infrastructure becomes more costly due to corruption in its selection, in its construction and in many cases, in its management, and its quality will have reduced incurring higher maintenance costs. This is in addition to its displacement of other built services higher on a scale of social need (Collier and Hoeffler 2005). Numerous 'monuments to corruption', environmentally destructive as well as financially corrupt, dubiously selected, corruptibly constructed and high-cost, stand as impotently non-productive projects of aggrandisement (Bosshard 2005).

These features conspire to create opportunities for corrupt management to flourish in all construction stages: proprietary information infringements; collusive tendering/bidding; cash inducements (i.e. bribery) for over-valuing work done (or not done) and covering up of inadequate materials or workmanship. Payment of bribes creates the need to inflate costs so as to absorb that additional overhead. In turn, at the lower end of the scale so to speak, such conditions increase on-site pressures to cut costs, to make short-cuts in materials and time, and to skimp on workmanship, it being in the nature of the construction process for one trade to materially cover another, facilitating the hiding of such practices.

In its construction, each stage of a building is concealed – from foundations under the ground and steel reinforcement in concrete, through to wall, floor and ceiling finishes, external cladding and the last coat of paint. Mistakes and omissions – accidental and intentional – have to be identified and rectified within each stage or they are covered up for ever – or until exposed by building failure. Pressures on builders to complete on time, increased by financial incentives, impeded by weather and by delayed deliveries, create circumstances in which temptations are rife for expediency and shortcuts. Given these endemic 'built-in' opportunities for deviation, it is not surprising that the construction industry is the hot-bed of illicit practice; 'cover-up' no doubt deriving from the construction industry (Lewis 2003, 2005). Driven and induced by corrupt transactions at management levels, corrupt practices on site, whether intentionally devious or caused by laziness, ignorance or apathy, are facilitated by similarly inadequate site management that, all in all, contributes most directly to eventual building failure and possible collapse.

Corruption in construction is not confined, however, to managers or practitioners. Professionals handling development projects, ahead of their start on site, may be complicit in corrupt practice. A quantity surveyor, for example, engaged in preliminary project costing before the final stage of going to tender/bid is reached, is in a position to leak that information to a prospective tenderer (Shakantu 2003), as well as being able to influence the listing of tenderers. Appointed as payments controller, as well as designer, an architect might seek to avoid certifying additional work caused by design or other documentation error (Stansbury 2005). In New York City, 31 architects, real estate brokers and managers as well as 24 other companies, pleaded guilty in 1999 to bribery charges related to bid-rigging on large construction projects (Osborn 2000).

Across earthquake-prone Japan in 2005, 78 buildings including 36 hotels were declared unsafe, their specification to falsified earthquake resistance data having been submitted by their architect during pre-construction stages (McCurry 2005). Apartment block residents were ordered to leave and hotels were forced to close to await demolition. In some cases, it was reported, steel structural members were one-quarter of their required size. The architect claimed he was under pressure from the developer-contractor, by whom he was engaged, who wanted the buildings finished quickly at minimal cost. Had he resisted, he said, he would have lost his valuable client. A national inquiry was instructed.

Corruption in construction affects entire national economies, impacts upon the well-being of the industry, is destructive of public participation and interest (Shakantu 2003) and results in higher costs and less construction overall. In housing construction, corrupt practices become increasingly devastating, as urban populations grow, as urban areas expand, opportunities for profit-making are imposed upon urban and rural self-build capacity. Expansion of would-be urban populations, in turn displaces former occupants of land taken for development. Land management, road building, water treatment and supply, and other preceding activities in which corruption is similarly rife, may require induced or forced displacement and relocation to marginal sites of entire or desiccated communities. Often illegally removed from their traditional resource base, communities may become more vulnerable to landslides, water shortage and drought, storms and flooding, and may be unable effectively to exercise their traditional responses (Lewis 1999).

Millions of people for whom there are no options about the places they inhabit, are obliged to respond to policies and activities in the control of others – or to often corrupt external pressures in the interests of others (Lewis 1999). The places in which people are obliged to build, or where buildings are built which they then occupy, have as much to do with vulnerability to natural hazards as does building construction and occupation.

Corruption and catastrophe

Turkey and Italy are the two most earthquake-prone countries in Europe (aspiring and Europe15).[4] They are two of the three countries with an incidence of 'more than 10' in 1974–2003 earthquake occurrence (EM-DAT, 2006: the third is Greece with many fewer losses from similar incidence). Can it be simple coincidence that these two most earthquake-prone European countries are also the two most closely identified with endemic corruption? Italy, at 40th, is the lowest Europe15 country on the corruption perception index; Turkey is placed 65th (Greece is 47th).

Turkey

During the twentieth century, earthquakes in Turkey have caused the destruction of almost 650,000 buildings (Mitchell and Page 2005: 27). The earthquake that occurred in north-west Turkey in August 1999, affected an area of approximately 72,000 sq km (28,000 sq miles) and killed more than 17,000 people. Forty-four thousand were injured and 600,000 were made homeless; damage was estimated at US$8.5 million. Places named as seriously damaged were Istanbul (periphery), Tekirdag, Yalova and Bursa to the east of the epicentral area, Izmit (Kocaeli) and Gölçük on the Izmit Estuary, Eskisehir to the south, Sakarya and Zonguldak on the Black Sea coast, and Bolu inland from them. Bolu was damaged again in another earthquake three months later in November (EM-DAT 2006).

The epicentre of Turkey's August 1999 earthquake occurred offshore from its north-western Izmit region, concentrating earthquake damage in this area of highest population density. The province of Kocaeli, with a density of 260 people per sq km, contains more than 20 per cent of a national population that has increased exponentially since 1945. A centralised government has attracted migration to the capital and to other cities now so devastatingly destroyed, migration from economically deprived eastern regions into informal squatter communities (*gecekondu*) on the edges of already large cities (Lewis 2003; Mitchell and Page 2005). Largely due to their siting, many of these communities are at high earthquake risk; continuing migration, in part triggered by earthquakes elsewhere and continuing Kurdish conflict, increases numbers and exacerbates an already severe problem.

Also at high risk are mid-rise apartment blocks constructed of reinforced concrete, in which occurred about 90 per cent of casualties of the 1999 Izmit (Kocaeli) and Bolu/Düzce earthquakes (Erdik and Aydinoglu 2000; Lewis 2003). Erdik states that in a range of earthquake intensities of nine (high), eight and seven, percentages of mid-rise reinforced concrete residential buildings that were damaged beyond repair were 40, 15 and 5 respectively. Victims of these earthquakes are described as 'urban middle class people'; corrupt construction practice was identified as the reason why building stock was exposed as being of such poor quality.

In 2003, the Bingöl earthquake in eastern Turkey caused a school dormitory building to collapse in which 85 people lost their lives (Gülkan *et al.* 2003). Polat Gülkan of the Middle East Technical University states that, as routine practice, government service buildings are built to 'template designs', the same building of each type being replicated 'all over the country' for schools, hospitals, tax collection offices etc. Done for reasons of economy, any design or specification errors are automatically transmitted from location to location; modifications in response to varying seismic risk levels, usually to steel reinforcement, are too easily forgotten or ignored where constructional and administrative integrity is variable or uncertain. As Gülkan concludes: 'Surely the minor expense in construction costs would more than make up for constantly recurring replacement costs and the accompanying social trauma'.

Ninety-two per cent of Turkey's overall area is stated as being 'at risk of a ground tremor' (Yuzer 1999); it is evident, therefore, that construction anywhere, for whatever purpose, has to take appropriate account of earthquake risk if it is not to fail in earthquakes of medium and even minor intensities. Additionally, high concentrations of population are to be discouraged in a country so severely prone to earthquakes – until such time as existing mapped zones of earthquake risk are geologically, geographically and socially identified, and have become publicly absorbed and understood.

Vulnerability to earthquake risk needs to be ameliorated by development planned and constructed to take risk fully into account, construction codes doing the same, with application of the codes enforced and inspected in all new construction – processes which cannot effectively show reduction in earthquake losses where corrupt practice in construction continues.

Turkey's first seismic code was issued in 1944, updated in 1975 and again in 1997, based upon five seismic zones (Ellul and D'Ayala 2003). Zone One of highest risk includes western and north-western Turkey, north, north-east and eastern Turkey, a central area, and bands through parts of the south-west and south-east. Seismic zoning is a major step but, for construction purposes, has to be accompanied by a building code, adequately but straightforwardly related to geographically defined zones of risk. An accompanying planning function is essential so as to indicate areas where new building could proceed away from ground conditions where, for example, earthquakes could cause subsoil liquifaction – a factor that exacerbated earthquake damage in Adapazari, Gölçük, Avcilar and Sapanca in 1999 (Green 2005). Overall, at present, as the report of the 'relatively moderate event' of the 2003 Bingöl earthquake emphasised, 'the current Turkish Code underestimates the effects on the majority of building stock in the region' (Ellul and D'Ayala 2003: 16). At present, say Turkish earthquake engineers, less than 25 per cent of all buildings in Turkey conform at all to the 1997 code, Turkey having failed to relate mapped zones of earthquake risk with an administration capable of assessing drawings and calculations, visiting buildings under construction, exercising punishments for non-compliance and preventing inadequate structures from being erected (Coburn 1995).

Extensive earthquake damage in Turkey results from the collapse of buildings inadequately constructed, and buildings in inappropriate places. It is insufficient however for state authorities to allocate blame solely to a failed system of building control, allegedly to deflect blame away from facilitation and condonation of corrupt practices within their own management systems.

Italy

Corrupt princes, popes and politicians have been known in Italy since the fifteenth century; Alexander VI, Rodrigo Borgia, and Pope Leo X who financed the construction of St Peter's Basilica by the sale of indulgences triggering Christianity's Protestant Reformation. Today, in Sicily for example, stories of the mafia and corruption are a tourist attraction (Best of Sicily 2007) and construction is synonymous with historically endemic corrupt systems and practices.

Sicily is one of the two islands of the earthquake-prone Mezzogiorno region of southern Italy, comprising also the island of Sardinia and the mainland regions of Abruzzi, Campania, Molise, Puglia, Basilicata and

Calabria. Italian for 'midday', the Mezzogiorno is hot; it is also relatively impoverished. Steep Apennine slopes of eroded non-arable soil mean that agriculture, employing most of the working population, is an inadequate source of support in an area of Italy where incomes and standards of living are generally lower, and illiteracy higher, than in the north. Naples and Bari on the west and east coasts, are two large industrial port-cities. Large-scale land reforms were instituted in 1946, much later than in the prosperous north. In 1950, the *Cassa per il Mezzogiorno* (Fund for the South) was initiated by central government to stimulate social and economic development, being superseded in 1986 by the *Agenzia del Mezzogiorno* (Barca 2001).

All of the four mafias are based in this Mezzogiorno environment (Alexander 2005); the Cosa Nostra in Sicily, the Camorra in Naples, the Ndrangheta in Calabria and the Sacra Corona Unita in Puglia. Entire economic sectors, for example of banking, transportation, construction and the buildings it creates, such as hotels, are controlled by this mafiosi in socio-economic contexts of condonation by which its activities are perpetrated and perpetuated. Since the post-World War II Marshall Plan, billions of dollars intended by the Italian and United States governments, the World Bank, and the European Commission as support for Sicilian and other economies, have 'disappeared'; as a result, it has become impossible to separate the mafia from political corruption in this region of Italy (Best of Sicily 2007).[5]

Earthquakes are larger and more numerous in the Mezzogiorno than in other parts of Italy (Alexander 2005). The 1980 Naples/Avellino/Potenza earthquake, one of the largest in Italy of recent times, affected the city of Naples, the towns of Avellino and Caserta in the Apennine foothills, and Potenza, a hundred kilometres to their south-east in the mountains; more than 4,500 people died. In the regions of Campania and Basilicata, 637 settlements reported damage in an affected area of 23,000 sq km, causing US$ 20 billion damage in total (EM-DAT 2006; Geipel 1982).

Even before the earthquake, in Naples, there were 80,000 homeless people, many of whom feigned to be earthquake victims for perceived forthcoming advantages. Numbers of homeless increased further, as many inadequately constructed new buildings were declared dangerously uninhabitable while aftershocks continued. Developers and their architects, and construction supervisors who had sanctioned inadequate construction, were indicted; but the process revealed more buildings, similar to those that had incurred damage, that were also then evacuated – only to be re-occupied by the homeless while the authorities 'turned a blind eye' (Geipel 1982: 186).

Looters and profiteers were tried in summary courts but the Camorra was known to take advantage of the catastrophe: 'In a situation of incipient anarchy, the Camorra offered itself as a kind of "*sottogoverno*" or government below government, and was sometimes more effective than the authorities' (Geipel 1982: 187). An example of overt self-interest was the technically well-equipped relief team that arrived on site from Sienna in the north, but was refused permission to commence rescue operations because the contract had already been given to a demolition firm yet to come from Naples (Geipel 1982).

In Avellino, east of Naples, the maternity wing of a recently built six-storey hospital collapsed, killing most of its occupants. Subsequent investigation indicated that drawings and specifications had been adequate but that 'substantial economies', involving foundations to inadequate depths and serious omissions in reinforced concrete structure, had been made by contractors during the construction process in which inspection had been absent or ineffective. Similar inadequacies were revealed in the wreckage of many other recent buildings of modern materials and construction, that failed to withstand earthquake motion in an area known to be earthquake prone (Alexander 2005).

North of the Mezzogiorno, and in the north of Italy, the eastern Alpine region of Fruili suffered earthquakes in May 1976 which killed 939 people, injured 2,400, destroyed 17,000 dwellings and made an initial 32,000 people homeless; another 157,000 were later made homeless by aftershocks and heavy rain. One hundred communes were affected in an overall area of 48,000 sq km; an area of 1176 sq km, 25 kilometres across, was 'totally levelled'. A second earthquake, four months later in September, caused the by then decreased numbers of homeless to increase again to more than 70,000 (Geipel 1982). There are no allegations recorded by Robert Geipel of corrupt construction practices having exacerbated earthquake damage in Fruili.

In the Belice Valley of western Sicily, after a small earthquake in 1968 caused significant damage and killed 200 people, very little had been done by 1983 to reconstruct buildings. Large sums of government money, intended for reconstruction, had 'disappeared' in poor accounting, opaque public administration and 'the interests of the underlying black economy' (Alexander 2005: 26). Reconstruction did eventually take place but only after concerted attempts had been made to break the power of the mafia, attempts which led to the assassination of an army general and senior investigative judges (Alexander 2005).

When buildings are not appropriately sited and adequately constructed to take full account of earthquakes and other hazards, they represent disasters waiting to happen. Thousands of people living in many medium- and high-rise apartment buildings, are destined to die when those buildings fail and collapse in inevitable future earthquakes. These lethal legacies are the landmines for ill-named 'natural' disasters, more potently lethal than the legacies of conflict. That all illegally constructed dwellings should be demolished is not cynically ridiculous: Japan has commenced the process (McCurry 2005), as also has Spain's building fraud and corruption prosecutor (Tremlet 2006).

Revelations, allegations and indictments in Italy

Italy's position as 40th on the 2005 Corruption Perceptions Index, places it way below all its fellow members of Europe15 and on a par with poorer countries of Malaysia, Hungary, South Korea and Tunisia. Summary measurements of national data, however, capture only part of the story (Golden and Picci 2005) in large and diverse countries where corruption, in common with many other characteristics, is likely to vary widely from region to region.

A new measurement of corruption to investigate 'the geographic dispersion of cumulative fraud and malfeasance' affecting public works construction across Italy's twenty provinces, was proposed in 2005 (Golden and Picci 2005: 4) of the universities of California at Los Angeles and of Bologna respectively. Public works construction was selected because of its vulnerability to collusion between levels of administration, elected officials, bureaucrats and private contractors. The authors observe that for the abuse of public office for personal gain to persist country-wide, elected officials are necessarily and regularly involved. As they also conclude, extensive and persistent corruption in public works, or any other sector, cannot be regarded as a phenomenon isolated from its broader political environment: 'In such a framework, corruption involves a non-benevolent principal rather than bureaucratic or institutional slippage from a benevolent one' (Golden and Picci 2005: 5). Revealed as intended and premeditated, is the extent and enormous scale throughout Italy of criminal fraud and corruption in the management of public works construction.

Using meticulous data gathered over a period of thirty years[6] (Golden and Picci 2005: 12), public works expenditure by province was compared against the cost of actual construction undertaken. It was well known that southern Italy was relatively underdeveloped, but it was considered necessary to attempt to assess by how much southern Italy was less well serviced by public infrastructure than the north. The common assumption was confirmed: the south is substantially less well provided, southern regions having barely two-thirds of the national average of infrastructure cost – and in 2005 it was declining in spite of declared national policy to achieve the contrary.

This comparison of built infrastructure values against government public works expenditure per region, shows southern Italy as having received more public works finance over the years, even though it has less infrastructure. The difference is interpreted as a measure of corruption: the regions that did not get what was paid for are those where politicians and bureaucrats were siphoning off public money during the construction process.

Golden and Picci (2005) concede that their measure does not reveal where unaccounted moneys went to, only that in some regions public expenditure failed to provide the same amounts of infrastructure as in others, with the most corrupt region spending four times more per infrastructure unit than the least corrupt. What the measure does do, they assert, is to convey hard information about what is widely known or believed to be: considerably greater corruption in the southern half of Italy, commencing south of Rome, than in the north, as a result of massive fraud as well as inefficiency.

In contexts of this kind, where endemic criminal fraud and embezzlement have become entrenched, almost traditional and in some areas, a tourist attraction, it would be difficult for any relatively small-scale building contractor to behave honestly – if after paying his kick-backs, such a contractor could afford to do so.

Earthquakes and media reaction in Italy and Turkey

Media reporting expresses degrees of press freedom as well as public interest. Italian media appear comparatively quiet on the subject of corruption except when it affects the prospects of their senior politicians. Before his resignation as Prime Minister in May 2006, Silvio Berlusconi, 'Italy's richest man' with a business empire spanning media, advertising, insurance, sport, food and construction, had fought off repeated allegations of corruption, being accused of embezzlement, tax fraud, false accounting and attempting to bribe a judge. Currently facing two trials for corruption and fraud, he is likely to benefit from a government pardon reducing by three years 'all but the most serious offences' (Hooper 2006). 'For some Italians, Mr Berlusconi's success as a business tycoon is evidence of his capabilities – a reason why he should run the country' (reported by BBC News 2006).

In October 2002, a moderate earthquake caused the collapse of a two-storey reinforced concrete school building in San Giuliano di Puglia, a hill-top village of 1,195 people, 80 km (50 miles) north-east of Naples near Campobasso in the Molise region (*Education Guardian* 2002a–e; BBC News 2002). A total of 29 deaths were confirmed: 26 children, one teacher and two other adults in nearby buildings. A second earthquake of similar magnitude followed in November. The school was reported as built in the 1950s and substantially modified since, with a second floor recently added. Seventy per cent of homes were reported damaged but the school, closely surrounded by older buildings which remained standing, was reported as having been alone in its collapse.

Italian press reports of this exceptionally emotive earthquake event, reviewed in English-language media, are appropriately incisive and hard-hitting. *La Republica* reported authoritative verbal allegations of inadequate construction in an area known to be earthquake prone. *Corriere della Sera* reported changes in seismic zoning for the village, from medium to high risk, made nationally in 1998 but not approved locally; negligence and incompetence of local administrators was also alleged. *La Republica* also reported that 57 per cent of Italian schools did not have 'certificates of safety' and that 5,500 school buildings did not conform to building regulations (*Education Guardian* 2002).

L'Unita in Rome headlined 'The earthquake and its accomplices' and the *Puglia Corriere del Giorno* reported a long history of infrastructure neglect in the poorer south, the school system being 'the Cinderella in terms of investment'. Referring to the national public spending debt, *Il Matino* asked how the money had been spent: 'Not on schools it seems – and even less on schools in the south'. Investigators were reported to be considering charges of manslaughter or criminal negligence (BBC Internet News 2002a–c). Corruption of whatever kind or application was not mentioned by name.[7]

Given the pervasive power of the mafia throughout southern Italy, across all commercial sectors inclusive of the media, such reports are encouraging. In Turkey, where earthquakes have been on a massive scale and no less emotive, press comment reported by English language media has been muted by comparison, focusing lower down the corruption hierarchy on developers, contractors and local administrators. Corrupt dealings at higher levels are not mentioned, nor even implied. Turkish media in print is, however, outclassed by well-gathered authoritative verbal statements, appearing in English-language media, from university professors, research institutions and, when it suits them, some senior politicians.

Similar sources indicate the media in Turkey to be less vociferous than its counterpart in Italy; reported allegations or indictments of corruption in construction are hard to find and, until recently, appearing not to exist. Whilst informed verbal statements from well-identified authorities and institutions are quoted, it is rare for newspapers to be named, even by non-Turkish sources; only one television station is quoted, with acknowledgment of its private ownership (*Guardian Unlimited* 1999a). The Turkish Newspaper *Hurriyet* was named by the World Socialist Web Site (1999), an exception that indicates a stifled norm. The state-run Anatolian news agency later reported that the 'Turkish broadcasting watchdog' had shut down a national television station for one week as punishment for 'provocative' coverage of the 'disorganised response to the disaster', the government having 'been stung by heavy criticism' (*Guardian Unlimited* 1999b). In the United Kingdom, *The Guardian* reported 'Rage at the greedy contractors who make a killing (*sic*) out of building cheap slums on what might as well be sand. Despair at the corrupt bureaucrats who turn a blind eye to construction laws in return for the builders' kickbacks' (Traynor 1999).

In these constrained contexts, statements made by some Turkish academics are invaluable. For example: a lecture on the 1999 Kocaeli and Düzce earthquakes, given in London to the Institute of Civil Engineers by Mustapha Erdick of Boğaziçi University, Istanbul (Erdik and Aydinoglu 2000); Alpaslan Özerdem's (1999) plea for better public understanding of, and involvement in, housing construction in Turkey; and the preliminary engineering report on buildings damaged in the Bingöl earthquake by Polat Gülkan and others (2003) at the Middle East Technical University.

Corruption thrives where it has subjugated the media, an act of corruption itself. For the eradication of corruption to succeed a free press is the prime medium; to rid itself of corruption, a country has to first free its media from control by the corrupt. In 2003, a previously rare *Turkish Daily News* editorial by Ilnur Cevíc (2003), forecast that an imminent report by a special parliamentary commission would seek the prosecution of two former Turkish politicians behind 'massive corruption cases in Turkey in the past decade'. Two years later, the most recent report on Turkey by Reporters Without Borders (2005a) comments that, as Turkey prepares for its possible acceptance into the European Union, legal reforms are beginning to take effect with positive effects for journalists.[8, 9]

Corruption and earthquake casualties

In Turkey, since 1903, 88,538 people have been killed in 71 earthquakes; of this total, 17,127 died in the earthquakes of August 1999. In the same earthquakes, 92,866 people were injured, 1,160,880 people were made homeless, and estimated costs of damage totalled US$ 16,096 million (EM-DAT 2006 and see Table 12.1). In Turkey, the cost of earthquake damage has been almost eight times higher than the next highest loss of flood damage during the same overall period.

In Italy since 1905, 115,324 people have been killed in 28 earthquakes, in which 3,749 people were injured and 197,300 were made homeless; the total estimated cost of damage has been US$ 30,484 million – almost exactly twice that of the cost of damage from flooding (EM-DAT 2006 and see Table 12.1; the second highest figure of deaths from natural hazards in Italy is 20,019 from extreme temperatures, 20,000 of whom died in 2003). Earthquakes are very significantly the most severe natural hazard for both Turkey and Italy.

In countries where development indices are higher, the cost of disasters is generally highest and numbers of casualties fewer, whereas in countries where development is less, the numbers of casualties are highest and the

Table 12.1 Turkey, Italy (and Greece): GDP per capita (2004) and earthquake incidence since 1903–5

	GDP per capita (2004: US$)	Earthquake total	Dead	Injured	Homeless	Cost of damage (millions US$)
Turkey	7,687	71	88,538	92,866	1,160,880	16,096
Italy	27,700	28	115,324	3,749	197,300	30,484
Greece	21,689	29	997	3,818	10,008	7,336

Source: Adapted from EM-DAT 2006

costs of damage are less (Disaster Research Unit 1975; Wisner *et al.* 2004).[10] Italy's GDP per capita (2004) is more than three-and-a-half times greater than that of Turkey but, with less than half the number of earthquakes in a period of one hundred years (approx), the overall cost of damage *and* number of deaths in Italy are each appreciably higher than those of Turkey (where numbers of injured and homeless have been significantly higher, a reflection of a lower level of development; see Table 12.1).

Italy's highest cost of earthquake damage is therefore, in these terms, an anomaly. High costs of earthquake damage ensue from high destruction incidence of buildings and infrastructure – larger than should be expected in a developed country. Pervasive and complicit corruption in construction can be suggested to have resulted in earthquake damage in Italy to the extent of causing a significant converse to a long-established phenomenon.

The relationship between GDP and disaster losses accounts also for related poverty and loss, poorer people incurring greater numbers of casualties. It could be argued, therefore, that relative poverty in southern Italy, the Mezzogiorno, is more susceptible to higher deaths in earthquakes. But poverty in southern Italy is, as has been described, closely related to regional impoverishment due to development funding having been massively, criminally and corruptibly 'milked'. In Italy, higher costs of damage *and* higher numbers of deaths can both be considered as ensuing from the same corrupt system.

Numbers and densities of population must also be taken into account. Italy's overall national population per sq km is 191.08, that of Turkey being 91.97 (OECD 2006b) but in Italy's earthquake-prone Mezzogiorno,[11] though less in number, population is higher per square kilometre at 210.6 (1999), than that of the north; Kocaeli's population in north-eastern Turkey is 260 per sq km (1999). Italy's principal earthquake-prone population is at a density less than that of Turkey's and cannot therefore be regarded as a reason for higher earthquake losses.

It has to be concluded that such a significant anomaly in southern Italy could be due to the embodiment of failed governance, such as a high incidence of corruption. Likewise, similar pervasive corruption in Turkey, as indicated by reported evidence of failed construction, has to be assumed to be a serious contributor to disproportionately high earthquake losses; by how much can only be conjectured. Corrupt practices do not always result in building failure, building failures are not always due to shoddy or corrupt construction practice; the consequences of corruption in construction nevertheless include the failure and collapse of buildings, especially in earthquakes, with increased numbers of casualties. Italy's almost formalised, certainly endemic and pervasive, insidious criminal system in the most earthquake-prone southern provinces, is unlikely not to be a reason for disproportionately high earthquake deaths and cost of damage. Similar conclusions will apply to all earthquake-prone countries where corruption prevails.

A conjectural estimate of future losses due to corruption in construction could be based, in a rare example, on the figure of 57 per cent of schools inadequately built in Italy's southern provinces (*Education Guardian* 2002a– e). If, for the sake of conjecture, the same percentage is assumed for all other public buildings in the south and if, towards the same purpose, the same figure is applied to commercially built dwellings, then over the period of another century, 57 per cent of those buildings will collapse or be seriously damaged in future earthquakes. On the one hand, not all buildings will collapse and not all of those that do will be due to corrupt practice; on the other hand, development will increase, especially in the hitherto impoverished south. It is collapsing buildings that kill people, not earthquakes themselves; in the next century, therefore, if corruption were to continue at the same scale, 57 per cent of all casualties in southern Italy could justifiably be ascribed to corruption in construction. Given the same earthquake incidence in Italy in the next one hundred years, it can be anticipated (see Table 12.1) that 65,735 Italians will die, who would not die in earthquakes if corruption in construction were to cease.

Italy provides a prime example of how corruption in construction can deprive and take away people's lives in earthquake-prone regions.

Counter measures

Vociferous allegations following earthquake damage, regularly directed at builders and contractors and often made by politicians, in many cases obscure the real contextual issue of corrupt practice higher up the scale of construction dealing, authorisation, and administration. Contractors, small builders and sub-contractors cannot, nevertheless, be exonerated; building failure and collapse are directly due to shortcomings in their management, practice and workmanship. What has to be recognised, as well as site practices during its execution, are the pressures and effects of entrapment, and temptations to emulate, those whose influence affects access for others to new projects and availability of present and future work.

Counter measures are here arranged in five categories:

- Contextual governance
- The wider construction industry and its related administrations, including bidding opportunities
- Construction on site, project and site management and its related procedures
- Media exposure and public participation
- Research and training.

Governance

Governance is the manner and function of governing; regarding various social and economic objectives, there can be good governance and malfunctioning governance. Corruption is both a cause and a symptom of bad governance; where corruption exists, it 'flouts norms of social behaviour and subverts hard-won democratic decisions' (UN-Habitat 2006). In the furtherance of good governance, corrupt influences that facilitate bad governance are exposed internationally, bribery and fraud often being dependent on money laundering and other concealment of proceeds for criminal and corrupt practices.

Governance to counter corrupt practices in construction, can be improved by: separating branches of government; installation of checks and balances; fostering competition; clear definition of roles and responsibilities; the undertaking of independent audits; compulsory disclosure of interests by public officials; furtherance of a sense of community, concern for the commons and ethical ambience (UN-Habitat 2006).

Other measures towards good governance against corruption in construction are:

- modernisation and strengthening of judiciary systems and law enforcement to encourage development of contractual relations;
- transparency, accountability, law making, policing and internal scrutiny and monitoring (adapted from Barca 2001; DFID 2006b: 28–34);
- radical modernisation of local administrations, with the introduction of accountable procedures for the selection of local projects;
- institution building, whereby private and public local actors are encouraged to come together and to create partnerships;
- incentives for networking, making full use of new information technologies, offering high-quality small firms the opportunity to market the diversity of their products with distant consumers;
- adequate planning, inclusive of the identification of areas at degrees of high risk, made public;
- improving pay and conditions of construction inspectors and other officials;
- strict accounting, procurement and auditing procedures;
- prevention of illegal trading in natural resources, for example, in timber.

To assist the prevention of corruption in construction, Transparency International have published a set of business principles for project owners – funders; construction, engineering and consulting companies – and a range of construction integrity pacts for independent monitoring and enforcement (Transparency International 2005).

The OECD (2006b) have published Guidelines for Multinational Enterprises, to set out what companies can do to meet standards on human rights, labour conditions, the environment and corruption (DFID 2006b). The third annual conference of the Asian Development Bank (ADB-OECD) of 2001, held in Japan, with the theme of Combating Corruption in the Asia-Pacific Region, included sections on developing efficient and transparent systems of public service, strengthening anti-bribery actions and promoting integrity in business, and supporting active public involvement (ADB 2002).

The first world-wide agreement on corruption, the United Nations Convention Against Corruption (UNCAC), came into force in December 2005. One-hundred-and-forty countries agreed to co-operate on all aspects of preventing, investigating and prosecuting corruption (DFID 2006b; UNCAC 2005).

Pre-contract tendering/bidding

Information brokers transmit detailed knowledge of competitor bids, acting as agents to those awarding contracts, or arranging for a pre-identified contractor to 'win' a tendering/bidding procedure, thereby rendering the process redundant. Measures against these practices are:

*   cessation and avoidance of cartels, collusion, bid-rigging, price-fixing rings and information brokering: indicators of these practices may be, for example, substitution of contractors, addition of new names to the list and repetition of pricing formats, systems and errors;
*   collusion in tendering/bidding to be stated as being unacceptable, with all signs of collusion stated as rendering suspect tenders/bids unacceptable (Crowley 2003).

Officials given to corruption are known to set up companies under their own management, or that of a relative, naturally related to the official's function. An official whose activity has to do with building projects may be the covert owner of a construction company with which he will place contracts (Palmier 2000: 2). Required is:

*   transparency of procedures installed for dealings by all appointed officers.

Complex administrative processes lend themselves to corruption by implying or requiring bribes as 'grease money' before sanctioning stages of their processes. Administrations that are not already complex, can be made so by officials impeding the progress of documentation until grease money has been paid by applicants. In some cases 'corruption is ... so institutionalised

... that pay-offs have become the lubricant that makes the bureaucracy run smoothly' (for those who can afford to pay; Chua 1999: 1–3). Required is:

- simplification of procedures and high penalties for corrupt bidding practices.

Principles for countering the offering or demanding of a bribe, have been established by a partnership of the World Economic Forum, Transparency International and the Basel Institute of Governance, as a contribution to improving business standards of integrity, transparency and accountability.

Project management

Required measures are:

- legislation, enforcement and provision of adequate, trained and empowered inspection of construction on progress, ensuring that both the opportunity and perceived need of bribery for the achievement of objectives are removed;
- re-evaluation and re-definition of control of local government building construction;
- encouragement of participation in earthquake insurance as a vehicle for independent certification of conformity with construction codes;
- re-examination of standardised design for government buildings;
- restriction of overcrowding and upper-storey extension of existing buildings;
- maintenance of old, damaged and poorly maintained buildings;
- facilitation of access to controlled provision and supply of (off-site) ready-mixed concrete;
- questioning of professional advisers on what they are doing about alerting, identifying and preventing corrupt practices in construction: 'the professions may be the first line of attack as well as defence' (Shakantu 2003).

Exposure: institutions, media and public participation

Corrupt practices rely upon popular tolerance or acquiescence; good democratic governance requires an informed and vigilant body politic to hold officials to high ethical account in the conduct of public business (UN-Habitat 2006).

'Corruption thrives in secrecy, and withers in the light': consequent shaming of its perpetrators by media exposure and public censure will cause corruption to diminish (Palmier, quoted in Quah 2004: 13). An independent media, therefore, is essential for the successful eradication of corruption.[12]

The World Bank investigates allegations of fraud, inside and outside the Bank; names of guilty companies and individuals are placed on the Bank's website (World Bank 2006b). Grassroots organisations and the media generate public debate on bribery and corruption standards (DFID 2006b).

Sectoral changes for the removal of ingrained quasi-cultural improbity are less feasible until people themselves take responsibility for the actions of their politicians: the International Declaration of Human Rights states that 'every citizen has the right to take part in the conduct of public affairs' (United Nations General Assembly 1948: Article 25).

Research and training

In the exercise of powerful media and public exposure of corruption, research and its publication is an equally powerful supplier of data and analysis. Initiated by Transparency International's 2005 Report on Corruption in Construction (e.g. Collier and Hoeffler 2005: 12, on the economic costs of corruption in infrastructure; Bosshard 2005: 19, on environmental risk from corruption), often courageous research into issues of corruption is now supported and strengthened by other institutions, journals and academics, some of whose work appears in the references consulted for the writing of this chapter. Other roles for training are:

- local manpower facilitation;
- improved and expanded facilities in earthquake-resistant construction for all building types and for architects and engineers practising in earthquake-prone countries.

It is more than thirty years in the United Kingdom, with research focus on developing countries (Disaster Research Unit), and more than fifty in the USA, since 'natural' disasters were identified as being due to the action and inaction of mankind. Emphasis then was on what was done innocently or inadvertently in everyday practices to cause disasters to be the manifestation of natural hazards, and what practical ameliorative adjustments could be made in hazard-prone situations.

International exposure of corruption during the past decade, and of its consequences in construction, has removed that once assumed innocence. Evident now is that deaths and destruction in disasters, especially in earthquakes, are exposed as being due to corruption on a criminal scale. In parallel, human rights law has become internationally established, based upon the International Declaration of Human Rights (United Nations General Assembly 1948) which proclaims *inter alia* the right to life, liberty and security of person, free from inhuman or degrading treatment, to 'public order and the general welfare in a democratic society' and, as a member of that society, to a standard of living adequate for health and well-being.

Article 30 emphasises that no state, group or person may 'engage in any activity or perform any act aimed at the destruction of any of the rights and freedoms of the Declaration' (United Nations General Assembly 1948).

It has been asserted (Green 2005), with reference to Turkish earthquakes but with relevance to earthquake-prone countries generally, that earthquake deaths due to collapse of inadequately constructed buildings are a violation of human rights law. In such contexts 'without addressing fundamental questions of political and social organisation, wider authoritarianism, clientelism and the corruption that flourishes within it' (Green 2005: 544), legislative reform on behalf of the protection of human rights will not succeed. As it is, the assertion continues, earthquake catastrophes are usually overlooked in international law as hazards of a purely geophysical kind to be 'responded to rather than approached preventively' and that for their recognition 'the ideological reconstruction of earthquake disaster in human rights terms may be necessary' (Green 2005: 544).

Where this is the case, such an ideological reconstruction is long overdue, if only to align itself with established advances in natural hazards research and development, and the policies that have ensued and that will continue to ensue.

Notes

1 William Shakespeare: *Twelfth Night*, Act 2, Scene IV, spoken by Viola to the Duke Orsino.
2 Corruption has been indicated as a significant cause of unhappiness: 'Lowest percentages of happiness, the most unhappy, are not always in the poorest countries, but among those "who lack human security, are plagued by corruption and civil rights abuse, or suffer from AIDS or alcoholism". GDP per capita does not ensure happiness or quality of life' (Bäckstrand and Ingelstam 2006); for comparisons between corruption and unhappiness scales, see Lewis (2006a).
3 During research for this chapter between October 2006 and January 2007, other countries in which corruption exposure has been reported have been: Australia, China, Equatorial Guinea, Iraq, Ireland, Lesotho, Russia, South Africa (where corruption in housing provision is reported to be a part of everyday life; McGreal 2007), Taiwan, United Kingdom, United Nations and USA; and countries in which corruption in construction is reportedly evident have been: Australia, Georgia, India, Japan, Kenya, and the USA.
4 The European Union of 1995 comprised the 15 countries of Austria, Belgium, Denmark, Finland, France, Germany, Greece, Holland, Ireland, Italy, Luxembourg, Portugal, Spain, Sweden and the United Kingdom.
5 After the mafia assassination in 1992 of two of its most prominent adversaries, a significant mafia boss was arrested. Between 1991 and 1993 many municipal governments were dissolved because of their ties with organised crime and, overall, the number of convictions increased; in the same year, electoral reform of municipal governments made it possible for city mayors to be elected and the responsibilities of local authorities were enhanced (Barca 2001: 3).
6 Research by Maurizio Di Palma and Claudio Mazziotta, as part of an investigation by the European Commission to evaluate which regions of member

countries were under-served and by which types of infrastructure. Measured and accounted were public works on roads, railways, airports, schools, hospitals and other public utilities and buildings.

7 In February 2007, 50 new buildings containing more than 300 apartments and 22 small villas completed on the outskirts of earthquake-prone Naples, were declared a crime scene. The investigating magistrate has stated: 'About €50–60 million (£33–40m) was invested here ... The money involved, the size of the site and the lack of permits leads us to believe a criminal organisation may be behind it, probably with political support.' The local mayor, elected as a candidate for Silvio Berlusconi's political party, was quoted by the *Corriere della Sera* as being 'unaware of the mammoth development' (Kington 2007).

8 The 2005 Press Freedom Index (Reporters Without Borders 2005b) shows Turkey at 98th on a scale of 167 countries. Italy is placed at 42nd, way below its other European partners. The United Kingdom is 24th and the United States is 44th; Denmark, Finland, Iceland, Ireland, Netherlands, Norway, and Switzerland all tie for first place; Burma, Iran, Turkmenistan, Eritrea and North Korea are placed at 163 to 167 respectively.

9 Pakistan provides other examples of bribery, violence, stifled media and earth-quakes. In the capital, Islamabad, 12 attacks were made against the media in the one month of October 2006, taking the total for that year to 100 including four journalists. Other attacks included injury, robbery, torture and police harassment, attacks on media property and forceful prevention from working. On the anniversary of the 2005 Pakistan earthquake 'hundreds of survivors' protested in Islamabad against embezzlement and bribery in the reconstruction programme. The report quoted Oxfam saying that corruption had compounded reconstruction problems (Observer, 2006). Pakistan ranks 150th on the 2005 Press Freedom Index. Intermedia appears to have since been discontinued.

10 A contrariety originally identified by the Disaster Research Unit (1975 pp. 7, 19 and 37), discussed in Lewis (1999 pp. 39–42), and in Bosher (2007) with reference to the World Disasters Report (IFRC 2002).

11 Inclusive of the eight regions of Abruzzi, Basilicata, Calabria, Campania, Lazio, Molise, Puglia and Sicily.

12 For comparisons per selected earthquake-prone country of press freedom and corruption scales, see Lewis (2006b).

References

ADB (2002) *Taking Action Against Corruption in Asia and the Pacific*, Manila: Asian Development Bank.

Alexander, D. (2005) 'The Italian mafia's legacy of high-rise death traps', *The Global Corruption Report 2005 Special Focus: Corruption in Construction and Post-conflict Reconstruction*, Berlin: Transparency International.

Bäckstrand, G. and Ingelstam, L. (2006) 'Enough! Global challenges and responsible lifestyles', *What Next*, Vol. 1., Development Dialogue, Upsalla: Dag Hammarskjöld Centre.

Barca, F. (2001) 'New trends and the policy shift in the Italian Mezzogiorno', *Daedalus*, 130(2): 93–114.

BBC News (2006) 'Berlusconi resigns as Italian PM', BBC News, London.

BBC Internet News (2002a) 'Questions over school building', 1 November, BBC website, Available HTTP: <http://news.bbc.co.uk/1/hi/world/europe/2385951. stm> (accessed 17 October 2006).

BBC Internet News (2002b) 'Italy's papers reflect mounting anger', 2 November, BBC website, Available HTTP: <http://news.bbc.co.uk/1/hi/not_in_website/ syndication/monitoring/media_reports/2391531.stm> (accessed 17 October 2006).

BBC Internet News (2002c) 'Prosecutors probe Italian school collapse', 2 November, BBC website, Available HTTP: <http://news.bbc.co.uk/1/hi/world/ europe/2391127.stm> (accessed 17 October 2006).

Best of Sicily (2007) 'The Mafia', *A guide to Sicilian travel and all things Italian* Available HTTP: <http://www.bestofsicily.com/mafia.htm> (accessed 13 October 2006).

Bosher, L.S. (2007) *Social and Institutional Elements of Disaster Vulnerability: The Case of South India*, Bethesda: Academica Press.

Bosshard, P. (2005) 'The environment at risk from monuments of corruption', *Global Corruption Report*, Berlin: Transparency International.

Cevíc, I. (2003) 'Yilmaz and Ciller have to explain: editorial', *Turkish Daily News*, 22 May.

Chief Technical Examiner's Organisation (2002) *Problem Areas of Corruption in Construction*, A Preventive Vigilance Publication of the Central Vigilance Commission, New Delhi: Government of India.

Chua, Y.T. (1999) *Robbed: An Investigation of Corruption in Philippine Education*, Metro Manila: Philippine Center for Investigative Journalism.

CIOB (2006) *Corruption in the UK Construction Industry: Survey 2006*, Ascot: Chartered Institute of Building.

Coburn, A.W. (1995) 'Disaster prevention and mitigation in metropolitan areas: Reducing urban vulnerability in Turkey', in Parker, R., Kremer, A. and Munasinghe, M. (eds) *Informal Settlements, Environmental Degradation and Disaster Vulnerability*, International Decade for Disaster Reduction (IDNDR), Washington, DC: World Bank.

Coburn, A. and Spence, R. (2002) *Earthquake Protection Second Edition* (first published 1992), Chichester: Wiley.

Collier, P. and Hoeffler, A. (2005) 'The economic costs of corruption in infrastructure', *Global Corruption Report*, Berlin: Transparency International.

Crowley, T. (2003) 'Construction and fraud', *Insurance International*, March/April. Available HTTP: <http://www.maxima-group.com/a-0509.shtml> (accessed 29 July 2004).

DFID (2006a) *Eliminating World Poverty: Making Governance Work for the Poor*, Government White Paper, London: Department for International Development.

DFID (2006b) 'Improving governance, fighting corruption', Speech by the Secretary of State for International Development, London: Department for International Development.

Disaster Research Unit (1975) 'Towards an explanation and reduction of disaster proneness', *Occasional Paper No 11*, Bradford: University of Bradford.

Education Guardian (2002a) 'Six killed in Italian earthquake', 31 October, *Education Guardian*, Available HTTP: http://education.guardian.co.uk/schools/ story/0,,823406,00.html (accessed 17 October 2006).

Education Guardian (2002b) 'Fresh quake hits village after school tragedy', 1 November, *Education Guardian*, Available HTTP: <http://education.guardian. co.uk/schools/story/0,,823932,00.html> (accessed 17 October 2006).

Education Guardian (2002c) 'Earthquake school "built on the cheap"', 2 November, *Education Guardian*, Available HTTP: <http://environment.guardian.co.uk/climatechange/story/0,,1850396,00.html> (accessed 17 October 2006).

Education Guardian (2002d) 'Quake warning was ignored', 3 November, *Education Guardian*, Available HTTP: <http://www.guardian.co.uk/italy/story/0,,825129,00.html> (accessed 17 October 2006).

Education Guardian (2002e) 'Quake village buries its lost children', 4 November, *Education Guardian*, Available HTTP: <http://education.guardian.co.uk/schools/story/0,,825580,00.html> (accessed 17 October 2006).

Ellul, F. and D'Ayala, D. (2003) *The Bingöl, Turkey Earthquake of the 1st of May 2003*, University of Bath Architecture and Civil Engineering Department and the Earthquake Engineering Field Investigation Team (EEFIT), Bath: University of Bath.

EM-DAT (2006), 'Disaster profiles – Italy: Earthquakes 1905–2006 and Turkey: Earthquakes 1903–2006', Brussels: OFDA/CRED International Disaster Database, Université Catholique de Louvain, Belgium. Online. Available: HTTP: www.em-dat.net (accessed 1 June 2006).

Erdik, M. and Aydinoglu, M. (2000) 'Rehabilitation, recovery and preparedness after 1999 Kocaeli and Düzce earthquakes', Lecture at the United Kingdom National Conference on the Development of Disaster Risk Reduction. London: Institute of Civil Engineers.

Fletcher, N. (2006) 'Bribery costs one in five UK companies business', *The Guardian*. 9 October, p. 26, London.

Geipel, R. (1982) *Disaster and Reconstruction: The Fruili (Italy) earthquakes of 1976* (Translation: Philip Wagner), London: George Allen & Unwin.

Glanz, J. (2006) 'Idle contractors add millions to Iraq rebuilding', *New York Times*, 25 October, New York.

Golden, M. and Picci, L. (2005) 'Proposals for a new measure of corruption, illustrated with Italian data', *Economics and Politics*, 17(1): 37–75.

Green P. (2005) 'Disaster by design: Corruption, construction and catastrophe', *British Journal of Criminology*, Vol.45, 528–46.

Guardian Unlimited (1999a) 'Thousands killed and injured as earthquake shakes', *Guardian Unlimited*, Mark Tran, 17 August, London.

Guardian Unlimited (1999b) 'Prime Minister defends rescue operation', *Guardian Unlimited*, John Arlidge, 25 August, London.

Guardian Unlimited (2006) Available HTTP: <http://www.guardian.co.uk/> (accessed 30 October 2006).

Gülkan, P., Akkar, S. and Yazgan, U. (2003) 'A preliminary engineering report on the Bingöl earthquake of May 1, 2003', Department of Civil Engineering and Disaster Management Research Center, Middle East Technical University, Turkey.

Haas, J.E., Kates, R.W. and Bowden, M. (1977) *Reconstruction Following Disaster*, Cambridge, MA: MIT Press.

Hooper, J. (2006) 'Pardon makes a mockery of 90% of Italian trials', *The Guardian*, 8 November, p.19, London.

IFRC (2002) *World Disasters Report: Focusing on Reducing Risk*, Geneva: International Federation of Red Cross and Red Crescent Societies.

KingTON, T. (2007) 'Police impound illegal "mafia town" built on broccoli fields', *The Guardian*, 13 February, p. 19, London.

Lambsdorff, J.G. (2006a) 'Corruption Perceptions Index 2005', *Global Corruption Report*, pp. 298–303, Berlin: Transparency International.

Lambsdorff, J.G. (2006b) 'Ten years of the CPI: determining trends', *Global Corruption Report*, pp. 292–7, Berlin: Transparency International.

Lewis, J. (1999) *Development in Disaster-prone Places: Studies of Vulnerability*, London: Intermediate Technology Publications.

Lewis, J. (2003) 'Housing construction in earthquake-prone places: perspectives, priorities and projections for development', *Australian Journal of Emergency Management*, 18(2): 35–44.

Lewis, J. (2005) 'Earthquake destruction: corruption on the fault line', *Global Corruption Report 2005 Special Focus: Corruption in Construction and Post-conflict Reconstruction*, Berlin: Transparency International, 23–30.

Lewis, J. (2006a) 'Corruption and unhappiness', Corruption Note 2., December, Mimeo.

Lewis, J. (2006b) 'Corruption and press subjugation in earthquake-prone countries', Corruption Note 1, November, Mimeo.

McCurry, J. (2005) 'Shaken confidence Tokyo', *Guardian Unlimited*, 22 December, London.

McGreal, C. (2007) 'Report attacks South African crime and corruption', *The Guardian*, 29 January, p. 16, London.

Mitchell, W.A. and Page, J. (2005) 'Turkish homeowners demand an end to earthquake devastation', *Global Corruption Report 2005 Special Focus: Corruption in Construction and Post-conflict Reconstruction*, Berlin: Transparency International, 27–9.

Observer (2006) 'Quake victims stage "bribes" demo', *The Observer*, 8 October, p. 38. London.

OECD (2006a), *The OECD Fights Corruption*, Paris: Organisation for Economic Co-operation and Development.

OECD (2006b) *Guidelines for Multidisciplinary Enterprises*, Paris: Organisation for Economic Co-operation and Development.

Osborn, J.E. (2000) 'Protecting against corruption in construction and renovation: Corporate counsel's essential role in making integrity pay: problems and solutions', New York City. Available HTTP: http://www.osbornlaw.com/jeopc_presents/publications/publication39.html (accessed 4 October 2006).

Özerdem, A. (1999) 'Tiles, taps and earthquake-proofing: lessons for disaster management in Turkey', *Environment and Urbanisation*, 11(2): 177–9.

Palmier, L. (2000) 'Corruption and probity', *Asian Journal of Political Science*, 8(1): 1–12.

Quah, J.S.T. (2004) *From Assessment to Action: Implementing Anti-corruption Measures in Asian Countries*, Washington, DC: World Bank.

Reporters Without Borders (2005a) *Turkey – 2005 Annual Report*, Paris: Reporters Without Borders.

Reporters Without Borders (2005b) *The Ranking – Turkey*, Paris: Reporters Without Borders.

Shakantu, W.M.W. (2003) 'Corruption in the construction industry: forms, susceptability and possible solutions', *CIDB 1st Postgraduate Conference*, Port Elizabeth, South Africa, 12–14 October.

Stansbury, N. (2005) 'Exposing the foundations of corruption in construction', in *Global Corruption Report 2005 Special Focus: Corruption in Construction and Post-conflict Reconstruction*, Berlin: Transparency International.

Transparency International (2005) *Global Corruption Report 2005 Special Focus: Corruption in Construction and Post-conflict Reconstruction*, Berlin: Transparency International.

Transparency International (2006a) *Construction Integrity Pacts*, Berlin: Transparency International.

Transparency International (2006b) *Transparency Watch Newsletter*, Berlin: Transparency International.

Transparency International (2007) 'UK actions on Saudi defence contract are a blow to anti-bribery convention', Press release 13 June, Berlin: Transparency International.

Traynor, I. (1999) 'First the terror, now the recriminations', *The Guardian*, 20 August, London.

Tremlet, G. (2006) 'Threat to 1,000,000 illegal Spanish houses', *The Guardian*, 31 October, p. 17, London.

UNCAC (2005) 'Conference of the States Parties to the United Nations Convention against Corruption', *United Nations Convention against Corruption*, Vienna: United Nations Office on Drugs and Crime, Available HTTP: <www.unodc.org/unodc/crime_convention_corruption.html> (accessed 5 March 2007).

UN-Habitat (2006) *Urban Governance: Corruption*, UN-HABITAT, Nairobi: United Nations Human Settlements Programme.

United Nations General Assembly (1948) *Universal Declaration of Human Rights* (Articles 3, 22, 28 & 30), Geneva: Office of the United Nations High Commissioner for Human Rights, United Nations.

Wisner, B., Blaikie, P., Cannon, T. and Davis, I. (2004) *At Risk: Natural Hazards, People's Vulnerability, and Disasters*, 2nd edition, London: Routledge.

World Bank (2003) *World Bank Sanctions Committee Procedures*, Washington, DC: World Bank.

World Bank (2006a) *Distinguished Panelists Say Corruption Fight is Central to Improving Lives of the Poor*, Press Release, Washington, DC: World Bank.

World Bank (2006b) *Governance and Anti-corruption*, Washington, DC: World Bank.

World Socialist Web Site (1999) 'Thousands die in Turkey earthquake', Shannon Jones, 19 August, Available HTTP: <http://www.wsws.org> (accessed 29 July 2004).

Yuzer, E. (1999) 'Ankara failed to heed expert's warnings', *The Guardian*, 26 August, London.

Chapter 13

Building resilience by focusing on legal and contractual frameworks for disaster reconstruction

Jason Le Masurier, Suzanne Wilkinson, Kelvin Zuo and James Rotimi

Introduction

The legal and contractual systems post-disaster can have a significant impact on the long-term recovery from an event. One of the requirements for reconstruction is the establishment of a comprehensive procurement framework for reconstruction. Following a disaster, there also needs to be an understanding of the effects the disaster has on the legislative and regulatory system of the country experiencing the disaster. Without developed frameworks, reconstruction and new development will be carried out on an ad-hoc basis with little regard for the needs of the society. Current normal procurement mechanisms used in the construction industry need to be assessed pre-disaster for their suitability to deliver the best economic outcome in the event of a disaster. Together with the procurement systems, it is important to manage the legislative and regulatory environment, paying particular attention to any changes that need to be put in place to assist reconstruction efforts. Addressing both the contractual and legislative approaches to post-disaster reconstruction prior to any event will help to build in resilience for a community. This chapter first examines the recovery and reconstruction environment post disaster with a focus on the legislative and regulatory problems faced in the post-disaster reconstruction phase. The chapter then examines current common construction procurement systems to assess their suitability for reconstruction following a disaster. Particular focus is made on the question of which is the best procurement option for reconstruction, with a view to delivering more resilient built assets and better built assets for the communities involved.

Contractual, legislative and regulatory considerations post disaster

Disaster management and the need to develop a resilient community capable of recovering from disasters is of increasing concern in many countries. The recovery process may present an opportunity for improvement in the functioning of the community, so that risks from future events can be reduced while the community becomes more resilient. The effectiveness of the process will depend on how much planning has been carried out and what contingencies are put in place prior to the disaster.

In preparation for disasters there is often an emphasis on readiness and response, with poor understanding and little consideration given to the implications of recovery (Angus 2005). Experience has shown that recovery is often carried out by modifying routine construction processes on an ad-hoc basis. Whilst this can work reasonably well for small-scale disasters, the effectiveness of reconstruction could be improved by modifying the legislative and regulatory framework in advance of a disaster and planning for adequate reconstruction procurement systems. For larger-scale disasters there is a greater imperative to have appropriate systems in place in advance, to allow effective coordination and delivery of reconstruction works.

Recovery is an integral part of the comprehensive emergency management process (Sullivan 2003). It refers to all activities that are carried out immediately after the initial response to a disaster situation. This will usually extend until the community's capacity for self-help has been restored. In other words, the end-state is when the assisted community reaches a level of functioning where it is able to sustain itself in the absence of further external intervention (Sullivan 2003). Recovery is defined as 'the coordinated efforts and processes to effect the immediate, medium and long-term holistic regeneration of a community following a disaster' (MCDEM 2005a: 5). Recovery requires a concerted approach that will support the foundations of community sustainability and capacity building and which will eventually reduce risks and vulnerabilities to future disasters. Jigyasu (2004) describes an increase in vulnerability of local communities after the Latur earthquake in India, where sustainable recovery interventions were poorly planned and implemented. Pre-planning for reconstruction will avoid the disaster becoming protracted and assist communities to more readily return to a pre-disaster state. However, the communities should aim to deliver a state that is an improvement on the pre-disaster state, in particular, incorporating hazard mitigation and risk reduction issues into the reconstruction decision making processes. The effectiveness of the process will depend on how much planning has been carried out and what contingencies are provided for in preparing for the disaster. It is expected that recovery and reconstruction works will restore the affected community in all aspects of its natural, built, social and economic environment. The recovery process may present an opportunity for improvement in the functioning of the community, so that risk from future events can be reduced while the community becomes more resilient (i.e. able to mitigate any further disasters and recovery rapidly from any disaster).

Improving recovery through improved legislation

Legislation that applies to routine construction provides for the safe development of infrastructure, capital improvements and land use, ensuring preservation and environmental protection, however there is often little provision in legislation to facilitate reconstruction projects. Much existing legislation was not drafted to cope with an emergency situation and was not developed to operate under the conditions that will inevitably prevail in the aftermath of a severe disaster. If well articulated and implemented, the regulations should provide not only an effective means of reducing and containing vulnerabilities (disaster mitigation), but also a means of facilitating better thought-out and designed reconstruction projects.

In comparison to routine construction, there is often little provision in legislation to cater for post-disaster reconstruction processes as part of recovery. When an official state of emergency is declared following a major disaster, special powers become available and routine statutory processes can be circumvented. However, once the state of emergency has been lifted the routine statutory processes become applicable, which can create sluggishness in the recovery process. The recovery stage can last several years and eventually transitions back to the point when construction processes can be considered routine.

To ensure robustness in the process, the rational starting point is the setting up of an institutional infrastructure for emergency management, which will formulate public policies for mitigation, response and recovery (Comerio 2004). These recovery policies should then be integrated into other emergency management areas as well as policies of sustainability and community capacity building (Coghlan 2004). The Ministry of Civil Defence and Emergency Management (MCDEM) in New Zealand encourages a holistic approach to the issue of recovery planning and believes this will be most effective if it is integrated with the remaining 3Rs of reduction, readiness and response (MCDEM 2005a).

Recovery is delivered through a continuum of central, regional, community and personal structures (Angus 2004). Responsibility for coordination of recovery will be determined by a number of factors including the scale of the disaster. In New Zealand the recovery planning and management arrangements are contained in the National Civil Defence Emergency Management Strategy (MCDEM 2004). The MCDEM, together with cluster groups of agencies, coordinate planning at the central level. Regional and territorial authorities are encouraged to produce group plans that will suit peculiar conditions of their local areas. However, unless lines of responsibility are made clear, management of recovery may involve an element of competition between central, regional and local levels of government for control of the process (Rolfe and Britton 1995).

Unless provision is made for recovery in regulations and legislation that apply to routine construction, the coordination and management of a major programme of reconstruction could become cumbersome and inefficient. For example it is unlikely that coordinating authorities and regulatory bodies would be able to cope with the volume of work, due to shortfalls in experienced personnel.

Legislation cannot be used for purposes other than those for which it is intended and there appears to be little provision in construction industry legislation for post-disaster situations. These polices need to be revised beforehand as hasty revisions during the course of reconstruction works do not provide the best solution to major disaster problems. Should the routine regulatory and legislative processes be followed after a major disaster it is unlikely that regulatory bodies would be able to cope with the volume of work. The conflicts in the interpretation of the different pieces of legislation need to be harmonised, whilst the roles and responsibilities of the various emergency and civil defence agencies and other stakeholders need to be made clear. The apparent division between those who, in practice, take responsibility for reconstruction and those who set policy and legislation, create barriers that need to be overcome.

Recovery is an enabling and supportive process, thus the heart of recovery is community participation. Consultation and communication is encouraged especially in identifying community needs and for collective decision making amongst all stakeholders. This way all stakeholders understand the process and their commitment towards agreed objectives is ensured. Typical reconstruction stakeholders will include:

- asset owners (may be private or public and the business community);
- lifeline agencies (i.e. transportation, utilities, telecommunications organisations);
- civil defence and emergency management groups (national, territorial and local government departments, police, fire brigade, relief and welfare agencies, health and safety personnel etc);
- insurance companies;
- non-governmental agencies (charities, funding organisations etc.);
- construction and reinstatement organisations.

All these stakeholders will need to understand their individual and collective responsibilities during the post-disaster reconstruction phase.

The recovery process will typically follow five key stages (Brunsdon and Smith 2004); impact assessment; restoration proposal; funding arrangements; regulatory process; and physical construction. In the *impact assessment* phase information is gathered in the recovery process aimed at gaining knowledge on the impact of the disaster event on individuals, community and the environment. It involves all stakeholders as it is at this stage that the necessary inspections and surveys (needs assessment) are carried out that will form the basis for all reinstatements activities. The needs assessments will include building inspections, insurances, and health and safety assessments. The success of this stage will depend on the levels of communication, consultation and planning between all stakeholders. The process must lend itself to reviews and updating to take account of new information at later stages.

In the *restoration proposal* stage, decisions are made on whether to repair, replace or abandon affected properties. These decisions are reached based on the input of the impact assessment activities. Realistic proposals for meeting the anticipated recovery task are presented for funding organisations' consideration.

For *funding arrangements*, affected parties may have access to different types of funds: from private insurance companies and from government. For instance in New Zealand, residential property owners are insured by the Earthquake Commission (EQC), New Zealand's primary provider of natural disaster insurance. EQC insures against damages caused by earthquake, natural landslips, volcanic eruption, hydrothermal activity, and tsunami. Secondary funding may come from charity organisations and external donor agencies.

At the *regulatory process* phase, design and regulatory approvals are sought for the reinstatement of damaged facilities. Processing of resource consents is usually painstaking and the target of approving authorities is to ensure that a considerable level of resilience is incorporated in all developments. New knowledge gained on risk from hazards after the disaster will assist approving authorities to correct former design concepts to mitigate future disaster risk.

Finally, at the *physical construction* phase the expectation is that this is the regeneration stage in the recovery process where every aspect of the community and its environment (natural, built, social and economic environments) returns to 'normalcy' or preferably to a state more able to withstand future disasters by building in resilience. Experience has shown that it is difficult to return to the pre-event status quo but effort is made to restore the functions of the affected community.

The processing of building consents at the early stages of reconstruction and recovery after an event has been identified as a potential bottleneck. Access to normal resource levels will be unlikely and inevitably there will be shortages of qualified people to handle impact assessments and consent processing. A more flexible approach to the standard consent process would be necessary to expedite the process and help cope with the high volume of consent applications after a major disaster. However, in New Zealand, the new Building Act (2004) requires that Territorial Authorities must not grant building consents on land subjected to natural hazards unless they can be protected from the hazard and, where waivers are granted, it requires that notices be placed on the land to indicate the risk of natural hazards they are exposed to. Implementing this Act will have far-reaching implications on insurance claims as the Earthquake Commission Act indicates that the EQC is not liable to settle any claim where there is an identified large risk. Current revisions to the mapping of vulnerable natural hazard zones may prevent existing properties from being compensated at all, although this may not be the case if the homeowner mitigates disaster events for their property.

Legislative requirements for disaster recovery and reconstruction need to be carefully considered so that in an event they are flexible enough to assist, not hinder, recovery and reconstruction. One of the key questions to be considered is how the construction industry and other stakeholders will facilitate the rebuilding. In order to rebuild an understanding and effective management of the construction contractual procurement systems for post-disaster reconstruction needs to be made.

Procurement options for post-disaster reconstruction

Procurement systems are the organisational structure adopted by the client for the implementation of the project process and eventual operation of the project (Griffith *et al*. 2000). Procurement is important in the reconstruction process after a natural disaster, but generally considered, it can be seen as a strategy designed to satisfy the client's development needs (Moore 2002). In a disaster recovery situation, the 'client' is most likely to be the government bodies coordinating the reconstruction process. A well-developed protocol or stipulated procedure should be available and clearly understood by the involved government agencies and appointed coordinators in such an event (Moore 2002; Wilkinson *et al*. 2004) but this is seldom the case in disasters, leading to reconstruction being carried out in an ad-hoc, uncoordinated way. In New Zealand, the MCDEM Director's Guidelines (2005b) proposes a management structure for coordinating recovery and it recommends the setting up of various task groups to achieve recovery objectives. Under the 'Built Environment Task Group' are sub-task groups for various parts of the built environment. For example, the 'Residential Housing Subtask Group' will act as a client and would be responsible to 'repair, reconstruct or relocate buildings – obtaining fast-track building and other consents, sufficient builders and materials, coordinating skilled trades and their work standards', thus expectation of the government acting as client for reconstruction is established.

Procurement strategies such as traditional, design-build, partnering and alliancing can be judged against, amongst other things, time, cost, quality, industry familiarity, communication and management. These factors have significance for reconstruction. Those managing disaster events will require guidance on the most useful strategy for reconstruction. The common purpose for all the procurement systems and the inherent part of a procurement system is to achieve the clients' objectives (Love *et al.* 1998). In an emergency situation reconstruction is required rapidly, and this should be at the forefront of the client's objectives. Cheung (2001) suggests procurement is critical as it determines the overall framework embracing the structure of responsibilities and authority for participants within the building process. In all projects, but particularly in a crisis situation, clear structures need to be determined and responsibilities and authority for rebuild established.

Internationally, recent changes in the construction industry such as changes in the type of client and development of new construction techniques have produced differentiation within construction processes and changes in organisational structures to satisfy a variety of clients' objectives (Love *et al.* 1998). This has led to the development of variety of procurement systems. The key to the management of construction project is the way in which the contributors are organised to use their skills effectively (Walker 2002). Effective use of skills is an essential element to rebuilding after a disaster. Where large loss of life occurs, skills are lost and it is important for the procurement system used to recognise the skills shortage and respond to it.

Commonly used procurement systems can be categorised as either traditional, integrated or management (Love *et al.* 1998). The traditional system is where the project process is separate and sequential in nature (construction follows design and tender) and is the oldest form of construction procurement (Moore 2002). An alternative integrated system, as described by Al Khalil (2002), is where a single organisation is responsible for design and construction of the project and the involvement of the client is at a single point. Common examples of these procurement systems are design and build, built-operate-transfer (BOT), built-operate-own-transfer (BOOT), turnkey and package deal. The management approach is where an additional role of construction manager (CM) or project manager (PM) is added in the organisation to look after the project objectives. These systems, as described by Walker and Hampson (2003), are a combination of the traditional system of procurement and an integrated approach because in this kind of system a separate entity, often called a project manager, acts in a management role and the project manager is responsible for all clients' objectives through one point of contact with the client. Other recently developed procurement systems are project alliancing and partnering which are largely based on the shared ethics and best for project approaches decided by project participants including collaborative relationships and trust between the parties (Broome 2002).

In a disaster, the ability of a governing organisation to quickly establish an adequate procurement system for rebuilding is crucial. For reconstruction following a disaster event the procurement framework needs to have the following (Wilkinson *et al.* 2006):

- short time for rebuilding;
- low cost;
- use of local material, labour and plant;
- well-developed communication links between the parties;
- local industry familiarity with the framework;
- well-developed relationships between the parties (including trust and respect between the parties).

Not all procurement systems will be suitable post disaster, so the governing organisations need to be advised of the best system to meet their needs.

The traditional approach to disaster reconstruction

The traditional procurement system is known as the system with separated design and construction organisation. Walker and Hampson (2003: 13) describe this: '... approach to procuring projects involves discrete design development, tender, contract award and construction phases. Each phase is, in theory, separate and distinct.' In this system the client appoints a designer such as an architect or design engineer who fully designs the project and helps the client in selecting the contractor by inviting the tenders.

In a disaster, the ability of a traditional framework to quickly respond to rebuild could be a disadvantage. If a short time for rebuilding is required then using the traditional system may be a disadvantage. As Wilkinson and Scofield (2003) suggest, the traditional system usually results in a longer time period for the whole construction project, mainly because the design is often fully completed before tendering and construction. They further point out that usually the longer the time period, the higher the cost, hence the traditional system is often associated with higher costs. A higher-cost product would be a disadvantage in a disaster rebuild situation, as financial resources will already be stretched. However, with full design documentation at tender stage, the total cost of the project should be known, which might introduce some clearer measure of reconstruction costs. The traditional system is associated with design by a design professional and a quality product. Administration of the project by a design professional usually means high quality as the design professional is often focused on quality. Care should be taken that, following a disaster, quality is not taken to mean importing expensive material at the expense of getting local infrastructure and material supply businesses functioning. The traditional system is a well-understood, tried and tested procurement system making it attractive from a familiarity viewpoint. If the parties have worked together with the framework on previous projects then there is likely to be well-developed relationships between the parties. However, the fragmented and sequential nature of the systems tends to mean that communication becomes complex. This is a disadvantage when working in a crisis situation where clear and easily understood communication systems are required. On balance, the main impediments to this system for reconstruction following a disaster are the time, cost and communication factors. The main advantages for reconstruction following a disaster are familiarity with the system and receiving a quality product.

The integrated approach for disaster reconstruction

In an integrated procurement system, design and construction are integrated and became the responsibility of one organisation. The design and build is an arrangement where one contracting organisation takes sole responsibility for the design and construction of the client's project (Masterman 2002). Design and build is the main member of the integrated system, other types include package deal, built-operate-transfer (BOT), built-operate-own-transfer (BOOT), novation and turnkey are also examples in the category. Cheung (2001: 429) describes the popularity of these systems in the construction industry: '... these options become fashionable when more and more contracting organisations are armed with a design function. In fact, for those with financial capability, their service can be extended to build-operate-transfer.' Masterman (2002: 68) suggests that design and build gained popularity because there were '... heavy demands upon the construction industry and shortage of construction resources, coupled with claims by contractors of greater efficiency and lower cost when using this method ...'. Wilkinson and Scofield (2003) suggest that these systems are good for typical construction where standardised techniques and materials are to be used. This system proves useful only when the client's brief and detailed design are properly communicated to the organisation responsible for the construction. Walker and Hampson (2003) further point out that systems like BOT and BOOT are more common for infrastructural projects than buildings because the concession allows for tolls or other payments to be made by end-users to cover the cost of both procuring the facility and its operation. Smith *et al.* (1994) describe that the BOT entity undertakes financing, design, and construction as well as operation and so the client is taking no direct cost risk other than the possibility that the facility does not meet its need or that the concession agreement is unsatisfactory. For design and build suitability in disaster reconstruction, consideration should be given to the reduction of time and cost which is of significance when rapid rebuild is required and financial resources are limited. As design and construction are overlapping activities, this can reduce the overall time of project completion. Design and build also permits the incorporation of constructability information during design, so following a disaster, local knowledge of material availability and other resources can be incorporated into the project.

Concern is expressed in the use of design and build because of the lack of checks and balances leading to lower quality (Beard *et al.* 2001). Quality may not be such a priority in a crisis situation, as recovery needs are more likely to focus on time and cost, although arguably this approach could lead to the increased vulnerability of a poorly planned and designed built environment to future disasters. Too often there are stresses on national and local administrations to rebuild key lifelines that have been lost; Menoni (2001: 105) notes that, 'Market forces put pressures to reconstruct as quickly as possible transportation networks to long distances and commercial and office buildings, hampering efforts to implement lessons learnt from the disaster in the attempt to reduce pre-earthquake vulnerability'. Extra quality and embedded forethought can help the reconstructed built assets and community to be more resilient, but there is inevitably a trade-off between time, cost and quality which communities need to make. Mulvey (1997) summarises that the design and build approach is especially successful in the case where the scope of the project is clearly defined, the design is a standard, repetitive design, and the schedule is tight. In a disaster, design and build procurement could be suitable because of this short time and low cost focus. This coupled with a local contractor's likely knowledge of the local construction industry such as fast access to local material, labour and plant makes it an attractive option. The design and build system is uncomplicated and therefore it is easy to establish well-developed communication links between the main parties, contractor and client. In many countries internationally, design and build also scores on local industry familiarity with the framework. The only caution to be noted is in the reputation of the design and build framework to be focused on repetitive, simple constructions and some concerns about the reliance on one company to be able to undertake all the work effectively – hence trust becomes a major factor.

The management approach for disaster reconstruction

The common characteristics of all management-oriented procurement systems are that in these systems a person or an organisation is contracted to manage the design and construction. Masterman (2002) suggests that these systems can more readily respond to the client's needs, especially accelerated commencement and completion. In the case of a pure project management arrangement the project manager has management authority for both design team and construction team. The project manager acts as the executor and is the only entity to interact directly with the client (Walker 2002), whereas in the construction management system where the project manager is usually a contractor, the expectation is that the construction expertise of the contractor is encouraged in the design. In this system, the construction manager is a consultant hired by the owner to oversee the project process on the behalf of the owner. Walker and Hampson (2003: 20) state about this expertise that '... this build-ability or constructability advice is crucial to the development of design solutions that maintain value in terms of the quality of product as well as providing elegant solutions to production problems'.

Having a management tier in a project could be advantageous during disaster reconstruction as the sole focus of the project management or construction management organisation is management of the project. The skill of the managing organisation is critical to its success, more so in a disaster scenario where quick and efficient decision making is required. Wilkinson and Scofield (2003) discuss the advantages and disadvantages of the project management system. They suggest that when a project manager is employed, the programme may be shortened due to the project manager using increased knowledge of project planning. By concentrating on management, the project manager can focus on reducing the overall time. This could be good for disaster reconstruction management. In New Zealand the Earth Quake Commission trialled a coordinated response to the Te Anau earthquake of 2003, using a large single contractor to coordinate and manage the recovery works on its behalf. Coordination was clearly an improvement on the situation where individual property owners competed for the services of a limited number of building contractors.

Where skilled local project managers can be used for managing reconstruction, the expectations of project success are high. As with design and build, the project manager or construction manager, using existing communication links between the parties, would be able to facilitate rapid rebuild – especially if the project manager has a well-developed sense of the local material, plant and labour situation. However, there is a need to check for local industry familiarity with the system and in particular the value of having another management tier in the project, increasing the lines of communication between the parties.

The partnering approach for disaster reconstruction

Partnering is not technically a procurement system, rather it is a framework which overlays any of the procurement frameworks in operation. 'Partnering does not supersede the process used by the procurement system chosen to implement the project but rather acts as a framework within which the selected system operates more beneficially' (Masterman 2002: 137). The definition provided by the Construction Industry Board (1997: 1) describes partnering as '... a structured management approach to facilitate team working across contractual boundaries'. In partnering there are three essential features: mutual objectives, continuous improvement and problem resolution (Bennett and Jayes 1995). *Mutual objectives* are the objectives that establish for everyone that their interests are best served by concentrating on the overall success of the project (Bennett and Jayes 1995). The second aspect, *continuous improvement*, involves the team and individuals feeling safe in measuring performance to learn from their experiences. This requires openness and honesty, hence trust and commitment are major issues for improvement. In continuous improvement, performance is measured and analysed to provide knowledge about how improvement can be achieved continuously. There must be a commitment to learn from experience and to apply this knowledge to improve performance (Bennett and Jayes 1995).

The third aspect, *problem resolution* requires the resolution of problems with an escalation strategy to solve them at the lowest organisational level possible. Partnering is a voluntary arrangement made between all of the project participants and has no legal standing and imposes no contractual obligations upon any of the parties (Broome 2002). But, as Egan (1998) and Latham (1994) discuss, the partnering approach can prove useful in that construction process and in improving construction practices. For disaster reconstruction, partnering may be a suitable option to overlay other frameworks. As a focus on relationships and trust are likely to be crucial in such a situation, partnering might provide a vehicle for reconstruction. If local companies in a reconstruction zone have an established long-term relationship, as a system based on trust requires, then the use of partnering might be viable. Internationally, familiarity with partnering concepts may still not be high, but could be quickly learnt, especially given the needs of the parties to face the common objective of rebuilding communities.

The alliancing approach for disaster reconstruction

Project alliancing is typically used on larger and more complex projects where there is a large amount of uncertainty, so would be useful in an emergency situation. The size and duration of the project has to justify the investment in setting it up, both commercially and culturally (Broome 2002).

Lynch (1989) points out that companies frequently lack sufficient management skills and resources to tackle extremely large and complex tasks. Team managers need to develop and nurture a culture of collaboration throughout the organisation and between organisations in order to manage such projects. Hence the attractiveness of inter-organisations with developed trusting and collaborative approaches to business. Philips (2005) reports that evidence in procuring construction contracts from Los Angeles following the 1994 Northridge earthquake was that the contracts were finalised within two weeks rather than three months and that the contractor had developed long and lasting relationships with Caltrans (the California agency responsible for overseeing the freeway reconstruction). This bodes well for any future emergencies where companies have an appreciation of the merits of the other parties and can trust and collaborate well, essentially having an alliancing project relationship. The next step is the formalisation of this relationship.

A critical success factor is the selection of the appropriate partners and the right price target (Walker and Hampson 2003). Such targets have been used in reconstruction. For instance, Philips (2005) described the reconstruction of the Santa Monica Motorway following the January 1994 earthquake centred in the heart of Los Angeles, where each contract had bonus incentives for early delivery and costs for being late. In alliancing this pain/gain is shared across all parties in the project.

But alliancing goes beyond price targets. McGeorge and Palmer (2002) name the following attributes of the project alliance:

- mutual trust between team members and parties;
- honest and direct communication; generous listening to each other;
- mutual support and respect for others;
- an atmosphere of integrity;
- accepting and maintaining a level of stretch or discomfort in declared targets;
- focus on achieving results;
- working on the basis of no blame if someone fails;
- individuals taking ownership of their actions and inactions;
- working in an environment where problems are not seen as negative, but as avenues to new possibilities.

In alliancing there are core principles that are regarded as fundamental for alliance relationships which can be defined as:

- collective ownership of all risks associated with the delivery of the project;
- sharing of the 'pain' or 'gain' depending on how actual project outcomes compare with the pre-agreed targets which have been jointly committed to;
- all participants operate at the same level and have an equal say;
- decision making based on condition that is 'best-for-project';
- responsibilities are clearly defined and parties respect and support rather than blame each other;
- all parties have full access to the resources, skills and expertise of each other;
- all financial transactions are based on an 'open-book' concept;
- innovative thinking is encouraged with a commitment to achieve outstanding outcomes;
- visible and unconditional support from the top level of each participating organisation;
- open and honest communication with no hidden agendas.

Many of these core principles can be reflected in the desires of communities for reconstruction projects following a disaster.

Reconstruction following a disaster – which construction procurement option is best?

The choice of reconstruction system is dependent upon a number of factors such as whether there is pre-planning for the disaster reconstruction, the types of construction relationships already in place, the ability of a client to understand and manage complex construction processes. If local companies in a reconstruction zone have established long-term relationships then the use of alliancing will be easier to implement, whereas the somewhat fragmented nature of a traditional design-bid-build system tends to mean that communication becomes complex. This is a disadvantage when working in a crisis situation where clear and easily understood communication systems are required (Wilkinson *et al.* 2004). Well-developed communication links between the parties, coupled with good relationships will aid the reconstruction process. Some clients, owners and public organisations already have such systems in place (Le Masurier *et al.* 2006). These organisations can demonstrate features of alliancing, such as developed links and established communication systems, which can be used in a crisis. Large infrastructure owner organisations, or serial clients, will have an advantage over one-off clients for a number of reasons, such as: experienced in procurement and delivery of construction projects; procurement processes already established; existing relationships with construction companies; ongoing projects from which resources can be diverted if necessary. These clients will have formed partnerships, and sometimes formal alliances, with other participants in the construction supply chain. They are in a better position than new clients to understand how reconstruction events will require a different approach and be more willing to use a system featuring collaboration and open communication.

If reconstruction projects following a disaster are treated piecemeal then the use of alliancing would not be justified on the many smaller individual projects. If however a coordinated programme of work is envisaged to facilitate post-disaster reconstruction, this situation presents an opportunity to use an alliancing approach. Hence alliancing would be appropriate in this situation due to the following factors:

- large scale of work programme;
- large degree of uncertainty and complexity;
- need for a target-cost type of payment mechanism to allow for variation in the scope of work and promotion of innovation in the execution;
- generation of a cooperative culture due to the wider social incentives to work together for the benefit of the whole community;
- the use of alliancing does necessitate pre-planning and detailed preparation.

Parties need to be confident that they understand how the relationship will work and the benefits for each party will need to be transparent, which is key to increasing trust between the parties. Preparation for an alliance relationship can be done in anticipation of a future event and the role parties will be able to play in such an event.

Since alliancing requires significant input in the early stages to establish an appropriate commercial and cultural framework, this would work against the use of alliancing if there is no prior planning and attempts are made to establish such a relationship in a reactive way following a disaster. However, some preparation can be done in advance which would facilitate adoption of an alliance for post-disaster reconstruction. Preparations should be carried out by agencies that would be responsible for delivering reconstruction projects (such as major public utility and highway infrastructure owners). Such preparations need to include:

- identifying construction supply chain companies with similar values and commercial culture to that of the owner;
- identifying facilitators who could be called upon to accelerate and maintain the culture;
- developing alliancing agreements that are provisionally agreed to in advance by potential alliance parties, including commercial and legal frameworks and dispute-resolution procedures;
- developing procedures for establishing target costs and pain/gain share.

By having reconstruction plans in place in the form of procurement strategies, collaborative arrangements between key organisations involved in a disaster event will be able to facilitate quick and effective reconstruction.

Conclusions

The legal and contractual systems post disaster can have a significant impact on the long-term recovery from an event and the ongoing resilience of society and built assets. The task of reconstruction after a major event can be an onerous challenge. It requires deliberate and coordinated efforts of all stakeholders for effective and efficient recovery of the affected community. This chapter has shown that the issues surrounding the implementation of the pieces of legislation concerning reconstruction after a major disaster are complex and interrelated. Without developed frameworks for legislation and procurement, reconstruction and new development will be carried out on an ad-hoc basis with little regard for the needs of the society. It is important to manage the legislative and regulatory environment, paying particular attention to any changes that need to be put in place pre-disaster to assist recovery and reconstruction efforts. Addressing both the contractual and legislative approaches to post-disaster reconstruction prior to any event will help to build resilience for a community. How recovery and reconstruction are achieved depends upon how the various legislative and contractual factors interplay and what requirements and facilities exist in the local communities.

References

Al Khalil, M.I. (2002) 'Selecting the appropriate project delivery method using AHP', *International Journal of Project Management*, 20: 469–74.

Angus L. (2004) *The Direction for Recovery in New Zealand*, NZ Recovery Symposium, Napier, New Zealand, Ministry of Civil Defence and Emergency Management.

Angus, L. (2005) 'New Zealand's response to the 1994 Yokohama Strategy and Plan of Action for a Safer World', World Conference of Disaster Reduction, Kobe-Hyogo, Japan.

Beard, J., Wundram, E. and Loulakis, M. (2001) *Design-Build: Planning Through Development*, New York: McGraw Hill Professional.

Bennett, J. and Jayes, S. (1995) *Trusting the Team: The Best Practice Guide to Partnering in Construction*, Reading: Centre for Strategic Studies in Construction, University of Reading.

Broome, J. (2002) *Procurement Routes for Partnering: A Practical Guide*, London: Thomas Telford Publishing.

Brunsdon, D. and Smith, S. (2004) 'Summary notes from the Infrastructure Workshop', *NZ Recovery Symposium*, Napier, New Zealand, Ministry of Civil Defence & Emergency Management. 12–13 July.

Cheung, Sai-on (2001) 'An analytical hierarchy process based procurement selection method', *Construction Management and Economics*, 19(4): 427–37.

Coghlan, A. (2004) 'Recovery management in Australia: a community-based approach', *NZ Recovery Symposium*, Napier, New Zealand, Ministry of Civil Defence and Emergency Management, 12–13 July.

Comerio, M.C. (2004) 'Public policy for reducing earthquake risks: a US perspective', *Building Research and Information*, 32(5): 403–13.

Construction Industry Board (1997) *Partnering in the Team: A Report by Construction Industry Board (Great Britain), Working Group 12*, London: Thomas Telford Publishing.

Egan Report (1998) *Rethinking Construction: The Report of the Construction Task Force to the Deputy Prime Minister on the Scope for Improving the Quality and Efficiency of UK Construction*, London: Stationery Office.

Griffith, A., Stephenson, P., and Watson, P. (2000) *Management Systems for Construction*, New York: Pearson Education.

Jigyasu, R. (2004) 'Sustainable Post-disaster reconstruction through integrated risk management', Second International Conference on Post-Disaster Reconstruction: Planning for Reconstruction, Coventry University, Coventry Centre for Disaster Management, 22–23 April.

Latham, M. (1994) *Constructing the Team, Joint Review of Procurement and Contractual Arrangement in the United Kingdom Construction Industry*, London: HMSO Publications.

Le Masurier, J., Rotimi, J.O.B. and Wilkinson, S. (2006) 'A comparison between routine construction and post-disaster reconstruction with case studies from New Zealand', in Boyd, D. (ed.) *Proceedings of the 22nd Annual ARCOM Conference*, 4–6 September, Birmingham, UK.

Love, P.E.D., Skitmore, M. and Earl, G. (1998) 'Selecting a suitable procurement method for a building project', *Construction Management and Economics*, 16(2): 221–33.

Lynch, R.P. (1989) *The Practical Guide to Joint Ventures and Corporate Alliances*, New York: John Wiley & Sons.

Masterman, J.W.E. (2002) *Introduction to Building Procurement Systems*, London: Spon Press.

McGeorge, D. and Palmer, A. (2002) *Construction Management – New Directions*, 2nd edition, Oxford: Blackwell Science.

MCDEM (2004) *National Civil Defence Emergency Management Strategy 2003–2006*, Wellington: Ministry of Civil Defence and Emergency Management.

MCDEM (2005a) *Focus on Recovery: A Holistic Framework for Recovery in New Zealand. Information for the CDEM Sector*, Wellington: Ministry of Civil Defence and Emergency Management.

MCDEM (2005b) *Recovery Management: Director's Guidelines for CDEM Groups*, Wellington: Ministry of Civil Defence and Emergency Management.

Menoni, S. (2001) 'Chains of damages and failures in a metropolitan environment: some observations on the Kobe earthquake in 1995', *Journal of Hazardous Materials*, 86(1–3): 101–19.

Moore, D. (2002) *Project Management: Designing Effective Organisational Structures in Construction*, Oxford: Blackwell Science.

Mulvey, D. (1997) 'Trends in project delivery – a contractors assessment', *ASCE Construction Congress Proceedings, Managing Engineered Construction in Expanding Global Markets*, 5–7 October, pp. 627–33.

Philips, P. (2005) 'Lessons for post-Katrina reconstruction', Briefing paper, Washington, DC: Economic Policy Institute.

Rolfe, J. and Britton, N. (1995) 'Organisation, government and legislation: who coordinates recovery?', in *Wellington After the Quake*, EQC/CAE, New Zealand, pp. 23–33.

Smith, N.J., Merna, A. and Grimsey, D. (1994) 'The management of risk in BOT projects', Internet 94: 12th World Congress on Project Management, Oslo, Norway, International Project Management Association, 9–11 June.

Sullivan, M. (2003) 'Integrated recovery management: a new way of looking at a delicate process', *Australian Journal of Emergency Management*, 18(2): 4–27.

Walker, A. (2002) *Project Management in Construction*, Oxford: Blackwell Science.

Walker, D. and Hampson, K. (eds) (2003) *Procurement Strategies: A Relationship Based Approach*, Oxford: Blackwell Science.

Wilkinson, S. and Scofield, R. (2003) *Management for the New Zealand Construction Industry*, Auckland: Prentice Hall.

Wilkinson, S., Gupta S. and Le Masurier, J. (2004) 'The development of a contractual framework for disaster reconstruction', 100th Anniversary of the San Francisco Earthquake Conference, San Francisco, 16–20 April.

Wilkinson, S., Le Masurier, J. and Zuo, K. (2006) 'An analysis of the alliancing procurement method for reconstruction following an earthquake', San Francisco 100th Anniversary Earthquake Conference, 19–22 April, San Francisco.

Chapter 14

The implications of the Civil Contingencies Act (CCA) 2004 for engineers in the UK

Andrew Fox

Introduction

Following the 9/11 terrorist attacks in New York, the British Government began to think seriously about its arrangements for civil contingencies. This facet of governance has undergone a long slow shift in emphasis since the end of the Second World War, through the demise of the Soviet Union and after the cessation of violence in Northern Ireland. The focus was originally on civil defence against military and terrorist threats but in recent decades moved towards an 'all hazards' approach (Coles 1998; Coles and Smith 1997; Moore 1996; Rockett 2000; Sibson 1991).

The process of change ultimately manifests itself in a blurring of focus on civil defence matters and led to a poor assessment of provision by the Society of Local Authority Chief Executives (SOLACE) in their 1983 report on 'The State of Civil Defence' (cited by Moore 1991). By 2001 the situation had improved to some extent after the introduction of the Civil Protection in Peacetime Act 1986 (Rockett 2000) but the events of 9/11 provided a sharp reminder of the vulnerability faced by all nations to the on-going threat of terrorism. In the UK it also came after a series of other notable civil crises including the Paddington and Hatfield train crashes in 1995 and 2000, the fuel crisis in September 2000, the foot and mouth outbreak of 2001 and repeated flooding events across the country during the 1990s and into the new millennium.

In 2000 the British Government resolved to take action and remedy the situation by setting out a new agenda for dealing with civil contingencies. This work was to be coordinated by the Civil Contingencies Secretariat that was established in July 2001 (UK Resilience 2007) and the centrepiece for this new agenda was the Civil Contingencies Act 2004 (hereafter referred to simply as the Act). The Act created new duties for responder organisations and provided a framework within which these duties were to be implemented. The essence of the new agenda was one of multi-disciplinary and multi-hazard planning; engaging a broad range of structural and non-structural measures to mitigate potential hazards and reflected a paradigm shift in the management of UK civil defence.

The author is not alone in arguing that engineers of all disciplines should take note of the requirements of the Act as it not only sets out a framework for the management of complex and multi-hazard environments of which engineers should be aware, but the Act also pointedly sets out to achieve the goal of resilience in local communities, which as a concept embraces the principles of sustainability in which the potential for input from the engineering community is considerable (Bosher *et al.* 2007a, 2007b; Fox 2002, 2004a, 2004b; Fox *et al.* 2003).

This chapter therefore sets out to explore the civil contingency arrangements in the UK and to assess their implications in relation to the work of engineers. It will seek to identify opportunities opened to the engineering industry by the Civil Contingencies Act and suggest ways in which the industry could better align itself with the paradigm enshrined within the new agenda.

The civil contingencies agenda in the UK

The civil contingencies agenda in the UK is well summarised by the eight '*principles of effective response and recovery*' (HMG 2005b: 6):

- **Continuity** – of service before, during and after critical incidents
- **Preparedness** – for all hazards and degrees of complexity in crisis situations
- **Subsidiarity** – ensuring that decisions are taken at the lowest appropriate level
- **Direction** – by clarity in understanding response and recovery objectives
- **Integration** – of planning and response by a collective of responder groups
- **Cooperation** – between all relevant individuals and organisations at all levels
- **Communication** – without delay to those that need to know, including the public
- **Anticipation** – of ongoing risks and indirect consequences following incidents and actions.

When drafting the new Act, the draft Bill was described (HMG 2003) as 'enabling' legislation, seeking:

1 To create a statutory duty on the part of local bodies to develop contingency plans for dealing with a range of emergencies, and
2 To provide powers for the Government to make regulations to deal with proclaimed emergencies.

The Act itself is split into three parts (HMG 2004):

- **Part 1** (clauses 1 to 16) applies only in England and Wales and sets out local arrangements for civil protection. It imposes on certain local bodies, described as Category 1 responders, a statutory obligation to prepare plans for dealing with the wide range of civil emergencies. Organisations listed as Category 2 responders are statutorily obliged to cooperate with the emergency planning and response processes and to share information with Category 1 responders.
- **Part 2** (clauses 17 to 30) applies to the whole of the United Kingdom and replaces the Emergency Powers Act 1920 and the Emergency Powers Act (Northern Ireland) 1926. Under the Act, a declaration of emergency can be made either by proclamation by Her Majesty or by a Secretary of State if he or she believes that a proclamation may occasion 'serious delay'. Clause 20 confers the power on Her Majesty or the Secretary of State to make regulations to control, prevent or mitigate the effects of an emergency, while clause 21 outlines their scope. A declaration of emergency can be made at a regional rather than national level.
- **Part 3** deals with amendments, repeals and revocations of legislation.

Following the enactment of the Bill in 2004, two important guidance documents were published. The first part relates to *Preparedness Planning* (HMG 2005a) and is intended to be viewed as statutory Guidance (with a capital 'G'), the second part relates to *Response and Recovery* (HMG 2005b) and is viewed as discretionary guidance (with a small 'g').

Duties, roles and responsibilities

The Act creates what it terms as a '*duty to assess, plan and advise*' (HMG 2004: section 2) and establishes a two-tier grouping of responder organisations who are obliged to cooperate with each other in relation to this new duty.

Category 1 responders
- Local authorities
- Emergency services
- Health services
- The Environment Agency

Category 2 responders
- Utility companies
- Transport companies
- The Health and Safety Executive

(Source: HMG 2004, Schedule 1)

The Act also introduces the concept of a '*risk register*', stressing that the development of such a register requires multi-agency cooperation which will be undertaken in the form of '*Local Resilience Forums*' working in '*Local Resilience Areas*'. Category 1 responders must participate in the forums and Category 2 responders may be involved as appropriate. The Guidance (HMG 2005a) does stress that organisations not identified under the Act can, and should, be involved as fully as possible.

The Guidance also encourages bilateral and cross-border cooperation to be developed outside the forum framework, sets out a six-step process for the assessment and management of risks and outlines a three-tier planning framework – local, regional and national planning.

Six-step planning process

1 **Contextualisation** – consider the social community, the environment, infrastructure and hazardous sites
2 **Hazard review and allocation for assessment** – identify hazards of a generic and site-specific nature and assign the people and processes to be used for the review
3 **Risk analysis** – assess likelihood and impact of risks within a five-year time frame
4 **Risk evaluation** – combine likelihood and impact assessments in order to rate the significance of risks
5 **Risk treatment** – prioritise risk-reduction measures in accordance with the rating of risks
6 **Monitoring and review** – undertake a full and formal review at appropriate intervals to suit the level of the risk.

At the regional level, both local and central level representatives are obliged to work together to address larger scale civil protection issues. Again the forum framework is utilised and the guidance documents state that the forums at local, regional and central level are not subordinate or hierarchical in any way, but serve to assist each other by facilitating the flow of information and to provide support in all directions (HMG 2005a, 2005b).

The voluntary sector is identified as playing an important role and Category 1 responders are required to work with and include these organisations in the planning process. How such collaboration is to work in practice is allowed to be flexible but the voluntary organisations are required to set out exactly what kind of service they can provide and to ensure that they train their staff to fulfil such a role.

Both the guidance documents (HMG 2005a, 2005b) recognise that not every organisation involved in a response is identified or covered by the Act and they emphasise that it is not intended that such organisations are to be excluded, stating that '*Category 1 responders should encourage organisations which are not covered by Part 1 of the Act to co-operate in planning arrangements'* (HMG 2005a: 160).

A Minister of the Crown is given powers under the Act to instruct any responder to perform a duty under the Act and may even issue instructions in relation to the Act when there is insufficient time to make other legislation. Importantly, the Minister can request information from responders in relation to the performance of their duties under the Act (HMG 2004) with failure to comply with such instructions or with other duties under the Act made an imprisonable offence.

Preparedness planning

A major proportion of the Guidance (HMG 2005a) is dedicated to planning and it states clearly that *'emergency planning'* is at the heart of the new duties placed on responders. The plans in question are required to address the prevention, reduction, control and mitigation of emergencies and their effects. Plans must enable the determination of when an emergency has occurred, provide for the training of key staff and involve exercising and review of the plan. What *risks* to plan for are the responsibility of the local responders to decide and in this they are encouraged to work collectively. The risks may be dealt with using generic plans or by specific plans.

Both parts of the guidance (HMG 2005a, 2005b) specifically address regional level planning, stressing its different nature to local planning in that its focus is for *'Crisis Management'*. Regional level plans are likely to be more generic in nature and supported by plans from National Government offices. The region is seen as providing support for coordination of local responders and, in the spirit of cooperation, is required to allow local responders access to their plans. The Act also requires Category 1 responders to maintain plans that will ensure they can continue to exercise their function in the event of an emergency. These so-called *'Business Continuity Plans'* are required to conform to all the same requirements as *'Emergency Plans'*.

Information sharing and advice

Information sharing and advice is the second major aspect of the Act. Responders have a duty to share information with each other and this information sharing is seen as an important part in developing the culture of cooperation intended by the Act. But the Act stops short of full participation and disclosure by admitting *'some information should be controlled if its release would be counterproductive or damaging in some other way'* (HMG 2005a: 25). In this regard, emphasis is placed on categorising types of information and obtaining consent for its disclosure.

The Act also places a duty on responders to *communicate* with the public. This duty is split into two distinct forms – warning and informing – and applies before, during and after an emergency event. Category 1 responders are also required to publish aspects of their Business Continuity Plans in so far as it is necessary or desirable for dealing with emergencies (HMG 2005a).

Local Authorities are targeted for specific duties in relation to providing advice and assistance to businesses and the voluntary sector. Such advice relates mainly to '*Business Continuity Planning*', but must also takes into account the local risk register when dealing with such programmes (HMG 2005a).

Response and recovery

The range of agencies involved in response and recovery activities is recognised as being more extensive than for planning and preparedness and as such, the tight duties in relation to Category 1 and 2 responders are not so evident. The guidance (HMG 2005b) reinforces the established system of response coordination using the Bronze, Silver and Gold command and control structure but is enhanced with additional guidance for localised, wide-area, terror and maritime incidents.

The main focus of response and recovery activities are listed as:

- saving and protecting life
- relieving suffering
- containing the emergency
- providing warnings, advice and information
- protecting personnel
- safeguarding the environment
- protecting property
- restoring critical services
- maintaining services
- promoting and facilitating self-help
- facilitating investigations
- facilitating socio-economic recovery
- evaluating response.

Part 2 of the guidance (HMG 2005b) explains that '*Regional Civil Contingency Committees*' will be convened when they can provide added value to a response. Their role will be to collate and maintain a strategic picture, assess what issues can be resolved at local level, facilitate mutual aid and ensure an effective flow of information between central and local level.

When central government support is required in a response, a '*Lead Government Department*' (LGD) will be designated to head the Government response (HMG 2005b). The LGD can decide on the use of emergency powers, mobilise military and other national assets, enact counter security measures and manage international relations and any public information strategy

The response and recovery guidance (HMG 2005b) also includes advice on the focus for humanitarian assistance by identifying what such programmes should cater for – injured and uninjured survivors, families and friends, the deceased, rescuers and response workers, children, young people, religions, cultures, minority groups, the elderly and the disabled.

What does the civil contingencies agenda mean for engineering?

Having outlined the essential elements of the civil contingencies agenda within the UK, this chapter now turns to consider the engineering industry. According to the Engineering Council (EC 2003) 'engineers' are defined as those individuals who use specialist engineering knowledge to optimise the application of new and existing technologies in the analysis and solution of engineering problems. Engineering solutions include products, processes, systems or services and engineers are seen as providing technical and commercial leadership in relation to such matters. As professionals, engineers are required to recognise and commit to obligations relating to the maintenance of standards in society, the environment and their profession.

The Royal Academy of Engineering (RAEng 2000) defines the engineering process as a bridge between science and technology and one that converts scientific knowledge into useful products wherein technology is both the means and the output from the application of the process. Engineering and technology are inextricably linked and the importance of technology to the economies of the world is one that has recently been recognised by economists who have added it to the shortlist of three primary inputs to all economic activity, the others being – labour, capital and materials (RAEng 2000).

A perennial problem faced by the engineering industry is its poor public perception and understanding (RAEng 2000; ETB 2005). This problem is compounded by the fact that by its nature successful engineering 'engineers out its visibility' (RAEng 2000: 12), or in other words, a successful engineering outcome would be the removal of the need for an engineering interface between the user and the product produced by the engineering process. The implications are that the benefits of engaging engineering expertise, advice and opinion are not always recognised and thus the onus is on engineers and the engineering profession to proactively market their services and to justify their engagement.

The engineering industry as a whole is vast and diverse; recent statistics published by the Engineering and Technology Board (ETB 2005) revealed that the UK engineering sector employs more than 2 million workers and accounted for 27 per cent of GDP in 2003. The report also shed light onto the distribution of engineers throughout the UK economy (Table 14.1).

Table 14.1 Distribution of engineers throughout the UK economy

Sector	Proportion
Manufacturing	40%
Finance and business	20%
Utilities	8%
Construction	7%
Transport and communications	6%
Other	18%

Source: ETB (2005).

As a body of individuals, those engaged in engineering are generally classified as chartered engineers, incorporated engineers or engineering technicians. There are an estimated 425,000 practicing engineers in the UK although only 158,000 were registered with the EC (ETB 2005) and the EC itself lists 35 member organisations with the ability to enrol engineers onto its register (EC 2003):

Institute of Acoustics
Royal Aeronautical Society
Institution of Agricultural Engineers
Chartered Institution of Building Service Engineers
Institute of Cast Metal Engineers
Institution of Chemical Engineers
Institution of Civil Engineers
British Computer Society
Institution of Electrical Engineers
Energy Institute
Institution of Engineering Designers
Society of Environmental Engineers
Institution of Fire Engineers
Institution of Gas Engineers and Managers
Institute of Healthcare Engineering and Estate Management
Institute of Highway Incorporated Engineers
Institution of Highways and Transportation
Institution of Incorporated Engineers
Institution of Lighting Engineers
Institute of Marine Engineering, Science and Technology
Institute of Materials, Minerals and Mining
Institute of Measurement and Control
Institution of Mechanical Engineering
Royal Institution of Naval Architects

British Institute of Non-Destructive Testing
Institution of Nuclear Engineers
Society of Operations Engineers
Institute of Physics
Institute of Physics and Engineering in Medicine
Institute of Plumbing
Institution of Railway Signal Engineers
Institution of Structural Engineers
Chartered Institute of Water and Environmental Management
Institution of Water Officers
Welding Institute

In addition to the above list, there are 14 affiliated organisations that contribute to the enhancement of engineering knowledge (EC 2003):

Institute of Asphalt Technology
Institute of Automotive Engineer Assessors
Institute of Concrete Technology
Institute of Corrosion
Association of Cost Engineers
Society of Engineers
Institute of Mathematics and Its Applications
Institute of Metal Finishing
Association for Project Managers
Institute of Quality Assurance
Institute of Refrigeration
Institution of Royal Engineers
Safety and Reliability Society
Institute of Vehicle Engineers

To this list should be added the Royal Academy of Engineers (RAEng), which promotes research and publishes reports on a wide range of engineering issues. One such report explored the 'universe of engineering' (RAEng 2000) and concluded that 50 per cent of the top 1,500 companies listed in the UK *Financial Times* depend on engineering to create, design and operate their products and services. The report also concluded that engineering plays a substantial role in eleven fields of the UK economy: agriculture and food; built environment; commerce, trade and finance; communications and IT; defence and security; education; energy and natural resources; engineered materials; health and social care; leisure and entertainment; and transport.

What the above summary reveals is that engineering is firmly embedded into the fabric of society. It can and does play a fundamental role in achieving sustainability, mitigating the effects of disasters and enhancing the ability of a community to recover following a crisis or emergency, which together define the concept of resilience as enshrined within the civil contingencies agenda (Fox 2004b).

It should be pointed out that in the field of civil contingencies planning, engineers are by no means devoid of experience, as a report looking into the operations of the engineering department of Teignbridge District Council (Teignbridge DC 2004) illustrates to good effect. In Teignbridge, the local engineering department was made responsible for emergency planning in 2000 and managed to change its role from one of a reactive provider of design services for capital projects to a proactive group giving engineering advice on all aspects of development. The report also provides an interesting insight into the range of legislation placing duties on local authorities and which may draw engineers into the emergency planning process, namely:

The Coastal Protection Act 1949
The Control of Pollution Act 1974
The European Water Framework Directive 2000
The Land Drainage Act 1991
The Local Government Act 1994
The Public Health Act 1936
The Water Act 2003
The Water Industries Act 1991

Another study commissioned by the Department for Environment, Food and Rural Affairs (HBR 2005) provides an illustration of how engineering advice may be engaged with the emergency planning process. The report specifically addressed engineering input to emergency plans for reservoirs and usefully highlights a number of other established regulations that are supplemental to the civil contingency agenda and with which engineers may well be familiar:

• Control of Major Accident Hazards Regulations (COMAH) 1999
• Radiation (Emergency Preparedness and Public Information) Regulations (REPPIR) 2001
• Further guidance on emergency plans for major accident hazard pipelines, Health and Safety Executive (HSE) 1997.

From a professional and ethical perspective it is widely accepted by the profession that engineers have a duty to uphold the safety of the public and it has been argued that this includes the issue of warnings of preventable disasters (Uff 2002). The Fellowship of Engineering (now the Royal Academy of Engineering) is widely quoted by supporters of this view who reference the draft guidelines for engineers they published in 1991 that sought to provide advice to engineers so that they may react in a responsible, prompt and disciplined manner when faced by potentially disastrous situations. The advice also stressed that engineers are frequently in positions to identify risks and emergency situations and are also frequently in positions of having to manage emergency situations (FOE 1991).

Another facet of the civil contingency agenda, which plays to the strengths of the engineering profession, is multi-hazard and multi-disciplinary planning. Engineers have had to cope with ever-increasing complexity in the projects they implement and in the risks that they manage (RAEng 2003a, 2004a, 2004b). This experience alone would be highly valued by the forums and responder communities commissioned to enact duties under the Act.

In light of the above summary it is perhaps inconceivable that the new civil contingencies responders would not call upon engineers in some shape or form when developing emergency plans. Similarly, response and recovery programmes are highly unlikely to escape the need for engineering expertise and as such engineers need to be aware of how their services can be engaged to the fullest potential and to the maximum benefit of the community.

Aligning engineering and the civil contingencies agenda

The civil contingency agenda may be summarised as a transparent and interactive framework of coordinated planning and response arrangements with a focus on risks and the promotion of key responders with a duty to develop resilience within society.

Engineering may be summarised as an industry that is deeply embedded into society, a profession that has significant expertise and experience in the solution of problems of a technical nature, a discipline that continues to develop an extensive array of useful products, processes, systems and services and an industry that can provide leadership and management in complex, multi-agency, multi-hazard environments.

Based on these two summaries the alignment of engineering to the civil contingency agenda and vice versa at a strategic level can begin to be realised.

Aligning the agenda to engineering

- To achieve resilience, the agenda needs to be embedded into the society that it seeks to serve. On this score the engineering profession can offer a significant gateway to opportunities and established networks within society.
- In order to achieve a coordinated framework of planning and response arrangements, the agenda will require a range of products, processes, systems and services. It is likely that those developed by the engineering industry would be the tools of choice adopted by the agenda.

- The risks associated with modern societies are often complex and socio-technological in nature. Through its detailed understanding of risk and technology the engineering industry has much to offer in this regard.
- The agenda aspires to transparency and full interaction in the delivery of the agenda through its key responders. The leadership and management capability provided by the engineering industry and its coverage in terms of key areas of the economy would provide a useful ally in this central aspiration of the agenda.

Aligning engineering to the agenda

- The engineering sector is a diverse and professional body that is constantly seeking opportunities to improve its service to society. In this respect the agenda's aim to establish resilience within the fabric of society should prove an irresistible challenge to engineers.
- The outputs from the engineering process are founded on the premise of their useful application. The agenda provides not only a market for existing outputs but also scope for the refinement and development of new outputs.
- As a professional body that prides itself in its leadership and management capabilities, and in particular its ability to deal with complex, multi-disciplinary and multi-hazard environments, the agenda should present a situation where engineers would feel their input is not only desirable but is absolutely necessary.

These are some of the strategic concepts that need to be recognised and accepted if any practical suggestions are to be taken seriously. What now follows is a more detailed consideration of areas for improving the alignment of the civil contingencies agenda and the engineering profession and for which a seven-stage structure has been adopted for analysis, namely:

1 Risk assessment
2 Responder agencies
3 Emergency planning
4 Coordination and control
5 Training
6 Warning and informing
7 Emergency powers.

1. Risk assessment

Engineers are often engaged as experts to make judgements on risk and safety. In this role their effectiveness is measured by the extent to which their audience believes their judgements. The RAEng (2003c) determined that the public prefer to leave decision making to experts and politicians, happy in the knowledge that they have the opportunity to protest should they judge any decisions to be unsatisfactory. It is important therefore that the experts have a good understanding of the public's concerns and priorities, the key to which is to engage with the social dimensions of risk.

In recent times, engineers have been realigning the understanding of their role in society and central to this role is the management of risk. Traditionally, engineers have treated risk as a technical issue but increasingly recognise that risk is as much about perception as it is about probability and impact (RAEng 2003b, 2003c). Risk perception involves a complex interaction between technical and social issues and requires a deep understanding of acceptability factors.

According to the RAEng (2003c), the acceptability of risks relies on a number of factors:

- Risks associated with natural hazards are more acceptable than human induced hazards.
- Ownership of the decision-making or control process makes risks more acceptable.
- Acceptance of risk must be accompanied by some benefit to the individual or the wider society.
- Familiarity and understanding of hazards makes risks more acceptable.
- Extensive or diffuse impacts are more acceptable than point impacts.
- Recurrent incidents are less acceptable than single incidents.
- A modest failure in a complex and poorly understood system is less acceptable than a major event in a simple and well-understood system.
- The response to an incident affects its acceptability and denial of responsibility can be less acceptable than the incident.

These ideas blend well with the philosophy adopted in the civil contingencies paradigm and its holistic view of risk. The social dimensions of risk are more engrained in the emergency management community than in the engineering community, primarily because emergency management is a relatively new discipline and has emerged from a truly multidisciplinary base, drawing knowledge from physical and social sciences as well as engineering theory and practice. The civil contingencies paradigm thus provides a window to the engineering community through which they can view a framework for integrating the social dimensions of risk into their own philosophy.

2. Responder agencies

Central to the new agenda paradigm is the identification of key responder agencies with the duty to plan and prepare for the response to emergency events. It thrusts to the fore a select group of organisations, creating a clear demarcation of their role within the civil contingency community and the wider public. Any belief in an individualist approach to the protection of communities and the response to critical incidents is firmly quashed as the Act embraces the principles of partnership and collective responsibility.

What the engineering industry may take from this is the vital importance of assigning clear roles and responsibilities in times of crisis and in preparations that pre-empt any response to critical incidents. Within the construction sector the recently revised Construction (Design and Management) Regulations (HSE 2007a) have made significant strides in this direction, addressing many concerns that have pervaded the industry in recent years. To a significant extent the CDM 2007 regulations have mirrored the civil contingencies paradigm by making their focus planning and management rather than the plan and other paperwork and by strengthening the requirements regarding coordination and cooperation (HSE 2007b).

3. Emergency planning

One element of the civil contingencies agenda that does mark a significant shift in the established paradigm is the planning associated with business continuity. The motivation for this may have arisen from either of two factors. Firstly, from the adoption of the resilience concept which holds that societal resilience is but a conglomeration of smaller resilient components at the individual, family, community, organisational and national levels. Secondly is the growing recognition that in times of disaster the responder organisations often become victims themselves thus undermining a key element for effective recovery; the ability of the responder organisation to continue to deliver its services before, during and after an emergency event.

Business continuity planning (BCP) as a management concept is not long out of its infancy stage with the first British Standard (BS 25999) outlining its fundamental processes, principles and terminology being published in 2006 (BSI 2006) and as a result engineering organisations may not be very conversant with its concepts. This aspect of the agenda therefore provides a significant opportunity for the engineering sector both to learn the value of BCP and to better align itself to the civil contingency agenda.

4. Coordination and control

It is perhaps the breakdown in the established system of coordination and control that prompted the review of civil contingency arrangements in the UK and in this respect the agenda has re-established clear roles and responsibilities and is not so much of a paradigm shift as a tidying up established practice.

The agenda reiterates the imperative for clear lines of responsibility and embraces the culture of collective action with a focus on the protection of the victims and facilitating recovery rather than on arguing who was to blame for the disaster. This in itself is a recognition that not every organisation or individual may wish to work for the collective good in times of crisis, focusing perhaps more on self-interest, which in many instances is a valid response strategy.

From the engineering perspective, the industry would probably argue that it has well-established working systems that cope adequately with risk and adversity (Bosher *et al.* 2007b), but the systems frequently used differ markedly from the civil contingency approach. The imperative of forced collective responsibility when responding to a declared crisis goes beyond many engineering crisis response systems and the hierarchy of command and control is less structured in engineering. The premise of the civil contingency command and control system is subsidiarity and revolves around the Bronze, Silver and Gold framework with an upward cascade of engagement. This gives priority to those at the frontline but enrols tactical and strategic level support, if needed and when called upon, as the scope and scale of the crisis reveals itself.

5. Training

A weakness in the established system of civil contingency planning was the commitment to the creation and maintenance of a cadre of trained personnel with both the knowledge of preparedness and recovery plans, and the experience gained either through engagement in real-life crises or simulated scenarios used in the testing of plans. The agenda seeks to rectify this in several ways, by nominating organisations with prime responsibility and placing duties upon them, but also by directing those with an interest to engage with this agenda to make the commitment to maintain a core of staff with the appropriate level of training and experience.

The nature of the training and experience in this context is worthy of note for clearly it lies outside the norm for it to be raised as an issue. Emergency plan writing, review and testing lie at the heart of this requirement, but coupled with the ability to communicate, inform and work with a collective of other responders and the general public. Associated with such training comes the authority and responsibility to lead and manage the development and implementation of contingency plans.

When viewed in such a light it is clear that such individuals have escaped the routine training and development programmes established by the engineering industry and the civil contingency agenda provides an opportunity whereby the alignment of the industry to the agenda can be facilitated. It should also be recognised that the individuals who receive this particular training will also act to facilitate the alignment of the engineering profession with the civil contingency agenda.

6. Warning and informing

The warning and informing aspect of the agenda marks the emergence of the systems into the modern age of information freedom and open governance. The perceived tradition within the civil contingencies community was one of secrecy but over recent decades the nature of governance has changed and a spirit of openness and transparency prevails; it was perhaps inevitable that in the review of civil contingencies arrangements such a paradigm would wind its way into the system.

This degree of openness is thus new to the civil contingency community so it raises opportunities for those with something to offer. The engineering profession understands the technology of communications, its products and services and would thus prove valuable to this new market. In addition, the skills to maintain such systems through times of crisis are contained within the industry.

As previously mentioned, the engineering industry is also vast and integrally keyed into all facets of society. This integrated network could be harnessed by the civil contingencies community to muster resources in times of crisis and to act as a conduit for the flow of information and the collection of data. In this regard the engineering community needs to be more proactive and engaged with the preparedness planning process so that its full potential can be realised in any response.

7. Emergency powers

The capacity to adopt emergency powers by the government following civil crises has been thoroughly revised by the Act. The implications for the engineering industry are as much to be aware of such potential declarations as to consider if their contractual forms need amending in light of this legislation. In addition the profession has the opportunity to align itself to the central authorities so as to ensure their services are engaged if such powers are exercised.

Conclusion

This chapter has attempted to provide an overview of the civil contingencies agenda in the UK following the introduction of the Civil Contingencies Act 2004. In doing so it has drawn upon Government guidance published to aid interpretation of what was a significant paradigm shift in the management of civil contingency crises, a key element of which is the creation of resilience within communities. The author has argued that the resilience objective should act as a magnet to engineers who hold that the sustainability and safety of communities are not only a professional but also a moral duty intrinsic to their profession and as such engagement with the civil contingencies agenda will move them some way to fulfilling this duty.

By outlining the nature and diversity of the engineering industry the chapter sought to add credence to any claim by the profession for greater involvement with the civil contingencies agenda and to provide guidance as to how the profession and the agenda could begin to better align themselves. Considerations ranged from the strategic to the more practical with advice for industry leaders and the engineering practitioner. And, in line with the collective non-hierarchical philosophy of the civil contingencies paradigm, which is premised on the idea of subsidiarity, it is left to the reader to judge what actions may be most appropriate to foster engagement between the engineering profession and the civil contingencies agenda.

References

Bosher, L.S., Carrillo, P.M., Dainty, A.R.J., Glass, J. and Price, A.D.F. (2007a) 'Realising a resilient and sustainable built environment: towards a strategic agenda for the United Kingdom', *Disasters: The Journal of Disaster Studies, Policy & Management*, 31(3): 236–55.

Bosher, L.S., Dainty, A., Carrillo, P., Glass, J. and Price, A. (2007b) 'Integrating disaster risk management into construction: a UK perspective', *Building Research and Information*, 35(2): 163–77.

BSI (2006) *BS 25999–1:2006, Business Continuity Management Part 1: Code of Practice*, London: BSI.

Coles, E. (1998) 'What price emergency planning? Local authority civil protection in the UK', *Public Money and Management*, 18(4): 27–32.

Coles, E.L. and Smith, D. (1997) 'Turning a job into a profession: the case for professional emergency management in the UK', Presented at the 4th International Emergency Planning Conference, 20–22 October, Prague.

EC (2003) *A Guide to the Engineering Profession*, London: Engineering Council UK.

ETB (2005) *Engineering UK 2005*, London: Engineering and Technology Board.

FOE (Fellowship of Engineering) (1991) *Draft Guidelines for Warning of Preventable Disasters Offered to the Professional Institutions for Consideration*, Available HTTP: http://www.raeng.co.uk/policy/pdf/draft_guidlines.pdf (accessed May 2007).

Fox, A. (2002) 'A framework to improve resilience planning for urban communities', *Security Monitor*, 1(5): 8–10.

Fox, A. (2004a) 'Planning for improved resilience', *Proceedings of the 2nd International Conference on Post-Disaster Reconstruction*, April 22 –23, Coventry University, England.

Fox, A. (2004b) 'Civil contingency – the engineer's role', Presentation at the Institution of Civil Engineers Presidential Conference, Cambridge, July.

Fox, A., Johnson, C. and Lizzaralde, G. (2003) 'A framework for improving the sustainability of housing initiatives to reduce the risk of disaster', *Proceedings of* the *International Civil Engineering Conference on Sustainable Development in the 21st Century*, JKUAT University, Nairobi, Kenya, pp. 399–404.

HBR (Halliburton Brown and Root) (2005) *Engineering Guide to Emergency Planning for UK Reservoirs: An Inception Report, Rev. A02*, London: Department for Environment, Food and Rural Affairs.

HMG (2003) *Draft Civil Contingencies Bill, Report and Evidence, Session 2002–03*, London: The Stationery Office.

HMG (2004) *Civil Contingencies Act 2004*, London: The Stationery Office.

HMG (2005a) *Emergency Preparedness: Guidance on Part 1 of the Civil Contingencies Act 2004, Its Associated Regulations and Non-statutory Arrangements*, York: Emergency Planning College.

HMG (2005b) *Emergency Response and Recovery: Non-statutory Guidance to Complement Emergency Preparedness*, York: Emergency Planning College.

HSE (2007a) *Managing Health and Safety in Construction: Construction (Design and Management) Regulations 2007 (CDM) Approved Code of Practice*, Sudbury: HSE Books.

HSE (2007b) *Construction (Design and Management) Regulations 2007, Baseline Study*, Sudbury: HSE Books.

Moore, G. (1991) 'Evaluating emergency plans', *Emergency Planning in the 90's, Proceedings of the Second Conference*, September, University of Bradford, England.

Moore, G. (1996) 'My practical experience in emergency management', *Disaster Prevention and Management*, 5(4): 23–7.

RAEng (2003a) *Risks Posed by Humans in the Control Loop*, London: Royal Academy of Engineering.

RAEng (2003b) *The Social Aspects of Risk*, London: Royal Academy of Engineering.

RAEng (2003c) *Common Methodologies for Risk Assessment and Management*, London: Royal Academy of Engineering.

RAEng (2004a) *'Trust Me, I'm an Engineer'*, London: Royal Academy of Engineering.

RAEng (2004b) *Humans in Complex Engineering Systems*, London: Royal Academy of Engineering.

RAEng and EC (2000) *The Universe of Engineering: A UK Perspective*, London: Royal Academy of Engineering.

Rockett, J.P. (2000) 'Wither, emergency planning', *Bristol Business School Teaching and Research Review*, Issue 2, Spring, Available HTTP: http://www.uwe.ac.uk/bbs/trr/Issue2/Is2-4_5.htm (accessed May 2007).

Sibson, M. (1991) 'Local authority – a planned response to major incidents', *Emergency Planning in the 90's, Proceedings of the Second Conference*, September, University of Bradford, England.

Teignbridge District Council (2004) *Engineering & Emergency Planning Best Value Review*, Newton Abbot: Teignbridge DC.

Uff, J. (2002) *Engineering Ethics: Do Engineers Owe Duties to the Public?*, London: Royal Academy of Engineering.

UK Resilience (2007) 'Introduction to the Civil Contingencies Secretariat', Available HTTP: http://www.ukresilience.info/ccs/index.shtm (accessed May 2007).

Chapter 15

Security planning in the resilient city

Stimulating integrated emergency planning and management[1]

Jon Coaffee

Introduction

For many years discussions have occurred amongst built environment professionals, urban managers and the agencies of security (especially the police) regarding the costs and benefits for urban authorities adopting counter-terrorism measures in the face of real or perceived terrorist threats. Some of the most historically explicit examples of such measures were seen in Northern Ireland in the early 1970s and 1980s where the military and urban planners used principles of 'fortress urbanism' to territorially control designated areas (Boal and Murray 1977; Boal 1995). This occurred most notably around the central shopping area in Belfast where access was barred, first, by concrete blockers and barbed wire, and then later by a series of high metal gates which became known as 'the ring of steel' (Brown 1985; Jarman 1993; Coaffee 2003). Subsequently this 'ring of steel' metaphor was reapplied in the 1990s to central London. The term has subsequently in more recent times been commonly used, almost ubiquitously, to describe any high profile security operation involving the sealing off of particular areas against the risk of terrorist attack.

However, in the post-September 11th era it is generally accepted that the nature of potential terrorist threats has changed, requiring an alteration in counter-responses from governments and security agencies at all spatial scales of defence, from the local area to the development of global coalitions to fight the 'war on terrorism' (Coaffee 2004). Not only were the attacks against New York and Washington in September 2001 unprecedented in terms of style of attack, damage caused and insurance claims, they also brought to the fore wider concerns about different types of 'postmodern terrorism' (Laqueur 1996), highlighting the links between strategically targeted terrorism (especially those using chemical, biological or nuclear products), new forms of urbanism, and the ever-advancing forms of technology which can be retro-fitted onto the existing built environment under threat.

In the post-September 11th world these new terrorist realities have led, in some cases, to dramatic and reactionary urban counter-responses based on Belfast-style fortification as well as the increasing use of sophisticated military threat-response technology. In short, the events of September 11th have served to influence the technological and physical infrastructure of targeted cities so that 'urban flows can be scrutinized through military perspectives so that the inevitable fragilities and vulnerabilities they produce can be significantly reduced' (Graham 2002: 589).

As well as stimulating a rethinking in how urban areas might be re-designed, the response of urban authorities to the enlarged terrorist threat has equally been focused upon how potential terrorist incidents should be prepared for and managed. Since September 11th the conceptualisation of terrorism, and how governments should respond to the dangers it poses, has undergone significant changes. This chapter will argue that the search for appropriate counter-terrorism solutions has led to a new synthesis of several academic and practitioner traditions, as policy makers and emergency professionals attempt to construct more holistic and multidisciplinary notions of security and its governance. It is further argued that in this effort emergency planning specialists have increasingly adopted a new vocabulary – centred on *resilience* – which is at once proactive and reactive, with an in-built adaptability to the fluid nature of the new security threats which challenge states and their urban areas in the new 'age of terrorism'.

New security challenges and the surge of multidisciplinary knowledge

Notions of risk and threat abound in contemporary debates about the future of the city. Although many of these debates have been ongoing for many decades, the events of September 11th, and subsequent acts of global terrorism, have led policy makers and academics to explore new ways to minimise risk, often through the addition of counter-terrorist security features into the design and management of the built environment. Such reflection has spawned various disciplinary studies, which have illuminated different aspects of the new security challenges faced. Such ideas in particular centre on the use of anti-terrorist urban design, often using advanced 'military style' strategies. They have also centred on the broadening of traditional risk and emergency planning responses, often focused on natural and environmental disasters, to encompass the different challenges thrown-up by the contemporary terrorist threats facing society and its economies.

Designing-out crime and terrorism

In recent decades a vast literature has developed around the concept of the 'defensive or fortress city' and how urban planners, police forces, private security, and policy makers can reduce the occurrence, and in particular the fear, of crime (see for example Crawford 2002; Marcuse and van Kempen 2002). Much of this contemporary work stemmed from the 1970s concepts of 'defensible space' and crime prevention through environmental design (see for example Newman 1972). From the 1990s these concepts have increasingly been re-appropriated in attempts to protect selective areas of cities from potentially destructive risks ranging from protests and riots to acts of terrorism (Jarman 1993; Azaryahu 2000; Coaffee 2000). These risks, until recently, have notably focused on the financial districts of major cities, with the starkest examples being in London where overt security measures have been utilised to design-out, or mitigate the impact of, terrorism, by constructing 'rings of steel', through the use of planning regulations and advanced technology in order to create 'security zones' where access is controlled and surveillance enhanced (Coaffee 2004).

Such defensive approaches were not restricted to territorial concerns but were also adopted at the individual building scale, placing great emphasis upon 'target hardening'. For example, pre-September 11th Martin Pawley (1998) argued that, as a result of an upsurge in urban terrorism, especially against 'the highly serviced and vulnerable built environment of the modern world', the new-wave of signature buildings could be replaced by an 'architecture of terror' as a result of security needs. Citing examples from Israel, Sri Lanka, North America, Spain and the UK he prophesied that building such structures could well take on an 'anonymous' design to make them less of an iconic target for terrorists. Pawley further emphasised that, once this security-oriented planning became established, it would be difficult to withdraw unless the terrorist threat significantly decreased.

In the aftermath of September 11th the construction of security cordons, exclusion zones, and ever-expanding CCTV networks around strategic sites has continued apace (Wood *et al.* 2003; Coaffee 2004). Today, urban governments and citizens alike question whether or not the plethora of post-September 11th security features that have subsequently been implemented at sites under threat actually improves security and reduces the fear of terrorist attacks, or indeed, whether we can adequately plan against such attacks in the future. For example, Graham noted that in the immediate aftermath of the September 11th attacks, such 'old defensive responses ... seem almost comically irrelevant in this new age of threat' (Graham 2001: 411). That said, many have argued that September 11th has merely signalled a surge towards an increasingly militarised city with temporary preventative measures becoming de facto solutions, more intense and legitimised, given the ongoing level of perceived threat (Warren 2002).

New socio-technical fixes

Since the 1970s and, more particularly, since the end of the Cold War in the late 1980s, a vast literature has grown around the idea of a 'Revolution in Military Affairs' (RMA). The term was initially used by Soviet academics in the 1970s to denote periods of 'military technical revolutions', when advances in technology precipitated new forms of weaponry and military tactics (Builder 1995). The reappraisal of threats in the new world order identified a large spectrum of potential threats (especially those from 'catastrophic terrorism'), from non-traditional sources (Ek 2000). As these threats have expanded, technologies and strategic thinking developed for military purposes have been increasingly transferred to the civic and urban realms and now intersect with everyday city life.

RMA advances have been used both *offensively* and *defensively*. They were used offensively, for example, in the network-centric warfare of the First Gulf War (1991) with the use of precision guided munitions to pinpoint targets from a distance, and in the current Gulf conflict to achieve 'full spectrum dominance' (Dillon and Reid 2001; Krepinevich 2003). Equally, they were used defensively to scan the 'homeland', mainly along territorial borders and within urban areas, for potential targets and threats. It is now clear that military-style technology is increasingly used in the civilian context of urban anticrime initiatives, as pointed out by Bishop and Phillips (2002: 94) who argued that military and civil defence have coalesced with the idea of risk, with threat and emergency being normalised:

> The military use has merged fully and completely with civil defence, protecting the populace from disasters. ... Emergency – being on alert, preparedness – has been our steady state for some half a century now; only now, we have more of it and the stakes get even higher.

This is not a new type of policy response; rather, it has been argued, events such as September 11th and the attacks, for example, in Bali (2002),[2] Riyadh (2003),[3] Madrid (2004),[4] London (2005),[5] Mumbai (2006)[6] and elsewhere, make the introduction of control technologies, which have been operationable for some time, easier (politically) to implement, particularly at the urban level.

In such a context, policy makers' analysis of the city through such a 'military gaze' offers insight into the new security risks cities face and in particular on how to protect not only their iconic buildings and public places but also their critical infrastructures – networks and information systems – long seen as vulnerable (Robinson *et al.* 1998). Appreciating the complex and interrelated nature of urban infrastructures that need protecting is particularly vital, as noted by Richard Little (2004: 52):

> Typically, hazard mitigation strategies for infrastructure have generally addressed first-order effect – designing robust systems to resist extreme loads imparted by natural events or malevolent acts such as sabotage and terrorism. However, because these systems do not operate independently, strengthening a single system is seldom effective at preventing outages.

September 11th also heralded many discussions regarding how major cities might look and function in the future if the threat of terrorism persists. In particular, there was a reassessment of the viability of building iconic skyscrapers, with some even predicting their demise (see for example Marcuse 2001; Kunstler and Salingaros 2001; Mills 2002). More broadly, others hypothesised about the nature of city life in the USA:

> The terrorist attack on the World Trade Center is propelling a civic debate over whether to change the way Americans experience and ultimately build upon urban public spaces. Are a city's assets – density, concentration, monumental structures – still alluring?
>
> (Vidler 2001: 6)

Likewise, after the Madrid train bombings in 2004 other commentators also questioned whether urban life has a future and whether a culture of fear and threat of catastrophic violence will end the urban age as the anxious flee the city (Klinenburg 2004).

Some commentators have more explicitly argued that technological advancement will become all-important in the battle against urban fear and terrorism. There is widespread evidence that the 'creep' of surveillance and other methods of social control in Western cities is beginning to 'surge' in response to the new terrorist threat (Wood *et al.* 2003). There is therefore growing concern that the mushrooming of automated and hi-tech systems will further erode civil liberties with democratic and ethical accountability giving way to defeating perceived risks of terrorism. In short, post-September 11th, there has been a commodification of surveillance with a technological drive to develop digital, automated and biometric systems, where boundary control and automated access become integral to risk management in an increasingly dangerous society (Lianos and Douglas 2000; Lyon 2003). For example, in the UK the Surveillance Studies Network[7] recently undertook work for the Information Commissioners Office leading to the publication of the *Report on the Surveillance Society* in November 2006.[8] This concluded that the UK is the most surveilled industrialised Western state with 4.2 million CCTV cameras, with looser laws on privacy and data protection, and that this Big Brother-esque state is expanding all the time, and is significantly driven, at least rhetorically, by the imposing and ever-expanding terrorist threat.

Environmental risk and insurance vulnerability

The growing fear of terrorist attack and the additional need for security planning, has also forced policy makers and emergency planners to reappraise pre-existing systems of managing disasters – most commonly associated with environmental risk. Social scientists have long studied natural hazards and the need to make contingency plans against them (Smith 1992) but until recently there have been few detailed studies of the impact of technological risk (Blowers 1999; Adam *et al.* 2000). Nearly thirty years ago in *The Environment as Hazard*, Burton *et al.* (1978: 1) argued that environmental hazards were having an increasingly disastrous financial impact and significantly influencing human settlement patterns:

> It may well be that the ways in which human kind deploys its resources and technology in attempts to cope with extreme events of nature are inducing greater rather than less damage and that the processes of rapid social change work in their own way to place more people at risk and make them more vulnerable.

They further argued that new ways of thinking about a more holistic form of environmental hazard management were necessary: 'Previous exploration of hazard proneness emphasizes either nature, society or technology. Our approach emphasizes the interaction amongst all three' (Burton *et al.* 1978: 18).

In more recent years such an integrated approach has been increasingly explored as 'risk theory' has developed as a significant academic tradition in the social sciences. Risk theorists have primarily been concerned with global environmental hazards which are now seen as defining characteristics in the organisation of contemporary society. It has also been intimately linked to insurance, which for centuries offered financial security against risk (see, for example, Beck 1992; Ewald 1993; Douglas 1994; Adams 1995). These hazards (most notably the release of nuclear material at Chernobyl and global climate change) have potentially severe environmental, social, economic and political implications, and through the media the public is becoming increasingly aware of their trans-national nature. This has led to a series of academic and media debates on the subject of risk and its effects on social relations, the interaction of local and global processes, the collapse of the idea of the nation-state, and the rise of pressure groups that challenge the existing social and political order, redefining the rules and principles of decision making.

As with significant environmental disasters of the past, the impact of September 11th, and of the recent catastrophic naturally induced disasters and pandemics (e.g. Severe Acute Respiratory Syndrome – SARS – and avian influenza), have shown the limits of insurance, in some cases even exceeding the limits of insurability. Reports indicate that many insurance companies are now trading in deficit following insurance industry losses of over $ 20 billion from September 11th, compounded by a series of natural disasters such as hurricanes and floods; some estimate that the three major US hurricanes in 2005 (Katrina, Rita and Wilma) will eventually cost the insurance industry up to $ 80 billion.[9]

In short, the nature of risk and its management through the insurance mechanism and associated issues of corporate liability has undergone something of a sea-change after September 11th with its unprecedented physical and financial impact. As a direct result, traditional disaster management and emergency planning provisions, in many cases remnants of World War Two and the Cold War, required modernising and rationalising in order to cope with the expanded nature of social risk and especially with the prevailing terrorist threat.

Bringing it all together: the concept of resilience

In the post-September 11th world not only were new or modified solutions to contemporary forms of terrorist risk sought but also new ways of managing these risks were initiated. As Molotch and McClain (2003: 679) argued, 'the attacks of September 11th indicate a new kind of threat to urban security and imply the need for new urban knowledges or at least fresh ways to apply older understandings'.

Traditionally, the study of defensive urban design, military innovation and environmentally focused emergency planning has enjoyed limited cross-fertilisation and has tended to be carried out in relative isolation. The aftermath of September 11th led to the understanding that more integrated approaches were required in order to cope with new security challenges, both internationally and within state boundaries. Consequently, the issues of crime, terrorism and contemporary warfare began to coalesce, their synthesis drawing on risk management, disaster recovery, and emergency planning, articulating a more holistic concept of security or combined multi-hazard management.

The language used in the counter-terrorist effort has similarly been modified and it is now common to talk of minimising the terrorist risk by developing *resilience*. The concept of resilience first emerged in research concerned with how ecological systems cope with stresses or disturbances caused by external factors (see for example Davic and Welsh 2004). More recently, the term has been applied to human social systems (Agar 2000; Pelling 2003), economic recovery (Rose 2004), and disaster recovery in cities and urban planning (Pickett *et al.* 2004; Vale and Campanella 2005). Post-September 11th metaphors of resilience have been increasingly used to describe how cities and nations attempt to 'bounce-back' from disaster, and to the embedding of security and contingency features into planning systems (Coaffee and Murakami Wood 2006). This implies the need to act within the conditions set by a 'new normality' and to develop a governance framework so that resilience is embedded effectively within emergency planning and management systems.

Today, resilience has undoubtedly become a relevant metaphor concept for politicians and policy makers alike, offering a new lexicon to make sense of the counter-terrorist challenge. It is a metaphor which can be applied in a variety of national and international contexts – a translation term – that allows, according to Gold and Revill (2000: 235), 'connections to be made between different strands of research with common terminology and consistent threads of analysis, without needing to make assumptions that the phenomena under investigation are the product of similar processes that apply regardless of cultural context'. There are thus a number of 'threads', or dimensions, around which the concept of resilience can be articulated, which differ from traditional notions of emergency planning and disaster recovery.

First, there has been a move towards *greater preparedness* rather than reactive disaster management, stimulated by new security challenges and natural catastrophes. The recent criticism of the US federal Government in the aftermath of Hurricane Katrina for the lack of preparatory planning and disaster response attests to the requirement of this move (see, for example, Burby 2006). Traditionally, most of the emergency planners' arsenal was linked to reacting to a disaster once it has happened and to developing appropriate plans to create 'a business as usual' situation as soon as possible. In contrast, more contemporary approaches view resilience as both reactive *and* proactive. As O'Brien and Read (2005: 345) note, 'the term resilience brings together the components of the disaster cycle – response, recovery, mitigation and preparedness'. Others support this idea in relation to the embedding of notions of a fluid form of resilience into the urban development process:

> Resilient cities are constructed to be strong and flexible rather than brittle and fragile ... their lifeline systems of roads, utilities and other support facilities are designed to continue functioning in the face of rising water, high winds, shaking ground and terrorist attacks.
>
> (Godschalk 2003: 137)

Second, there has been a *broadening of the emergency planning agenda*, to cover not just disaster recovery from natural and environmental hazards or accidents but to focus specifically on new security challenges. Most countries have only recently begun to appreciate the global nature of the threats posed by both natural hazards and technological accidents, and importantly in the post-September 11th era, incidents of catastrophic terrorism.

The contemporary terrorist threat has altered the established agendas of emergency planning. On the one hand, a far greater variety of threats now presents itself. Concerns about how terrorists might use deadly chemicals, nuclear material, biological agents or the internet in attack have promoted a review of 'are we prepared?' Similarly, the damage inflicted by September

11th has prompted questions about the adequacy of current contingency planning arrangements, insurance policies and business continuity cover. What is important here, is that whilst there are considerable concerns about the potential of terrorism to inflict massive damage, there are already considerable risks that emergency planners should not de-prioritise, including the impact of natural hazards, the spread and containment of diseases and the periodic threat of weather abnormalities. As O'Brien and Read (2005: 359) further note in relation to the UK, 'resilience in the face of international terrorism is an obvious current priority ... but wider considerations should not be consumed by this current single course of threat'. Equally, as Alexander (2002: 211) has noted, terrorism has a significant and wide-reaching influence on emergency planning practice:

> It is a curious paradox that terrorism now dominates world thinking despite the fact that natural and technological disasters of the more usual kind have not become less serious and remain vastly dominant in terms of size and frequency of their collective impact.

Third, *the key role of institutional resilience* and multi-level governance has been promoted in different tiers of government; this emphasises holistic solutions to risk management to be undertaken by governments, service sectors and other emergency agency managers to create strong and transparent institutional arrangements to protect complex infrastructural systems. Given the high financial and potential political cost of enhancing 'resilience', it now makes sense to include not only emergency planners and the emergency services, but also architects, planners, policy specialists, insurance experts, engineers, legal experts, and so forth in decision making processes about emergency management and security planning. As Little (2004: 55) points out, resilience is not just about physical robustness or designing out risk:

> Resilience is often provided by means of increased robustness which increases failure-resistance through design and/or construction techniques ... [but] these characteristics are just as critical for institutions as for the physical systems themselves.

He further argues that a fully inclusive governance system be enacted for dealing with resilience issues:

> Developing a successful strategy for urban security requires that these interactions be understood and enabled by all involved stakeholders. Security will be neither holistic nor effective if it is restricted to narrow professional or disciplinary stovepipes or if interactions among government officials, security professionals, program and financial staff, and emergency responders occurs only on a product by product basis.
>
> (Little 2004: 55)

This, however, is not just about coordinated thinking and multi-agency working *per se* but also about such 'joined-up thinking' occurring at the appropriate spatial scale. Whereas traditionally, localities, more often than not, have been responsible for both pinpointing potential target risks and pulling together the appropriate response, new larger scale risk events have necessitated a rethink about the appropriateness of this institutional infrastructure leading to *the rescaling of the response* – from locally coordinated systems to centralised and sub-national organisation.

The reorganisation of the UK resilience response provides a concrete example of this change, ongoing before September 11th but hastened as a result of it. Historically, response was locally organised, with 'central government quite willing to let local agencies deal with emergencies' (O'Brien and Read 2005: 353). In 2000, strategically targeted nationwide protests on the transport network (blockades of oil refineries, go-slow convoys on motorways) regarding the price of fuel, led to significant impacts on the national economy. These protests also led to critical questions regarding 'who was in charge' for coordinating the response within the petrochemical industry and emergency services. It became clear that recent rounds of privatisation, reorganisation and a general 'hollowing-out of the state' had left a disorganised 'chain of command' without authoritative leadership, thus indicating that a reform of the emergency planning procedures was long overdue. Such a reform was also prompted by the outbreak of foot and mouth disease and a number of serious flooding incidents in 2000–1. These disruptions – fuel protests, foot and mouth disease, and flooding – are collectively referred to as the 3Fs by UK emergency planners. Just after these incidents the events of September 11th and the concern that key sites in and around larger cities, especially London, would be targeted by terrorists, led to a significant speeding up of this ongoing reform process.

In the UK reform has been swift and has led to wholesale change. What were perceived as outdated emergency planning procedures have been replaced, most notably under the Civil Contingencies Act (CCA 2004) by a new resilience governance structure which aims to coordinate a multi-agency and multi-level response. In this scenario, the need to improve coherence in the established lines of control and communication and action between the local, regional and national tiers of government in the event of disaster or emergency was addressed in a number of ways as the CCA which sought to establish a consistent degree of civil protection and emergency preparedness across the UK. In particular, the CCA established multi-agency partnerships – *Local Resilience Forums* – in sub-regional areas, where neighbouring local authorities and emergency services could coordinate emergency responses. Since 2003, *Regional Resilience Teams and Forums* have coordinated the actions of the local forums, managing key relationships with responders, other UK regions and central government, and ensuring effective communication and resource flow between and across different tiers of government. Completing the resilience governance pyramid is the national level *Civil*

Contingencies Secretariat, who assess potential risks, making sure different tiers of government can respond effectively, providing strong leadership and guidance to the resilience community. This national response also involves other national departments such has the Home Office, the Office of the Deputy Prime Minister (renamed the Department of Communities and Local Government in 2006), and the Regional Co-ordination Unit.

At the local government level, having a fit-for-purpose resilience infrastructure is now a statutory responsibility across all key public services. This can be seen as a formal attempt to embed resilience not just in policy rhetoric but also in the core practice of government institutions and agencies. This has led to: greater performance management pressures for meeting national minimum standards for emergency arrangements; improved multi-agency working and cross-institutional training; the promotion of business continuity to local organisations; and the statutory requirements to develop enhanced scenario building and test emergency and contingency plans where local government has, in conjunction with regional and national partners, begun to hold regular table-top exercises and 'live' simulated procedural tests. There has also been a statutory responsibility to develop systems of communication for 'warning and informing' the public about risks faced, although this key priority has, to date, been downplayed, with importance given to developing the resilience governance system.

In order to achieve urban resilience in the UK, a monitoring system that attempts to pinpoint any potential vulnerability or weakness, has also been implemented at different spatial levels. At a national level the currently ongoing *National Capabilities Programme* provides the main evaluative framework by which the government can ensure comparative levels of resilience across all parts of the UK. From 2004, Regional Resilience Forums have produced *Regional Risk Assessments*, which evaluate the probability and likely impact of potential emergencies. From 2005, such judgments have been informed by *Community Risk Registers* compiled by Local Resilience Forums that pinpoint local risk priorities and can assist in identifying gaps in organisational ability and inform the planning process regarding the scale of response required. At the local government level, risk management evaluation has now been embedded within every Local Authority's *Comprehensive Performance Assessment* and has been pushed to the forefront of the local government modernisation agenda. Traditionally, whilst most authorities have had a 'stand-alone' risk management strategy in place, the emphasis is now on embedding risk management throughout local government as part of good corporate governance (Coaffee and Murakami Wood 2006). These surveys are important as they provide the strategic drivers of reform for resilient planning, dictating priority work-packages and timetables of work. Finally, protocols are now increasingly tested through a series of local, regional and national level exercises to ensure all organisations are fully prepared for all types of emergencies (not just terrorism) and fully conversant with the detail and chain of command within emergency plans. Exercises have tended to be of three types: discussion forums to allow plans to be finalised, table-top simulations, and live rehearsals. Exercises are timetabled nationally and occur at all spatial scales, coordinated by the various resilience forums, local and regional authorities and lead Government departments. Increasingly such exercises are also taking place with international partners (ibid.). Examples of such exercises include the well-documented *Osiris II* (the simulation of a chemical attack on the London underground in September 2003), *Merlin Aware* (to practice multi-agency strategic command in the event of a CBRN incident in North East England in October 2005) and *Atlantic Blue* (involving the UK, US and Canada simulating a major international terrorist incident in April 2005).

The fourth key change from traditional disaster management has been the *reappropriating emergency planning-type policies for everyday civil use* which has continued apace and for a number of reasons. On the one hand, the opportunities offered by the terrorist threat for governments to adopt social control technologies in the 'national interest' and for 'homeland security', amidst the ongoing climate of fear, are almost unprecedented. As Graham (2004: 11) notes, 'in the wake of 9/11 and other catastrophic terrorist acts of the last few years, the design of buildings, the management of traffic, the physical planning of cities and neighbourhoods, are being brought under the widening umbrellas of national security.' On the other hand, the necessity of creating safe and secure business environments to maximise the attractiveness of urban areas to footloose capital has become paramount. A number of commentators now emphasise the links between city growth strategies and security planning with the retrofitting of defensive security often seen as a vital selling point for urban competitiveness (Murakami Wood and Coaffee 2007).

The threat of terrorism has now embedded resilience implicitly and explicitly in numerous social and planning policy discourses. For example, Wekerle and Jackson (2005) have argued that in North America numerous policies have 'hitchhiked' on the anti-terrorism agenda, including extensive use of control technologies in public and private space; promotion of low-density urban sprawl as a safer alternative to high-density but easily targeted high-rise cities; and the more ephemeral connections such as social movements linked to environmentalism being labelled 'terrorist', and community crime-reduction schemes such as neighbourhood watch being adopted to aid 'community resilience'.

The conjoining of national counter-terrorism policy and local level police safety initiatives is now widespread and often referred to in resilience policy. This has been aided by the use of military technology in the civil sphere. For example, by mid-2006 the UK had a network of advanced tracking security cameras on all main roads and motorways, based on smaller schemes developed from First Gulf War technology. The technology was first installed in the financial districts of London in the 1990s as a response to terrorist threats, and then as part of a wider congestion charging zone over much of central London (Coaffee 2004). This latest importation of 'panoptical technology' for targeting crime reduction, traffic management and anti-terrorism, highlights the pervasive interconnections of the military and policy makers – long referred to as the military–industrial complex.[10]

Conclusion

In today's so-called 'age of terrorism' the security challenges facing states and urban agglomerations continue to be in a state of flux. Given the unpredictable nature of this threat and the inherent vulnerabilities it creates, we should not be surprised by the increased attention given to both preventative and recovery strategies. The response to terrorist risk, especially after September 11th, usually poses the question, 'Are we prepared?' rather than 'Can we prevent it?' Given the vast array of targets, strategies and technologies available to would-be terrorists, traditional, and often static approaches focused on planning against terrorism are no longer suitable. As Little (2004: 57) notes:

> Threats are unpredictable and the full range of threats probably unknowable. We will never be able to anticipate all possible threats and even if we could, there is not enough money to deploy technologies to address them. Security in this situation needs to be flexible and agile and capable of addressing new threats as they emerge. Protective technologies have a key role to play in making our cities safer but only if supported by organizations and people who can develop preattack security strategies, manage the response to an attack, and hasten recovery from it.

Thus resilience is quickly becoming a key factor in shaping how society is structured and, most notably, how civic leaders respond to security challenges. As Godschalk (2003: 42) notes, 'if we are to take the achievement of urban resilience seriously, we need to build the goal of the resilient city into the everyday practice of city planners, engineers, architects, emergency managers, developers and other urban professionals. This will require a long-term collaborative effort to increase knowledge and awareness about resilient city planning and design'. Such knowledge is likely to come from several areas of inquiry all of which will form a new synthesis for planning resilient environments. However, at present in many countries developing resilience against terrorist threats is an agenda which has been developed almost *exclusively* by politicians and emergency planning professionals with little if any discussion with citizens, the business community, town planners, urban designers and other built environment professionals.

Likewise, developing resilient urban responses to terrorism is a fluid process and one which must be able to adapt to changing types of terrorist threat, whist balancing the effectiveness of resilience response with the acceptability of such actions to the public and wider stakeholders. In particular, greater attention and government resources are now being given to the changing nature of terrorist threats which has forced a rethinking of traditional emergency planning and counter-terrorism tactics given the increased magnitude of the threats faced, especially those from CBRN (chemical, biological, radiological and nuclear) sources which many terrorist groups have expressed significant interest in utilising in attacks. Equally, attacks conducted by suicide attackers and tactically aimed at soft targets such as hospitals, schools, shopping promenades, and more generally crowded places, are posing new questions for those tasked with emergency management and embedding resilience into the design and management of urban areas.[11]

Finally it is important to note that the emergency planning and disaster management agenda in most countries is much wider than just terrorism, although it would be true to say that in recent years the ongoing threat of terrorist attack has been the key catalyst for emergency planning reform. What many countries are now seeking to achieve is a more resilient integrated emergency planning and management system which can better help prevent, prepare, respond and recover from a variety of risk incidents including terrorism, flooding, severe weather abnormalities such as heat waves and cold spells, pandemic influenza, and infectious animal diseases. At present the UK in particular is at the forefront of developing resilience policy and practice in all its guises – through the development of more holistic and integrated emergency management – and provides a template that might in the future be adapted by other countries. Here it is important that policy and knowledge transfer from existing programmes is developed within a particular political and social context, is carried out in a pragmatic and thoughtful way in line with current or predicted institutional realities, threat levels and funding streams, and is not transferred uncritically or hastily (see for example Dolowitz *et al.* 2000: 3).

Notes

1 This paper is based in part on a paper by the author (Coaffee 2006), published by Taylor and Francis. It is also the result of a UK Economic and Social Research Council's New Security Challenges programme, project (Ref: 228250034) that was focused upon new and creative ways of conceptualising post-September 11th security agendas.

2 During October 2002 explosions destroyed two nightclubs in Bali, killing 202 people.

3 In May, four cars laden with explosives were driven into expatriate housing compounds, killing 34 and wounding over 200.

4 A series of bombs in March 2004 on the Madrid transport network killed 191 people and injured hundreds more.
5 Bombs exploded on one bus and three underground trains, killing 56 people and injuring over 700. This occurred the day after London was awarded the 2012 Olympic Games.
6 In July 2006, seven bombs exploded within 11 minutes on the Mumbai train system, killing 209 people and injuring over 700.
7 The Surveillance Studies Network are an independent not-for-profit organisation dedicated to the study of surveillance practices.
8 Available HTTP: <http://www.privacyconference2006.co.uk/files/discussion_eng.pdf>. This report was intended to begin a public debate on the pros and cons of the adoption of particular technologies such as ID cards and CCTV.
9 See for example 'US Hurricanes Cost Lloyds of London a Record £2.9bn', *The Guardian*, Thursday 1st December 2005.
10 The military–industrial complex was first described in 1961 by President Eisenhower in his Farewell address to the American nation.
11 Research into these processes on effectiveness and acceptability of approaches to counter-terrorism and resilient design is now a key government priority and the subject of ongoing government-funded academic research. For instance, the *'Resilient Design (RE-Design) for Counter-terrorism: Decision Support for Designing Effective and Acceptable Resilient Places' Project* is currently being undertaken by the author, the editor of this book and two colleagues and is being funded by the Engineering and Physical Sciences Research Funding Council in the UK. The aim of the 'RE-Design' Project is to ensure that effective and acceptable resilient public places are attained through the structured and considered integration of counter-terrorism initiatives into the decision-making processes of key stakeholders.

References

Adam, B., Beck, U. and van Loon, J. (2000) *The Risk Society and Beyond*, London: Sage.

Adams, J. (1995) *Risk*, London: UCL Press.

Agar, W.N. (2000) 'Social and ecological resilience: are they related?', *Progress in Human Geography*, 24(3): 47–64.

Alexander, D.E. (2002) *Principles of Emergency Planning and Management*, Harpenden: Terra Publishing and New York: Oxford University Press

Azaryahu, M. (2000) 'Israeli securityscapes', in J.R. Gold and G.E. Revill (eds) *Landscapes of Defence*, London: Prentice Hall, 102–13.

Beck, U. (1992) *Risk Society: Towards a New Modernity*, London: Sage.

Bishop, R. and Phillips, J. (2002) 'Manufacturing Emergencies', *Theory, Culture and Society*, 19(4): 91–102.

Blowers, A. (1999) 'Nuclear waste and landscapes of risk', *Landscape Research*, 24(3): 241–64.

Boal, F.W. (1995) *Shaping a City – Belfast in the Late Twentieth Century*, Belfast: Institute of Irish Studies.

Boal, F.W. and Murray, R.C. (1977) 'A city in conflict', *Geographical Magazine*, 49: 364–71.

Brown, S. (1985) 'Central Belfast's security segment – an urban phenomenon', *Area*, 17(1): 1–8.

Builder, C. (1995) 'Looking in all the wrong places? The real revolution in military affairs is staring us in the face', *Armed Forces Journal International*, May: 38–9.

Burby, R.J. (2006) 'Hurricane Katrina and the paradoxes of government disaster policy: bringing about wise governmental decisions for hazardous areas', *Annals of the American Academy of Political and Social Science*, 604(1): 171–91.

Burton, I., Kates, R. and White, G. (1978) *The Environment as Hazard*, Oxford: Oxford University Press.

Coaffee, J. (2000) 'Fortification, fragmentation and the threat of terrorism in the City of London in the 1990s', in J.R. Gold and G.E. Revill (eds) *Landscapes of Defence*, London: Prentice Hall, 114–29.

Coaffee, J. (2003) *Terrorism, Risk and the City*, Aldershot: Ashgate.

Coaffee, J. (2004) 'Rings of steel, rings of concrete and rings of confidence: designing out terrorism in Central London pre and post 9/11', *International Journal of Urban and Regional Research*, 28(1): 201–11.

Coaffee, J. (2006) 'From counter-terrorism to resilience', *European Legacy* – Journal of the International Society for the Study of European Ideas (ISSEI), 11(4): 389–403.

Coaffee, J. and Murakami Wood, D. (2006) 'Security is coming home – rethinking scale and constructing resilience in the global urban response to terrorist risk', *International Relations*, 20(4): 503–17.

Crawford, A. (ed.) (2002) *Crime and Insecurity: Governance and Safety in Europe*, Devon: Willan.

Davic, R. and. Welsh, H.H., Jr. (2004) 'On the ecological roles of salamanders', *Annual Review of Ecology, Evolution, and Systematics*, 35 (2004): 405–34.

Dillon, M. and Reid, J. (2001) 'Global liberal governance: biopolitics, security and war', *Millennium*, 30(1): 41–66.

Dolowitz, D.P., Hulme, R., Nellis, M. and O'Neil, F. (2000) *Policy Transfer and British Social Policy: Learning from the USA*, Buckingham: Open University Press.

Douglas, M. (1994) *Risk and Blame: Essays in Cultural Theory*, London: Routledge.

Ek, R. (2000) 'A Revolution in military geopolitics?', *Political Geography*, 19: 841–74.

Ewald, F. (1993) 'Two infinities of risk', in B. Massumi (ed.) *The Politics of Everyday Fear*, Minneapolis: University of Minnesota Press, 221–8.

Godschalk, D. (2003) 'Urban hazard mitigation: creating resilient cities', *Natural Hazards Review*, 4(3) (August): 136–43.

Gold, J.R. and Revill, G.E. (eds) (2000) *Landscapes of Defence*, London: Prentice Hall.

Graham, S. (2001) 'In a moment: on global mobilities and the terrorised city', *City*, 5(3): 411.

Graham, S. (2002) 'Special collection: reflections on cities, september 11th and the "war on terrorism" – one year on', *International Journal of Urban and Regional Research*, 26(3): 589–90.

Graham, S. (ed.) (2004) *Cities, War and Terrorism: Towards an Urban Geopolitics*, Oxford: Blackwell.

Jarman, N. (1993) 'Intersecting Belfast', in B. Bender (ed.) *Landscape: Politics and Perspectives*, Oxford: Berg, 107–38.

Klinenburg, E. (2004) 'After Madrid, does urban life have a future?', *New Statesman*, 22 March, Available HTTP: <http://www.newstatesman.com/200403220002>.

Krepinevich, A. (2003) 'The unfinished revolution in military affairs', *Issues in Science and Technology*, 19(4): 65–7.

Kunstler, J. and Salingaros, N. (2001) '*The End of Tall Buildings*', published by PLANetizen (online), Available HTTP: <http://www.planetizen.com/> (accessed 17 September 2001).

Laqueur, W. (1996) 'Post-modern terrorism', *Foreign Affairs*, 75(5): 24–36.

Lianos, M. and Douglas, M. (2000) 'Dangerization and the end of deviance: the institutional environment', *British Journal of Criminology*, 40: 261–78.

Little, R. (2004) 'Holistic strategy for urban security', *Journal of Infrastructure Systems*, 10(2) (June): 52–9.

Lyon, D. (2003) *Surveillance after September 11*, Cambridge: Polity Press.

Marcuse, P. (2001) 'Reflections on the events', *City*, 5(3): 394–7.

Marcuse, P. and van Kempen, R. (2002) *Of States and Cities: The Partitioning of Urban Space*, Oxford: Oxford University Press.

Mills, E.S. (2002) 'Terrorism and US real estate', *Journal of Urban Economics*, 51: 198–204.

Molotch, H. and McClain, N. (2003) 'Dealing with urban terror: heritages of control, varieties of intervention, strategies of research', *International Journal of Urban and Regional Research*, 27(3): 679–98.

Murakami Wood, D. and Coaffee, J. (2007) 'Lockdown! Resilience, resurgence and the stage-set city', in R. Atkinson (ed.) *Securing the Urban Renaissance*, Bristol: Policy Press (in press).

Newman, O. (1972) *Defensible Space – Crime Prevention through Urban Design*, New York: Macmillan.

O'Brien, G. and Read, P. (2005) 'Future UK emergency management: new wine, old skin?', *Disaster Prevention and Management*, 14(3): 353–61.

Pawley, M. (1998) *Terminal Architecture*, London: Reaktion.

Pelling, M. (2003) *The Vulnerability of Cities: Natural Disasters and Social Resilience*, London: Earthscan.

Pickett, S., Cadenasso, M. and Grove, J.M. (2004) 'Resilient cities: meaning, model, and metaphor for integrating the ecological, socio-economic, and planning realms', *Landscape and Urban Planning*, 69: 369–84.

Robinson, C.P., Woodard, J.B. and Varnado, S.G. (1998) 'Critical infrastructure: interlinked and vulnerable', *Issues in Science and Technology*, 15(1): 61–7.

Rose, A. (2004) 'Defining and measuring economic resilience to disasters', *Disaster Prevention and Management*, 13(4): 307–14.

Smith, K. (1992) *Environmental Hazards: Assessing Risk and Reducing Disaster*, London: Routledge.

Vale, L. and Campanella, T. (eds) (2005) *The Resilient City: How Modern Cities Recover from Disaster*, Oxford: Oxford University Press.

Vidler, A. (2001) 'The city transformed: designing "defensible space"', *New York Times*, 23 September.

Warren, R. (2002) 'Situating the city and September 11th: military urban doctrine, "pop-up" armies and spatial chess', *International Journal of Urban and Regional Research*, 26(3): 614–19.

Wekerle, G.R. and Jackson, P.S. (2005) 'Urbanizing the security agenda: anti-terrorism, urban sprawl and social movements', *City*, 9(1): 34–49.

Wood, D., Konvitz, E. and Ball, K. (2003) 'The constant state of emergency? Surveillance after 9/11', in K. Ball and F. Webster (eds) *The Intensification of Surveillance: Crime, Terrorism and Warfare in the Information Age*, London: Pluto Press, 137–50.

Chapter 16

'Planning ahead'

Adapting settlements before disasters strike

Christine Wamsler

Introduction

Civil engineers, architects, builders and urban planners[1] have the task of developing secure and sustainable settlements. However, they are unconscious – but significant – contributors to the fourfold increase in the number of disasters that has taken place during the last 30 years (UNISDR 2006). This relates to the fact that they commonly view the interlinkages between disasters and the built environment (and related planning practices) as a simple one-way, cause-and-effect relationship (Figure 16.1). In fact, the limited perception that disasters are the (uncontrollable) cause and the destruction of the built environment is the effect, is widespread amongst those professionals, who consequently have a tendency to see disaster risk management[2] in a purely physical way. Their responses are thus very limited and mainly focused on the post-disaster context. Moreover, even preventive tools, such as building codes or land-use zoning, are currently of low relevance to the urban poor whose lives are most at risk (Figure 16.2). With

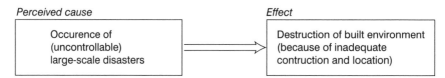

Figure 16.1 Common, 'erroneous' view of the interlinkages between disasters and the built environment (and related planning practices) as being a simple one-way, cause-and-effect relationship

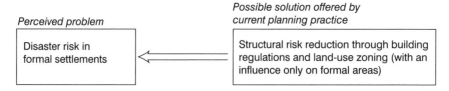

Figure 16.2 Common, 'erroneous' view of the interlinkages between the potential of planning practice for disaster risk management as being a simple one-way problem-and-solution relationship with limited efficiency and area of influence

more than one billion 'slum' dwellers worldwide, who often have no choice but to live in precarious and life-threatening conditions (UN-HABITAT 2003), planners have to urgently re-evaluate their work to provide more adequate solutions.

Responding adequately to disaster risk is inherently complex. Disasters occur when a hazardous event strikes a vulnerable human settlement, with the coping capacity of its inhabitants further influencing the extent and severity of damages received.[3] Unfortunately, at present, planners often negatively influence *all three* risk components (i.e. hazard, vulnerability and coping capacity). Hence, the task of developing secure settlements cannot be achieved unless planners thoroughly understand the interlinkages between disasters and the built environment (and related planning practices) and – based on this – integrate disaster risk management into their everyday work. In fact, incorporating knowledge about how to make houses safer into their work is just one of many issues that they need to address. Against this background, the objective of this chapter is threefold:

a. Preparing the ground

To demonstrate the complex interlinkages between disasters and the built environment (and related planning practices). It will be argued that the reality is much more complex than the one-way, cause-and-effect relationship mentioned above. It is, in effect, a reciprocal two-way and multifaceted relationship that, to date, has not been well understood and theorised. A new analytical framework for viewing this relationship is presented.

b. Reality versus current planning practices

To show how the identified interlinkages between disasters and the built environment (and related planning practices) are currently addressed. This will be mainly discussed in the context of aid programming[4] in the fields of disaster risk management and human settlement development planning (including social housing, settlement upgrading, new settlement development and urban governance programming). The analysis covers related programmes and stakeholders, institutional structures, the discourses of experts and practitioners, their working priorities, concepts, terminology and tools, as well as the historical development of both fields of work. The challenges identified at the global, national, municipal and household levels are illustrated, as is the gap between reality and current practices, both of which can lead to increased disaster risk.

c. A way forward

To show how disaster risk management could be better integrated into planning practices. A strategic and conceptual model is presented that provides guidance on how international, national and municipal (aid) organisations working in settlement development planning in so-called developing countries[5] could adopt a more proactive approach towards disaster risk management.

The following sections reflect the three objectives and follow an inherent logic: the two-step analysis of the interlinkages between disasters and the built environment and how it is tackled in practice is the necessary basis and input for the third step, namely, the development of the model mentioned above. Its aim is to overcome identified challenges and modify current planning practices so that they match better to reality (Figure 16.3).

This chapter summarises the results of research, undertaken from 2003 to 2007, that have been presented in different publications (Wamsler 2004; 2006a–g; 2007b). Here, their outcomes are incorporated within a new and comprehensive model that addresses how disaster risk management could be better integrated into human settlement development planning and

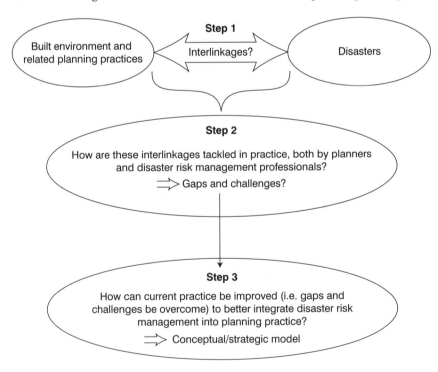

Figure 16.3 Inherent logic of research objectives presented in this chapter

programming. Sections 2 and 4 have been complemented by additional, more recent research results obtained during 2007.[6] Note that, in the following text, references are included only if the content is based on work other than the author's.

Preparing the ground

The following two sections show, first, the possible impacts of disasters on the built environment and related planning practices in cities and, second, the reverse interrelation (i.e. the influence of the built environment and related planning practices on risk and disasters). This two-way relationship has, to date, not been well understood and theorised. A framework for viewing it is thus presented as a first attempt to provide a comprehensive and exhaustive systematisation of the issues involved. The framework systematises the relationship by filtering out 12 key aspects. Tables 16.1 to 16.13 provide an in-depth analysis of each aspect.

The interlinkages: disasters ⇒ built environment

Natural hazards and/or disasters have widely varying impacts on the built environment and related planning practices. These impacts are not only physical but also socio-economic, environmental, organisational and institutional. In fact, disasters can:

- disrupt city functions;
- intensify urban hazards and create new ones;
- increase urban inequalities (producing 'poverty traps');
- create new challenges for future urban development;
- create barriers to sustainable urban development; and
- have a (negative) impact on the resources invested in the built environment.

Unfortunately, these negative effects, which are described in detail in Tables 16.1–16.6,[7] are not only extensive, but can – over decades – negatively impact the urban poor, as well as municipal and national development. The effects can generally be categorised as 'immediate and short-lived', 'immediate and long-lasting', 'delayed and short-lived' and 'delayed and long-lasting'.

Table 16.1 Disrupted city functions caused by disasters

Disaster impacts that can lead to the disruption of city functions

- Damage/destruction of housing stock.
- Damage, destruction or interruption of technical and social infrastructure (i.e. infrastructure for water supply, sanitation, energy, transport, communication, education and health services. Examples are electricity failure; blocked accesses to houses or settlements; damaged health care units and schools; contamination of drinking water wells; and destroyed bridges.
- Malfunctioning of technical and social infrastructure due to disaster impacts (e.g. accidents due to insecure pathways).
- Loss of architectural heritage (i.e. buildings and sites of cultural value), also undermining the collective quality of life, as well as national economies (e.g. because of fewer tourists).
- Destruction of whole cities (e.g. cities wiped out because of the rise in sea levels caused by climate change).

Table 16.2 Intensified and new urban hazards caused by disasters

Disaster impacts that can lead to creation or intensification of hazards

- Damage to the integrity of ecosystems creating future hazards (e.g. destruction of vegetation leading to landslides, or damages to mangroves leading to erosion and increased wave energy).
- Aggravated environmental degradation triggering secondary hazards (e.g. soil instability and erosion caused by earthquakes leading to landslides during 'normal' rain or through waste water flows).
- Modification of the landscape of settlements, thus reshaping their hazard patterns (e.g. through the change of the course of a river).
- Contamination of the environment through recovery and/or preparedness measures creating new hazards (e.g. plastic sheets – used for protecting slopes or temporary shelter – being blown away, contaminating the environment and blocking river flows or water channels).

Table 16.3 Increased urban inequalities (creating 'poverty traps') caused by disasters. As the poor are disproportionately affected, disaster impacts can intensify differences in status and the patterns of social inequality

Disaster impacts that can lead to increased urban inequalities which, in turn, could cause poverty traps

- Forced eviction of 'slum' dwellers affected by disaster.
- Direct and indirect post-disaster expenses, together with the disruption of local and household economies pushing already vulnerable groups further into poverty (e.g. loss of income earners through death or injury, interruption of production or access to markets, destruction of productive assets such as home-based workshops).
- Governance problems at different levels, resulting in aid budgets being skewed towards the recovery of one group or sector as opposed to another, resulting in increased urban inequalities.

Table 16.4 New challenges for future urban development caused by disasters

Disaster impacts that can lead to additional challenges and even barriers to future urban development

- Increased number of urban dwellers, due to immigration, affected by disaster or by decreased agricultural productivity caused by climate change.
- Increased number of homeless people (due to loss of housing and land belonging to people affected by disaster) and hence a need for living space.
- At the household level, erosion of livelihoods, savings and physical capital, increasing the number of people and settlements that depend on outside assistance, for instance, to access adequate rental housing or house ownership, house and infrastructure maintenance, etc.
- Modification of the landscape of cities, affecting past and future planning (e.g. infrastructure planning).
- Construction of temporary housing/settlements that over time need to be transformed or replaced to offer permanent solutions for disaster victims.

Table 16.5 New barriers to sustainable urban development caused by disasters. These barriers decelerate (positive) development processes

Disaster impacts that can lead to decelerated urban development

- Increased need for resources for specific (already planned) urban developments because of, for instance, contamination of the environment as a result of disaster impacts (contaminated soil, wells, etc.).
- Reduced capacity/functioning of housing/planning organisations that are directly or indirectly affected by disasters (e.g. national governmental and non-governmental aid organisations suffering through reduced reputation, damaged office buildings, disaster-affected programme measures and staff death, injuries, leave, etc.).
- Death, (temporary) disablement or migration of key persons (and workforce in general) at the national, municipal, local and household level, leading to an erosion of social capital for urban planning and governance.
- Aggravation of political stresses, leading to increased corruption, bureaucracy, political conflicts and rivalry at all levels, which affect developments at settlement, city or country level.
- Aggravation of social stresses and shocks such as disease and psychological shocks, which affect developments at settlement, city or country level (e.g. community distress; family disruptions; burglaries due to damaged houses and/or increased need, illnesses caused, for instance, by waste water entering houses; HIV/AIDS, trauma).
- Impacts on national fiscal and monetary performance, indebtedness, the distribution of income and scale and incidence of poverty, all negatively influencing the provision and financing of housing and infrastructure.

Table 16.6 Change of resources invested in the built environment caused by disasters. These effects are mostly negative in terms of achieving sustainable urban development

Disaster impacts which can lead to changed investments in the built environment

- Reduced support/assistance by planning authorities to affected 'slum' communities because of increased and unacceptable risk levels (e.g. stopping land legalisation processes).
- Disruption of national economies and related governance functions of planning authorities, for instance, due to post-disaster expenses and relocation of development investments (e.g. investments aiming to provide sustainable access to safe housing, drinking water and sanitation being disbursed on emergency issues).
- Lower output from damaged or destroyed public assets and infrastructure, resulting in fewer resources that can be reinvested in the built environment.
- Increased temporary investment and hence activity in the (formal and informal) construction industry due to rehabilitation and reconstruction efforts.
- Disruption of local and household economies, affecting people's investments in improving their living conditions (e.g. incremental housing, including infrastructure), which immediately increases their vulnerabilities (vicious circle).

The interlinkages: built environment ⇒ disasters

The reverse analysis indicates that the built environment and related planning practices constitute one of the main causes of disasters, not only in terms of generating increased vulnerabilities. In fact, they can:

- increase vulnerability;
- increase exposure to existing hazards;
- intensify/magnify urban hazards and create new ones;
- constantly change vulnerabilities and hazards (and thus make them hardly controllable);
- reduce the national and municipal coping capacities
  - because of inadequate disaster risk management systems, or
  - because of inadequate urban management/governance systems; and
- reduce the coping capacities of low-income households.

Tables 16.7–16.13[8] describe in detail how these effects, which can be classified in physical, socio-economic, environmental, organisational and institutional terms, can be generated. This detailed systematisation clearly illustrates that disasters are not one-off events caused solely by natural hazards but are generated by complex and interacting development processes in which planning practices play a major role.

In addition, the research undertaken at household level reveals that disasters are the outcome of a non-linear development process, with the key variables underlying the complex system of risk and disaster occurrence reinforcing each other. Hence, disasters not only make the already precarious conditions of 'slum' dwellers worse, but can also create vicious circles of increasing risk. 'Poverty traps' can be the outcome.

With growing urbanisation and climate change, the described reciprocal two-way and multifaceted relationship is becoming increasingly alarming.

Table 16.7 Increased vulnerability caused by the built environment and related urban planning practices. Increased vulnerability leads to reduced capacity to resist, absorb or recover from hazard impact. Hence, a condition is created where hazards easily create disasters

---

Aspects or activities that can lead to increased vulnerability

- High number of inhabitant living in cities, expressed in high population densities and surface areas of cities.
- High concentration of social networks, buildings and infrastructure in cities, including state governments and financial centres.
- High concentration and overcrowding of people in hazard-prone areas (especially the poor living in 'slum' areas), both in inter-city and peripheral communities.
- High concentration of highly defenceless population groups in cities (for example, weakened by HIV/AIDS or other diseases, conflict or malnutrition). Note: such groups are mainly found within the lower income groups; in turn, the poor's inadequate and unsanitary housing results in ill-health. Furthermore, space restrictions can influence transmission of disease (e.g. through violations).
- High and increasing number of poor and destitute persons living in areas that are socially excluded and politically marginalised, leading to limited access to information (and hence knowledge) about, and resources for, housing, infrastructure, risk reduction, etc.
- Poorly constructed residential and commercial buildings and infrastructure. Note that substandard buildings relate, amongst other things, to corruption in the construction sector, lack of control mechanisms and lack of financial resources and knowledge.
- Use of inadequate construction techniques because of lack of knowledge, together with rapidly changing environments (e.g. shelters being constructed of inflammable materials, or earthquake-resistant buildings being vulnerable to increased wind storms).
- Densely built settlements that, because of their layout (narrow paths for movement, many lanes and alleys with dead ends) can inhibit effective emergency services (e.g. evacuation).
- Densely built settlements, which allow damage to spread easily from one shelter to the next (e.g. fire spreading from one roof to the next, or a domino effect being created when earthquake affected houses fall on neighbouring buildings).
- Land used for residential, industrial and transport purposes at too close a proximity to each other.
- Closeness of shelter in risk areas to environmental hazards, intensifying the impacts of natural hazards, namely, the impact but not the hazard as such (e.g. leaking sewage pipes from better-off settlements passing through 'slum' areas to discharge into near-by rivers result, even during minor floods, in the immediate contamination of whole settlements).
- Construction of shelter on plots that are too small and have no space available for mitigation works (related to lack of living space, combined with inadequate financial resources and knowledge).
- Non-existence of infrastructure networks/services or inadequate capacity of existing ones (e.g. for waste collection, pedestrian and vehicle circulation, rain and waste water services). In 'slums', this results, for instance, in people living uphill allowing waste and storm water to flow down on to their neighbours' land, and people from inside and outside the settlement tipping solid waste down their neighbours' hills or into the nearby rivers.
- High dependency of people living in urban areas on infrastructure networks/services, with the result that the disruption of these can cause societies to collapse completely (e.g. transportation and banking systems).

Table 16.7 continued

- Full dependency of many poor urban households on housing as a productive asset for pursuing their livelihoods. In fact, economic activities are incrementally related to the housing of the poor (i.e. labour and room rental).
- Importance of the informal construction sector, on which the poor are mainly dependent, which provides limited social responsibility.
- Local livelihood practices that are not suited/adapted to densely built areas (i.e. very limited space can result in substandard modification of the built environment to permit economic activities, e.g. removal of supporting walls or creation of land fills).
- Limited access to clean water and sanitation (i.e. access to this basic need is denied to around one-quarter of urban households, which undermines health and hence also causes vulnerability to 'natural' disasters).
- Restricted access of 'slum' dwellers to regular income, influenced by the segregation/marginalisation of people from specific geographical living areas. In turn, unemployment and low income levels influence the quality of housing and infrastructure within these areas (vicious circle).
- 'Slum' dwellers' strategies/efforts to gain and expand their living space to cope not only with the growing number and size of 'slum' households but also with the lack of alternative living areas. Strategies include people living downhill felling trees or excavating the slopes below their neighbours' houses; people building latrines close to declivities; and/or claiming land from nearby rivers.
- Few mutual rights and obligations within 'slums' related to the settlements' maintenance and development (e.g. no rules as regards excavation of the slopes below houses or the construction of latrines close to declivities).
- Lack of knowledge at all levels on how to construct safe buildings and settlements (due to poor education system, limited professional training, marginalisation of 'slum' dwellers, etc.).
- Importance of status (expressed by the built environment) combined with the lack of knowledge resulting in the construction of modern-looking houses without technical safety features in risk areas.
- Conventional belief of 'slum' dwellers and partly also of representatives of planning authorities and aid organisations that disasters are purely 'divinely driven'.
- People's (false) perception of cities as secure places, influencing them not to invest money and effort in the built environment and related security measures.
- Planners' and builders' false perception that hazard-resistant design is too costly, while the implementation of hazard-proof measures in building and infrastructure design can be relatively inexpensive in terms of construction costs.
- Unwillingness of 'slum' dwellers to invest in security measures (related to insecure land tenure as well as to promises of outside help not being honoured).
- Many people living on land without having/access to secure tenure (amongst other things this also results in forced evictions after disaster occurs).
- Internal segregation within 'slum' areas expressed in the built environment (e.g. the poorest living on the ground floors, which are particularly vulnerable to flooding, or in inaccessible areas, which are virtually impossible to evacuate).

Table 16.8 Increased exposure to existing hazards caused by the built environment and related urban planning practices

Aspects or activities that can lead to increased exposure to existing hazards

- Geographic positioning of cities on disaster-prone sites (e.g. 8 of the 10 most populous cities in the world sit on or near earthquake faults). The location of many cities was chosen in the pre-colonial or colonial eras when mainly economic or other strategic factors were considered for site selection (e.g. proximity to mineral resources, close to the coast, etc.).
- Development or expansion of illegal settlements in/into marginal high-risk areas (e.g. near rivers or on steep slopes), because of the malfunctioning of land and property markets in cities and the inability of formal housing and planning sectors to cater for the priorities of the population (e.g. access to work opportunities). The latter is also related to urban dwellers' priorities – they frequently 'choose' to live in hazardous locations if it provides access to work.
- Spread of housing and infrastructure towards risky open land (because of fast growing urbanisation and the lack of inner-city land).
- Increased proximity of housing and infrastructure to environmental hazards (shelter close to industries, heavy equipment, pipelines, effluent drains, toxic disposal sites, etc.).

Table 16.9 Intensified, magnified and newly created urban hazards caused by the built environment and related urban planning practices. Hazards are being intensified or reshaped and new hazards introduced, thus increasing the number and magnitude of urban disasters

Aspects or activities which can lead to intensified or even new hazards

- Cities, while covering only 0.4% of the Earth's surface, produce the vast majority of the world's carbon dioxide emissions, thus contributing to climate change. With cities growing in population and wealth, increased production and consumption is an engine for climate change. Climate change, in turn, reshapes hazard occurrence, compounding global and local insecurity.
- Cities' 'heat island effect' (created, for instance, by concentration of heat and pollutants from power plants, industrial processes and vehicles) can cause and exacerbate heatwaves.
- Layout of streets (e.g. straight streets lined with tall buildings) can result in turbulence and wind gusts, hailstorms and localised rainfall.
- Some building features can create new hazards (e.g. antennas and electrical equipment on top of buildings that attract lightning).
- Transformation of cities' environment by urbanisation processes on inadequate land (e.g. developments of new urban areas on watersheds that modify hydraulic regimes and destabilise slopes, increasing the risk of floods and landslides; or colonisation of garbage landfills, which increases landslides and unplanned urbanisation of new areas).
- People's construction and livelihood practices, as well as urban expansion, result in overexploitation of natural assets and environmental degradation that demolishes natural protection and magnifies hazards. (Such practices include coral reef mining, sand dune grading, mangrove cutting, conversion of mangrove coasts into intensive shrimp-farming pools or development over mangrove swamps so that the natural coastal habitats can no longer protect against storm surges. As a result, erosion and wave energy are increased, and deforestation can cause a higher risk of landslides or drought.)

Table 16.9 continued

- Use of livelihood practices that are inadequate for densely built areas/housing, putting neighbours or whole settlements at risk (e.g. through dangerous production processes or cooking on an open fire).
- Lack of trees to purify air and stabilise soil, resulting in increased storm water runoff and erosion.
- Lack of open space to absorb storm water (and provide wildlife habitats).
- Lack of infrastructure, combined with inadequate use of existing infrastructure can create new hazards (e.g. fire through illegal electrical connections).
- A lack of infrastructure or unsuitable use of existing infrastructure can easily transform small hazards into disasters. (For instance, inadequate waste or water disposal can intensify flooding as a result of blocked water gutters and river flows; or can result in landslides because of uncontrolled water dispersal on unstable soil. Fire hazards can be created as secondary hazards after earthquakes, for instance, as a result of insecure electrical connections.)

Table 16.10 Constantly changing vulnerabilities and hazards caused by the built environment and related urban planning practices

Aspects or activities that can lead to constantly changing vulnerabilities and hazards

- Constantly changing extension of settlements, sometimes on inadequate land, and increased closeness to hazardous areas/elements (see also Table 16.11).
- Constantly changing local conditions of the built environment (e.g. through change of layout, landscape and density of settlements due to urbanisation processes, combined with the impacts of disasters and climate change – see foregoing tables).

Table 16.11 Reduced coping capacity due to non-adequate disaster (risk) management systems caused by the built environment and related urban planning practices

Aspects or activities which can lead to reduced coping capacity due to non-adequate disaster (risk) management systems

- Lack of emergency infrastructure and related back-up systems (e.g. emergency fire access and evacuation roads, adequate width of highways).
- Out-of-date and incompatible (paper) maps, as well as lack of maps/information on informal settlement, making effective disaster response impossible.
- Exclusion of the urban poor and/or other vulnerable groups (e.g. women) living in marginalised settlements from decision-making processes (resulting, for instance, in those groups being unwilling to use emergency shelters).
- Urban growth leading to disaster management agencies not having the capacity to provide basic supplies and assistance, especially for marginal settlements.
- Urban growth leading to cities growing together and merging without effectively integrating their disaster agencies; this results in confusion and in inability to coordinate disaster response and disaster risk management.

Table 16.12 Reduced coping capacity due to inadequate urban management/governance systems caused by the built environment and related urban planning practices

Aspects or activities that can lead to reduced coping capacity because of inadequate urban management/governance systems

- Use of imported and/or colonial building and planning regulations that do not consider local factors (e.g. local hazards).
- Use of inadequate enforcement schemes resulting in non-compliance of building and planning regulations.
- Use of 'best local practices' for the design and construction of infrastructure which ignore relevant considerations for hazard-resistance (for instance, as a result of local traditions, lack of knowledge/existence of building and planning codes or poor enforcement schemes).
- Absent or poor certification and licensing of planning professionals (who would be responsible for applying, enforcing or inspecting codes) because of, for instance, disciplinary traditions and corruption.
- Planning authorities' lack of political power to control the construction sector (e.g. corruption leading to the use of substandard bricks or other building materials by contractors, and inability to force developers and property holders to plan in secure areas and/or invest in security features).
- Out-of-date and incompatible (paper) maps, as well as lack of maps/information about informal settlements, resulting in poor planning capacities.
- Frequent non-compliance of planning ministries with regulations prescribing the use of environmental impact assessments (EIA) for new urban developments.
- Insufficient capacity of urban authorities and the private sector to supply adequate housing or basic infrastructure at the same speed as urbanisation processes (especially for poor and marginal settlements).
- Urban growth leading to cities growing together and merging without effectively integrating their planning agencies (resulting in confusion and inability to coordinate urban planning efforts and housing provision).
- Exclusion of the poor and other vulnerable groups (e.g. women) living in marginalised settlements from decision-making processes (resulting, for instance, in those groups being unwilling to invest in improved housing and a generally apathetic civil society). Note that 'slum' dwellers often see national and municipal governments as unhelpful, and even a hindrance, to their efforts to improve their situation. In fact, the actions taken by planning authorities and the information obtained by them with respect to the development and legalisation of planned settlements are often viewed as contradictory and unreliable.
- Centralised planning institutions that do not generally have institutional interlinkages with disaster agencies.
- Decentralisation of planning functions without decentralised technical and financial resources (also due to structural adjustment processes), resulting in erosion of living standards (e.g. through poor maintenance of rental property and funds being unavailable for housing and related risk reduction).
- Lack of communication/cooperation between national and municipal planning authorities.
- Few adequate mechanisms to finance/access safe land and housing for the poor.
- Unequal distribution of support by housing/planning organisations for incremental housing and infrastructure in 'slums' (this is related to individual 'slum' dwellers' relationships with planning authorities, corruption, and communities' organisation level).
- Politicisation of building and planning processes, as, for instance, illustrated by global bidding process between cities to attract investment, resulting in planning authorities being forced to lower environmental regulations (and security of workers).
- Corruption at all levels, unnecessary bureaucracy and political rivalry within and between different sectors and ministries.

Table 16.13 Reduced coping capacity of 'slum' dwellers caused by the built environment and related urban planning practices. In comparison with rural areas, this is related to reduced solidarity and reciprocity, as well as a lack of resources and knowledge on adequate coping

Aspects or activities that can lead to reduced coping capacity of 'slum' dwellers to manage risk and disasters

- Increased ease of mobility caused by urbanisation, which creates loose socio-economic community networks that enable dwellers to 'default' on obligations to relatives and neighbours (e.g. migrants lose traditional rural networks of family and neighbours that they could rely on during and after disasters).
- Vulnerable habitat, combined with high risk of suffering hazards, cause 'slum' dwellers within a settlement to frequently experience simultaneous and persistent impacts on their living conditions, a situation that negatively affects solidarity and reciprocity between neighbours.
- Loss of trust in hierarchical structures established by housing/planning agencies (as well as disaster risk management agencies), related to corruption, co-optation and political factionalism, which negatively affects solidarity and reciprocity (e.g. unequal assistance depending on individual's relation to housing/planning authorities).
- Concentration of highly defenceless population groups in urban areas (for example, weakened by HIV/AIDS or other diseases, conflict or malnutrition) results in reduced community cohesion and individual coping (e.g. with HIV/AIDS, many of the able-bodied, adult workforce who would normally engage in disaster coping activities are too weak from the disease. After these people die, households are composed of the elderly and very young, who often lack labour capacity or knowledge.)
- Urban lifestyle results in inhabitants being unwilling to get engaged in community organisation (e.g. because of (a) frequent changes in living place within the city that result in a lack of ties to the respective place; or (b) time restrictions, for instance, due to long commuting distances between home and work, or having to work at several different jobs, combined with unsocial working hours).
- 'Slum' dwellers have very few assets that can be sold to help themselves or others (e.g. because of limited income and small plot sizes that do not allow farming or the keeping of livestock).
- Rapidly changing living environments (due to urbanisation, disasters, climate change, etc.) which negatively affect people's coping knowledge and ability.
- Lack of knowledge of urban 'slum' dwellers on adequate coping due to rural–urban migration or frequent changes of dwelling within the city.

Reality versus current planning practices

The interlinkages presented in the last section indicate the powerful potential of the built environment and related planning practices for reducing (or increasing) risk and hence disaster occurrence. However, the comparison between reality and current planning practices shows that these interlinkages have not been effectively confronted by planners or by disaster risk management professionals. Furthermore, it is possible to identify an unfruitful gap – and even tension – between the related working fields, which finds expression in the respective:

- literature;
- stakeholders and institutional structures;
- discourses of experts and practitioners;
- working priorities, concepts, terminology and tools; and
- sector-specific programmes.

These aspects are described briefly in the following section.

The gap between urban planning[9] and disaster risk management

Technical literature

Literature analysis shows that only a small amount of systematic research has been carried out on the linkages between disasters and the built environment (and related planning practices). As a result, on the one hand there is a large amount of literature emerging from the planning field that deals with purely construction-related issues in the post-disaster scenario of mainly large-scale disasters. Only very few publications are based on a more proactive rather than reactive attitude that also include non-structural aspects and/or consider small-scale everyday disasters. An exception to this are publications on cities and general development issues which have an ecological and health-centred approach. However, these take account of, but do not specifically focus on, broader disaster risk reduction measures.

On the other hand, the analysis of the literature emerging from the disaster risk management field shows that general disaster studies tend to focus on the hazards themselves and hence mostly address related scientific aspects and solutions (e.g. high-tech prediction systems). However, there are also more socially oriented disaster studies that mainly look at (social) causes of vulnerability. In this respect, since the early 1990s a growing literature has emerged in Latin America and the Caribbean, Asia and Africa, born of disaster reduction research and applications carried out by developing country researchers and institutions. This literature forms the basis of many of the contemporary approaches to disaster risk management now being discussed and advocated at the international level (UNDP 2004). Nevertheless, most authors give secondary importance to the built environment and its related planning practices. In fact, more socially oriented disaster studies seem commonly to neglect planning (including social housing and infrastructure development) as a vitally important risk reduction measure, as it is perceived as a purely physical measure that only deals with the symptoms of the problem rather than the cause. Only some very recent publications fully recognise urban disasters and the importance of adequate housing and planning practices for sustainable risk reduction (e.g. UNDP 2004).

Stakeholders and institutional structures

To begin with, compared with other development sectors there are only a limited number of specialised networks, organisations and departments working on either settlement development planning or disaster risk management in developing countries. The reason for this is their marginal status at the global, national and municipal levels. Furthermore, cooperation between the few existing sector-specific stakeholders is mostly non-existent. The gap between them is expressed and further aggravated by: (a) their separate institutional and inter-institutional structures; (b) the lack of adequate channels to optimally support and coordinate their contribution to risk reduction and risk financing;[10] and (c) the separate budget lines for development and emergency relief (with the latter still being the main funding source for disaster risk management). This applies to both international donor organisations and national governmental and non-governmental implementing organisations.

In addition, at the national level in developing countries, one (by-)product of the promotion of disaster risk management on the part of donor organisations is the change to the implementing organisations' internal structures: new and separate structures for disaster risk management are often added on, without, however, being adequately integrated and/or consolidated. Increased and sustainable integration is therefore seldom achieved. Furthermore, at the household level, low-income households often perceive national and municipal planning authorities as being unhelpful, and even a hindrance, to their and other organisations' risk reduction efforts.

Discourses of experts and practitioners

The limited view of many planners as regards the correlation between disasters and the built environment was shown in Figure 16.1. This limited perception of (large-scale) disasters being the cause and the destruction of the built environment being the effect is often combined with the erroneous assumption that pro-poor urban development automatically reduces risk.

Consequently, planners do not generally perceive disaster risk management as being part of their sphere of activity.

Disaster risk management professionals, on their part, often share the perception presented in Figure 16.1 and consequently believe that settlement development planning has no real relevance to sustainable disaster risk management. This view is also related to their understanding that urban planning is a purely structural and formal tool – related to building regulations and conventional land-use zoning – which is incapable of tackling the problems of the urban poor whose lives are most at risk (Figure 16.2). Furthermore, the planning/construction sector is perceived as one of the most difficult development sectors with which to work, because, it is said, knowledgeable and experienced experts are rare.[11]

Working priorities, concepts, terminology and tools

The perceptions of planners and disaster risk management professionals, just described, are related to their different professional backgrounds which – due to the respective theoretical and practical training – influence the use of distinct working priorities, concepts, terminology and tools. Other approaches are met with criticism. The research indicates that together with a lack of coordination between different implementing stakeholders and the competition on the ground, this can result in the duplication of small-scale efforts (e.g. research efforts into hazard-proof construction) and higher investment costs, as well as the mutual incompatibility of their respective programme measures (e.g. risk reduction training, the plans and maps developed, and the hazard-proof construction standards promoted). As regards the latter, almost every organisation imposes different methods and approaches within their programmes (e.g. for training or the elaboration of plans and maps), thus hampering related local developments.

Sector-specific programmes

The incompatibilities between the different professional disciplines and related institutional and organisational structures impede the establishment of more integrated programmes that are needed to properly tackle urban risk. In fact, on the one hand, internationally promoted programmes in the field of disaster risk management do not seem to actively integrate planning-related issues. On the other hand, development agencies or departments, whose focus is urban settlement planning, seem mainly to overlook possible disaster occurrence in their programmes.

However, at the national, municipal and household levels, the occurrence of disasters and the resulting distress can – at least, temporarily – push forward an integration process. This was the case in El Salvador after Hurricane Mitch in 1998 and the 2001 earthquakes. Especially since 2001, relief, development and housing/planning organisations initiated a shift to include disaster risk reduction and related planning measures in their fields of action. However, because disaster risk management was promoted by most international agencies as a new and autonomous field of activity, mainly needing to be integrated into programme implementation, the actual integration within housing/planning organisations was for the most part limited to the adoption of new pilot programmes or specific programme components for disaster risk management.[12] Thus, irrespective of whether the organisations opted to ignore increasing disaster risk or to carry out direct disaster risk management work, they failed to consider the basic strategy of responding indirectly (i.e. through their core work), thus missing the opportunity to sustainably reduce risk. In addition, the following problems occurred to some extent:

- As many organisations and their staff were not well suited to undertaking such disaster risk management work, ineffective work (and even undesirable programme outcomes) resulted.
- Taking on direct disaster risk management work caused organisations' core work to suffer where they did not have sufficient human and organisational capacity to perform both tasks.
- Even if the direct work on disaster risk management was carried out effectively, there was an unproductive increase in competition with other organisations as well as duplication of effort. This was partly because most of the additional knowledge and institutional capacities required were built up independently and internally by each organisation, rather than through the creation of cooperative partnerships.
- Once the new programmes or programme components ended, the work in disaster risk management could not be continued, as it was usually not linked to the organisations' core work and not backed up by adequate operational, organisational, institutional and legal frameworks.

Research at the household level revealed further problems of partly integrated programmes being implemented by housing/planning organisations in high-risk areas. In fact, a gap was encountered between what households need and undertake to deal with disasters and risk, and how organisations support them, creating a barrier for effective disaster risk management. At the household level more than 100 coping strategies could be identified, with households spending on average 9.2 per cent of their income on reducing disaster risk and preparing for the following winter. However, while these household strategies to cope with disasters and risk include risk reduction, self-insurance and recovery mechanisms, the analysed housing/planning organisations looked mainly at how to reduce physical risk.[13] For instance, risk and loss financing is usually not integrated into housing finance mechanisms (i.e. government and non-government subsidies, microcredits and family savings, mutual or self-help). In addition, the risk reduction measures implemented were often unsustainable, as the organisations seldom analysed the key variables and causal loops underlying the complex system of risk and disaster occurrence in the programme areas, nor did they take into consideration the local risk reduction strategies that already existed.[14] Hence, after project implementation, the programme beneficiaries continued to cope – as before – without having obtained better structures for implementing and/or financing their own efforts. It has moreover emerged that some programme measures somewhat hinder future coping ability. For instance, families who wish to obtain loans for further risk reduction or general housing improvements are often not able to use their project houses as collateral, as assisted housing cannot become bank property in the event of a default in payment. Programme beneficiaries are therefore unable to use their assets effectively to reduce the risk they face. Another identified barrier for effective disaster risk management at household level is the lost trust of 'slum' dwellers in both community solidarity[15] and hierarchical structures, as well as the fear of being hoodwinked by the authorities.

Despite the described situation, the research at household level reveals that the organisational structures and mechanisms for social housing provision and financing offer a potentially powerful platform for tackling disasters and risk.

Root causes of the identified gap

The current separation between urban planning and disaster risk management, presented in the last section, does not match up with the identified reality (see section before last). The reasons can be found and are based in the roots and the subsequent historical development of the respective fields of work that is briefly described in the following.[16]

Urban planning theory and practice

Originally, one of the main functions of the city was considered to be to provide defence, not against natural hazards but against human threats from the 'outside', such as wars and armed conflicts. In this context, Meurman (1947) coined the term 'protective city planning' for fire and air protection, suggesting that vulnerable facilities should be 'deconcentrated' and isolated from the rest of the city.[17] Since the architectural Modern Movement, more inner-city (man-made) threats, such as assaults and accidents, have been factored into the vulnerability equation, with a move towards greater protection of cities through physical means and electronic surveillance.[18] In this regard, the term 'defensible space' was created in the 1970s by Newman (1972). In parallel, 'nature ecology' and 'urban ecology' studies gave consideration to planning that ensured compatibility between urban planning and the natural environment. However, the focus there is mainly on the conservation of the environment and climatic design features (i.e. not on aspects relating to natural hazards). More recently, there have also been some discourses on integrated and preventive urban planning, based on consideration of climate change and related hazards. Concrete achievements, however, are still an exception.

Planning schemes promoted by international agencies

The specific history of planning theory and practice promoted by international agencies provides further important background regarding the underlying reasons of the identified gap. In the 1960s and 1970s cooperating governments in developing countries received financial support to build (conventional) housing for the poor on a mass scale. As most of these efforts were declared unsuccessful, at the beginning of the 1970s donors started to support site and service programmes. From 1972, they also assisted squatter upgrading, and in the early 1980s the development of housing-finance institutions (World Bank 1993). In parallel, urban community-development workers have championed participatory methodologies at the settlement level since the 1960s. In line with this, and because of the failure of conventional and traditional urban planning and the lack of adequate responsiveness by planners to the fast-changing needs of developing cities, Otto Koenigsberger (1964) introduced the concept of 'Action Planning' (i.e. community-based schemes supported by government agencies). This approach was subsequently further developed by Hamdi into Community Action Planning or so-called MicroPlanning (Hamdi and Goethert 1997).

During the 1970s, planners started to involve themselves in discussions on disaster management as interest was growing in the design and implementation of ways to mitigate disaster losses through physical and structural measures (for example, through building levees and flood defences, or increasing the resistance of structures) (UNDP 2004). However, with the developing concept of disaster risk management, planners' role again diminished during the 1990s (see also below). Together with the shift from 'delivering' to 'enabling' housing and settlements since the 1970s, it became even more difficult to promote and implement disaster risk management measures. Indeed, the 'enabling' approach promoted can be viewed as an obstacle to integrated risk reduction and urban settlement planning.

At the national and city level, structural adjustment programmes, which were introduced by the World Bank and the International Monetary Fund (IMF) in the 1980s and 1990s, also strongly influenced the current challenges. They not only increased vulnerability but also marginalised urban planning by decreasing the influence and political role of planners and national planning units. Based on the Millennium Development Goals and the outcomes of the World Summit on Sustainable Development, held in Johannesburg, South Africa, in 2002, international donors now promote the private sector as a leading provider of urban infrastructure and services, including drinking water and sanitation. Unfortunately, this means that programmes with a focus on settlement development planning tend again to lean towards the more structural aspects, thus obstructing more holistic planning, which would include disaster risk management. However, some recent developments have given reason to hope for better integration (e.g. trends such as the 'strengthening of local governments', 'decentralisation', and so-called Sector-Wide Approaches).

Disaster risk management

Regarding the history of disaster (risk) management, it is important to point out that this is still a relatively new area of knowledge which is developing slowly and undergoing a multifaceted process of institutionalisation. Traditionally, discussions about disasters have taken place in the emergency relief arena. Until the 1970s the dominant view was that 'natural' disasters are synonymous with natural events/hazards such as earthquakes, flooding, etc. In other words, a natural hazard was, ipso facto, seen as a disaster. The magnitude of a disaster was hence considered to be a function of the magnitude of the hazard. Consequently, the emphasis of national governments and the international community was on pure disaster management (i.e. responding to the events and, in the best-case scenario, preparing in advance for disasters in order to improve existing response capacities). As mentioned above, from the 1970s onwards planners began to get involved in disaster discussions, focusing on the fact that the same natural hazard can have varying impacts on the built environment. A general trend evolved to associate disasters more with their physical impact than with their natural trigger, promoting conventional and traditional engineering or planning practices as an important mean of mitigating disasters. However, in many countries efforts to reduce risk by these means have been minimal (UNDP 2004).

Beginning quietly in the 1970s, but with an increased emphasis during the 1980s and 1990s, social sciences researchers triggered a shift in thinking, by pointing out that the impact of a natural hazard mostly depends on the capacity of people to cope (i.e. the ability to absorb the impact and quickly recover from loss or damage). With the advent of the term 'disaster risk management' (replacing the term disaster management), the focus of attention moved to social and economic vulnerability. This shift was further reinforced by the mounting evidence that natural hazards have widely varying impacts on different countries and different social groups within these countries (UNDP 2004). Spurred on by the International Decade for Natural Disaster Reduction (IDNDR), between 1990 and 1999, as well as by the occurrence of a number of highly destructive large-scale disasters at the end of the 1990s, many pilot programmes in the field of disaster risk management emerged in developing countries, with international agencies providing increased resources. However, the post-disaster context remained the focus of intervention. Growing experience gained within the mentioned pilot programmes, combined with ongoing development of the disaster risk management concept, meant that a common understanding gradually evolved (UNDP 2004). The causal factors of disasters are now understood to be directly linked to development processes, which generate different levels of vulnerability. The UN International Strategy for Disaster Reduction (ISDR), established in 2000, helped raise the profile of related discussions. In fact, it promoted the idea that reducing disaster risk requires a long-term engagement in development processes and, hence, an increased engagement of international organisations in this field.[19]

More recently there have been discourses to the effect that development processes are not only generating different patterns of vulnerability but also altering patterns of hazard – an argument that is causing increasing concern, especially as evidence mounts regarding the potential impact of global climate change (UNDP 2004).

Today, disaster risk management can be considered a constantly enhancing and altering paradigm that integrates the trends and perceptions mentioned above. However, while during the 1990s the – formerly promoted – purely structural planning measures were less and less seen as a solution and thus 'deleted' from the disaster risk management agenda, hardly any alternative planning strategies were developed to replace them.

A way forward: how to better match practice to reality?

The two previous sections have shown (a) the interrelation between disasters and the built environment (and related planning practices); and (b) that this interrelation is not given enough attention by international and national stakeholders working in either settlement development planning or disaster risk management. This situation can contribute to increasing risk and disasters in two ways: first, through the implementation of programmes that focus only on planning *or* disaster risk management; and second, through the lack of initiatives that integrate the two fields.

While, in the meantime, implementing (and donor) organisations working in settlement development planning increasingly demand guidance on how to sustainably integrate disaster risk management within their core work, no adequate sector-specific and praxis-oriented tools are available. This is a paradox, as at a global level there is a fast-increasing number of tools for assessing progress in disaster risk management, most of them developed as a result of top-down processes created by international (and national) organisations. To make matters worse, compared with other cross-cutting issues, such as gender or HIV/AIDS, the idea of mainstreaming disaster risk management is widely underdeveloped and/or misunderstood. As a result, existing tools and ongoing discussions confuse, and hence do not differentiate between, the terms and concepts of 'mainstreaming' and 'integrating', and thus, are often very limited in their scope.[20]

To counteract the situation just described, an 'Operational Framework for Integrating Risk Reduction for Aid Organisations Working in Human Settlement Development' was developed and first published in 2006 (Wamsler 2006d).[21] The framework aims to support aid organisations with concrete tools and guidance to:

- evaluate the relevance of integrating disaster risk management within their organisation;
- identify and prioritise the different possible strategies for integrating disaster risk management;
- formulate activities to implement the selected strategies;
- evaluate possibilities for financing these activities; and
- define an implementation plan.

Based on the ongoing research, on workshops held in Central America for operational staff and programme managers, as well as on the lessons learned from organisations currently using the tool in practice, the Operational Framework was further developed during 2006–07. The result is the conceptual and strategic integration model presented below.[22] It is based on seven complementary strategies that counter the currently incomplete approaches to integrating disaster risk management. These strategies are presented in the following sections; they are summarised in Figure 16.4 and Table 16.14, and are partly illustrated in Box 16.1.

Integration strategies at the household level

The first three strategies present possible ways of integrating disaster risk management, that is, risk reduction and risk financing, into programme implementation at the household level (Figure 16.4, left side).

Strategy I: Direct stand-alone disaster risk management

This is the implementation of specific programmes for disaster risk management that are explicitly and directly aimed at financing or reducing disaster risk. These stand-alone programmes are distinct, and they are implemented separately from other existing work carried out by the implementing housing/planning organisation. Examples would be programmes aiming at (a) establishing early-warning systems or organisational structures for risk reduction (e.g. specialised disaster risk management committees); (b) constructing physical disaster mitigation (e.g. embankments to reduce flooding); or (c) offering independent disaster insurance (i.e. insurance policies not included in housing financing schemes being offered to the poor).

Table 16.14 Overview of the proposed complementary strategies for integrating disaster risk management (DRM) into the efforts of (aid) organisations working in settlement development planning and programming

Strategies		Description	Main question to be answered by an organisation (working in settlement development planning)
I	Direct stand-alone DRM	DRM programming	What dedicated programmes can be implemented separately and additionally from the organisation's core work in order to specifically address risk and disaster occurrence?
II	Direct integrated DRM	Adding DRM programming elements to core activities	What dedicated programme measures can be added to the organisation's core work in order to specifically address risk and disaster occurrence within the programme areas?
III	Programmatic mainstreaming of DRM	DRM mainstreaming (in programme implementation)	What can be done within the core work of the organisation in order to (a) reduce risk and (b) increase the coping capacities of the programme beneficiaries? (Or at least to ensure to (a) not increase risk and (b) not reduce coping capacities.)
IV	Organisational mainstreaming of DRM	Institutionalisation of DRM mainstreaming (and programming)	What can be done to sustain and back up DRM mainstreaming (and programming)?
V	Internal mainstreaming of DRM	DRM to reduce the organisation's own risks	What measures can be taken so that the organisation, i.e. its offices and staff, become more disaster resilient?
VI	Synergy creation for DRM	Coordination and complementation for improved DRM integration	How can the DRM mainstreaming and programming activities of the organisation be coordinated and complemented with the work of other implementing (aid) organisations?
VII	Educational mainstreaming of DRM	Shift towards non-conventional settlement development planning to integrate DRM in the philosophy that drives urban planning	What has to be done so that universities and other training institutions facilitate the integration of DRM into urban actors' spheres of activity?

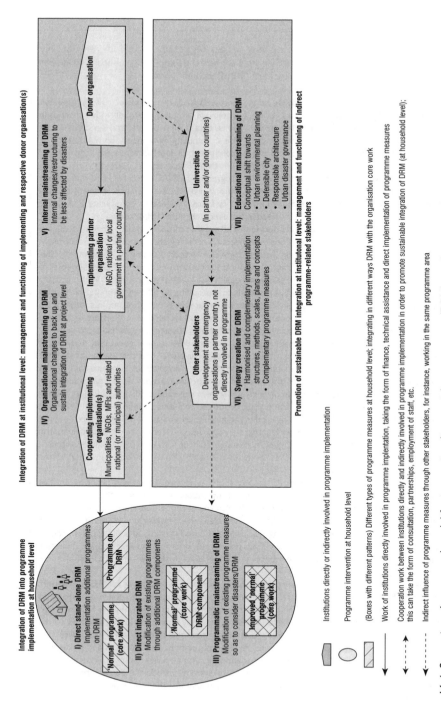

Integration of DRM into programme implementation at household level

I) Direct stand-alone DRM
Implementation additional programmes on DRM

'Normal' programme (core work) | Programme on DRM

II) Direct integrated DRM
Modification of existing programmes through additional DRM components

'Normal' programme (core work) | DRM component

III) Programmatic mainstreaming of DRM
Modification of existing programme measures so as to consider disasters/DRM

Improved 'normal' programme (core work)

Integration of DRM at institutional level: management and functioning of implementing and respective donor organisation(s)

IV) Organisational mainstreaming of DRM
Organisational changes to back up and sustain integration of DRM at project level

Cooperating implementing organisation(s)
Municipalities, NGOs, MFIs and related national (or municipal) authorities

Implementing partner organisation
NGO, national or local government in partner country

Donor organisation

V) Internal mainstreaming of DRM
Internal changes/restructuring to be less affected by disasters

Promotion of sustainable DRM integration at institutonal level: management and functioning of indirect programme-related stakeholders

Other stakeholders
Development and emergency organisations in partner country, not directly involved in programme

VI) Synergy creation for DRM
• Harmonised and complementary implementation structures, methods, scales, plans and concepts
• Complementary programme measures

Universities
(In partner and/or donor countries)

VII) Educational mainstreaming of DRM
Conceptual shift towards
• Urban environmental planning
• Defensible city
• Responsible architecture
• Urban disaster governance

Institutions directly or indirectly involved in programme implementation

Programme intervention at household level

(Boxes with different patterns) Different types of programme measures at household level; integrating in different ways DRM with the organisation core work

Work of institutions directly involved in programme implantation, taking the form of finance, technical assistance and direct implementation of programme measures

Cooperation work between institutions directly and indirectly involved in programme implementation in order to promote sustainable integration of DRM (at household level); this can take the form of consultation, partnerships, employment of staff, etc.

Indirect influence of programme measures through other stakeholders, for instance, working in the same programme area

Figure 16.4 Strategic and conceptual model for integrating disaster risk management (DRM) into human settlement development projects, at both household and institutional levels

Box 16.1 Hypothetical example of how an aid organisation – a Mexican housing/ planning organisation called UNAGI – was triggered to apply the different strategies for integrating disaster risk management (DRM) into its work.

After a recent disaster, and in response to the increased funding for disaster risk management being offered by international donors, UNAGI employs a new staff member with expertise in risk reduction and financing and designs and implements a pilot programme on disaster risk management. The pilot programme aims to raise community awareness about disaster risk through the distribution of leaflets and the establishment of local disaster risk management committees. Thus, UNAGI becomes engaged in the *stand-alone direct DRM strategy*.

With the experience gained from the pilot programme, UNAGI then starts to include risk reduction and financing activities in its ongoing housing projects. For instance, it begins to raise risk awareness and promotes community emergency funds alongside its community training for self-help housing. Thus, it becomes involved in the *direct integrated DRM strategy*.

One year later, UNAGI's managers decide that all programmes should take greater account of disasters and seek to maximise their positive effects on reducing and financing risk. Accordingly, UNAGI carries out research analysing the links between its social housing activities and disaster risk. In one project area, it finds that basing housing credits on income capacity makes it impossible for the people most vulnerable to disasters to qualify for UNAGI programmes. Without doing any direct risk reduction work, UNAGI responds to this finding by offering them partial housing subsidies and smaller credits for physical mitigation measures in existing houses. In another area, community research provides evidence that beneficiaries are vulnerable to disasters because of their dependency on informal vegetable trading and that past housing projects had increased their socio-economic vulnerabilities by resettling them far from their income-generating activities. It is also discovered that these housing projects used expensive roof tiles that were not durable. Acting on these findings, UNAGI sets up a local production workshop for concrete roofing tiles to provide a more disaster-resistant and cheaper construction material. At the same time, the workshop allows some households to diversify away from vegetable trading. In addition, in both project areas, advice on disaster-resistant construction techniques is provided, disaster insurance mechanisms are included in the housing credits and neighbourhood and women's associations are established which campaign for greater transparency in government and grassroots participation in urban planning decision-making. Thus they increasingly build up a stake in municipal development planning (e.g. as regards legalisation of land). In this way, UNAGI becomes involved in the *programmatic mainstreaming strategy*.

Over time, UNAGI realises that its various efforts in risk reduction and financing are not sustainable in the long term because they are not institutionalised and/or anchored within the organisation's general management and project planning cycle. It thus starts to engage in the *organisational mainstreaming strategy*. As an initial step, the organisation revises its policy to formalise its commitment to integrating risk reduction and financing, and develops a financial strategy to sustain this integration. In addition, risk assessments and capacity analyses (including the analysis of local coping strategies) become routine tasks in the planning phase of all social housing programmes.

Several months later, there is an earthquake in Mexico. Unexpectedly, UNAGI is affected: its head office is damaged, four staff members are severely injured and there are problems communicating with field offices. This forces the organisation to engage in the final strategy: *internal mainstreaming*. A team is formed to predict the likely impacts of future disasters on the organisation's finances and human resources, analysing potential direct and indirect losses (e.g. costs related to damaged buildings, vehicles, reduced reputation, staff absences and sick leave). Based on this work, UNAGI acquires an organisational insurance policy and improves its working structure by installing an enhanced communications system, introducing better processes for information sharing, and revising its workplace policy. In addition, the head office is retrofitted to become more disaster-resistant.

Strategy II: Direct integrated disaster risk management

This is the implementation of specific risk reduction and/or risk financing activities alongside, and as part of, other sector-specific programme work. The only difference from Strategy I is that this work is carried out in conjunction with, and linked to, other programme components. An example would be the establishment of a local disaster risk management committee or the offer of capacity building for socio-economic risk reduction within the framework of a self-help housing project. Another example could be the implementation of disaster awareness campaigns and simulations alongside a settlement upgrading project.

Strategy III: Programmatic mainstreaming

This is the modification of sector-specific programming in such a way as to reduce the likelihood of any programme measures actually increasing risk, and to maximise the programme's potential to actually reduce and/ or finance risk. Hence, the objective of programmatic mainstreaming is to ensure that the ongoing core work is relevant to the challenges presented by natural hazards and disasters. In contrast to the two strategies described above, the programme's main objective is not risk reduction or risk financing as such. The modifications can be of a physical/structural, environmental, institutional and organisational nature. An example of this strategy could be a settlement upgrading programme that adjusts its loan system to the specific needs of vulnerable households at risk (e.g. offering smaller credits with different conditions and integrated risk insurance, considering beneficiaries' limited capacity for payment). Programmatic mainstreaming can also result in the elaboration of new activities within the organisation's working field that are needed to take existing risk into account. An example of this would be a social housing organisation becoming engaged in land-use planning and urban governance issues for risk reduction or the offer of risk and loss financing schemes through their existing housing financing mechanisms.

Integration strategies at institutional levels of implementing organisations

Compared with Strategies I–III, presented in the last section, the following two strategies (i.e. IV and V) do not refer to the integration process at household level, but aim to counter the challenges identified at institutional levels (Figure 16.4, right side, middle). This section refers to both the donors' national counterparts and their implementing partners. The latter are often governmental authorities (e.g. municipalities and decentralised ministries of education and health), together with microfinancing institutions (MFIs).

Strategy IV has direct relevance for programme implementation at the household level; Strategy V is only indirectly related.

Strategy IV: Organisational mainstreaming

This means modification of the organisational management, policy, working structures and tools for programme implementation, in order to back up and sustain (direct and/or indirect) disaster risk management at project level, and to further institutionalise it. In fact, if integrating disaster risk management into programme work is to become a standard part of what an organisation does, then organisational systems and procedures need to be adjusted. The objective is to ensure that the (implementing) bodies are organised, managed and structured to guarantee that risk reduction and risk financing are sustainably integrated within their core programme work. This includes, for instance, the adaptation of institutional objectives as well as programme planning tools already used, such as logical and results-based frameworks or vulnerability and capacity analyses.[23] A summary of issues to be taken into account in the programming, identification and appraisal stages of construction projects is presented by Benson and Twigg (2007: 147–9) and Rossetto (2006: 9–14). In addition, organisational mainstreaming also means the adoption of new tools needed for adequately integrating disaster risk management into settlement development programming. Examples are risk mapping or causal loop diagrams for analysing the key variables – and their causal relations – underlying the complex system of risk and disaster occurrence.

In the case of governmental organisations, for instance, housing and planning ministries and municipalities, organisational mainstreaming importantly includes the following activities: (a) revision or creation of national or municipal legislation and policies; (b) the formal standardisation of methods and approaches to elaborating maps and plans for urban planning and disaster risk management; and (c) the creation of improved institutional structures between the national and municipal levels and the respective disaster risk management bodies.

Strategy V: Internal mainstreaming

This means modification of the organisation's functioning and internal policies so that it can reduce/finance its own risk to impacts created by disasters. The focus is on the occurrence of disasters and their effect on the organisation itself, including staff, head office and field offices. The objective is to ensure that the organisation can continue to operate effectively in the event of a disaster. In practice, internal mainstreaming has two elements: (a) direct risk reduction and risk financing activities both for staff and for the physical aspects of the organisation's offices, for instance, the establishment of emergency plans and retrofitting; and (b) modification of how the organisation is managed internally, for example, in terms of personnel planning and budgeting.

Integration strategies at institutional levels of donor organisations

Donor organisations that wish to promote the integration of disaster risk management through and within their partner organisations need, themselves, to be committed to disaster risk management and its integration. This is a precondition if they wish to prove effective in supporting their partners in doing the same. To that effect, not only the national partner and their cooperating implementing organisations, but also donor organisations, would have to integrate risk reduction and risk financing within their work. In sum, the organisational and internal mainstreaming Strategies (IV and V) also apply to international donor organisations (Figure 16.4, right side, top).

One important organisational change within a donor organisation, to be effected as part of the organisational mainstreaming process, would be the allocation of (primarily) development resources to push forward the integration of risk reduction and risk financing into urban planning and housing. Importantly, these resources would need to be channelled in such a way that they do not promote integration only in programme implementation at the household level. Indeed, it is equally crucial to promote integration at the institutional levels of implementing governmental and non-governmental organisations, which would affect related national and municipal legislation, operational instruments and internal structures (without separate ones necessarily being added).

To better illustrate the differences between the presented Strategies I–V in practice, a hypothetical example is presented in Box 16.1. It describes how a Mexican housing/planning organisation was pushed towards applying the different strategies to its work. In contrast to this hypothetical example, (aid) organisations could, and should, take a more proactive approach for the design of an adequate and sustainable integration strategy. In this context, the Operational Framework mentioned above can assist an organisation in 'planning ahead' before disasters strike, by guiding the selection and prioritisation of the appropriate strategies. Once the strategies are selected, the framework provides matrixes for the formulation of related programme measures. These matrixes include: (a) input and process indicators to get the integration process started; (b) input and process indicators in the form of benchmarks (i.e. the operational state that an organisation should seek to achieve); and (c) output indicators. The matrixes are organised into different subsections. Those for Strategies I–III include indicators related to human resources and capacity building; risk identification and community research; and physical, socio-economic, environmental, institutional and organisational programme components. The matrixes for organisational and internal mainstreaming (i.e. Strategies IV and V) include indicators regarding human resources and capacity building; risk identification and staff research; working structure and procedures, policy and strategy; financial management; and external relations.

Key changes for integrated housing and settlement development

For each of Strategies I–V, the Operational Framework offers sector-specific guidance for housing/planning organisations by providing specific reference activities and recommendations. Looking at programme implementation, while it is impossible to provide universal guidance that fits all types of programmes, most housing/planning organisations would probably need to modify their programming and functioning to, first, improve the content and scope of their (direct and indirect) risk reduction measures so as to better reduce potential disaster impacts at the household and institutional levels and, second, integrate adequate (self-)insurance and recovery mechanisms. The latter is crucial to improving the chances of people and institutions 'bouncing back' quickly, and to a reasonable level, after disasters strike. To achieve the described changes in practice, the following four key aspects would have to be considered:

1   Implementation of additional/modified sector-specific measures for supporting disaster risk management during, and also *after*,[24] programme implementation through:
    - the integration of risk and loss financing into the organisation's housing financing mechanisms (i.e. microcredits, government and non-government subsidies and family savings); and
    - the expansion of existing housing financing mechanisms to support the financing of risk reduction, (self-)insurance and recovery measures for the urban poor.
2   Improvement of programmes' sustainability by taking careful consideration of the perceptions, capacities and needs of the beneficiaries as regards risk reduction and risk financing. It is essential within programme implementation to consider encouraging and scaling up selected coping strategies, as well as offering better alternatives, where needed. Coping strategies can be divided into: (a) those that can increase the capacity of low-income communities to manage urban disasters and disaster risk in both the short and long terms; (b) those that increase capacities in the short term but decrease them in the long term; (c) those that decrease capacities in both the short and long terms; and (d) those that decrease capacities in the short term and increase them in the long

term. Naturally, those under (a) should be the main ones considered for support. In this context, careful attention should be given to the cost-effectiveness and sustainability of assistance.[25]

3   Reduction of barriers to coping. This is another important aspect related to people's coping strategies. An example would be giving permission to use project housing as collateral when applying for specific credits (e.g. for risk reduction).

4   Improvement of programmes' sustainability by ameliorating the social relations among the 'slum' dwellers, national and municipal authorities and local-level organisations, as well as within communities themselves. This is crucial because of the lack of trust and the tensions identified between and within the different levels. Measures related to this could be the improvement of communication and decision-making structures for integrated planning and the creation of related community rights and obligations. In fact, enabling the achievement of social cohesion, inclusiveness and open participation in decision-making is crucial to improving communities' coping capacities and to reducing urban vulnerability.

Promotion of sustainable disaster risk management: coordination with other implementing organisations

As mentioned above, the gap between the distinct working fields and related professional groups, together with a lack of coordination with other stakeholders carrying out programmes, can result in competition, the duplication of small-scale efforts, higher investment costs and mutual incompatibility of programme measures. Unsustainable disaster risk management is the outcome. Strategy VI aims to counter this situation.

Strategy VI: Synergy creation for disaster risk management (integration)

This is the promotion of 'harmonised' risk reduction and risk financing into the management and functioning of other implementing organisations, including both relief and development organisations. The idea is to create synergies instead of competition through coordinating and complementing each other's work (Figure 16.4, bottom, middle). Coordination among the work of different organisations could be achieved by: (a) working with unified implementation structures (e.g. municipal committees for local development along with political and operational focal points for programme implementation); (b) the standardisation and unification of methods, scales and contents for the development of specific maps and plans; (c) the standardisation or flexible adjustment of the concept of disaster risk management within the different organisations; and (d) the coordinated inclusion of activities for capacity building and socio-economic development for risk reduction and financing. Complementation and compatibility could be achieved by: (a) working through different municipal/local commissions (e.g. for relief, risk reduction, programme implementation); (b) the development of compatible products and services, such as maps and plans with different contents and scales; and (c) the implementation of additional sector-specific activities (that take risk indirectly into account).

Promotion of sustainable disaster risk management: cooperation with universities and other training institutions

The work on disaster risk management is a field of activity where interaction or cooperation between academia and practice can, and must, complement each other so that sustainable solutions for the urban poor can be developed. This can be by means of partnerships, by consultation, or by employing professional staff. Thus, in parallel to the integration processes described so far, a partly independent process needs to take place to generate a more proactive approach on the part of planners towards disaster risk management. This is crucial so that their work will match up with settlements' current challenges. Hence, the focus of Strategy VII are universities and other training institutions (Figure 16.4, lower right side).

Strategy VII: Educational mainstreaming

This is the development of a conceptual shift in the philosophy that drives urban planning towards non-conventional settlement development planning to allow disaster risk management to be incorporated into urban planners' spheres of activity. In fact, planners require a different knowledge base and radically different skills. This will assist in bringing planners and disaster risk management professionals closer together by helping them to move towards an understanding of the risk that urban dwellers face. The four concepts presented below can help to initiate the required shift. Donor organisations could promote such a shift directly by supporting, for instance, universities or ministries of education as their counterparts. Another more bottom-up approach would be the involvement of universities and training institutions in local programme implementation.

Urban environmental planning

This concept expresses the need for the interconnection between urban planning and broader environmental development aspects, thus incorporating both large-scale and small-scale everyday hazards/disasters. Examples of concrete measures could be: (a) the use of participative and broader environmental impact assessments as well as more adequate performance indicators for selecting and designing integrated planning measures; (b) the integration of legal frameworks and agendas related to urban planning and environment protection; and (c) the adaptation of planning codes based on climatic area-specific characteristics.

Defensible city

This concept expresses the need to integrate protection (against natural hazards and disasters) as a key aspect of integrated urban planning. This strategy includes innovation and the use of structural as well as non-structural planning measures. Examples could be: (a) the construction of firebreaks, flood defences, access and evacuation roads to and from specifically vulnerable areas, escape routes to emergency shelters, protected rooms in basements (for hurricanes) or top floors (for tsunamis); (b) the setting up of back-up facilities (such as transportation systems) when structural/physical measures fail; or (c) the creation of incentives to build in a safe manner (e.g. tax inducements, exchanging rights and insurance schemes).

Responsible architecture

This concept encapsulates the need for planners not only to engage in large-scale structural improvements of the formally built environment but also to directly target informal settlements, thereby combining large-scale structural improvements with structural and non-structural small-scale measures. Examples of the latter are: (a) the exchange of dwellings between low- and high-risk groups; (b) awareness raising and door-to-door advice offered by 'barefoot planners' regarding the design and use of buildings; (c) technical training of informal builders; and (d) the creation of local construction centres. The active use of small-scale measures could enable a better link to be forged with other development professionals as well as with disaster preparedness experts, which might generate further positive outcomes.

Urban disaster governance

This concept contains the idea of the combined domain, where disaster and urban planning are coordinated, mediated and altered through joint governance practices. The domain of urban disaster governance is hence the realm in which the interrelationship between disasters, urban planning and society becomes apparent. To facilitate timely, equitable and strategically coherent decisions in resource mobilisation and supply, it is important to identify those governance tools that will be likely to simultaneously benefit disaster risk management and settlement development planning by, amongst other things:

- fostering equality in participation in decision-making across genders, religious and ethnic groups, and castes and economic classes;
- engaging with the local knowledge of individuals and communities at risk;
- combining such knowledge with scientific information (for instance as regards hazards and disaster-resistant structures); and
- reforming governance practices that might inadvertently contribute to the generation of vulnerabilities. For instance, coordinating disaster risk management networks that are often in unproductive competition with one another.

Conclusions

Increased disaster risk is possibly one of the greatest threats to sustainable urban development that developing countries face today. Paradoxically, the built environment (and related planning practices) are not only affected by disasters; they can also constitute one of its main cause, creating:

- increased vulnerability to natural hazards;
- greater exposure to existing hazards;
- intensified and/or magnified hazards;
- newly created hazards;
- constantly changing vulnerabilities and hazards (thus making them quasi uncontrollable);
- reduced coping capacities of national and municipal institutions; and
- reduced coping capacities of urban low-income households.

Although growing urbanisation and climate change make these negative effects even more alarming, organisations working in settlement development planning have, as yet, not tapped into their full potential to address disaster risk. Even worse, these organisations are partly contributing to increasing disaster risk.

Merely developing and implementing hazard-proof measures is not tantamount to integrating disaster risk management into planning practice. In fact, structural adaptation of this kind needs to be combined with and backed by an integral 'take-up system' – at both the household and institutional levels – that integrates structural and non-structural, large- and small-scale measures. 'Institutional level' refers to the following institutions:

- governmental and non-governmental implementing organisations;
- donor organisations;
- other stakeholders working in programme implementation; and
- related universities and other training institutions.

The Integration Model presented here offers such a 'take-up system', providing a new concept for integrating disaster risk management into the efforts of organisations working in settlement development planning and programming. If implemented into practice, it could help to (a) overcome the constraints that these organisations currently face to get disaster risk reduction translated into their working practices; and (b) enable planners to take on the role of developing secure and sustainable communities.[26] They could thus considerably contribute to the reduction of post-disaster destruction and, hence, the forced evictions of the urban poor whose lives are most at risk.

Notes

1 In the following, the umbrella term 'planners' will be used for all the mentioned professional groups.
2 Disaster risk management includes risk reduction and risk financing. Risk reduction has become a popular term used to bring together those measures to minimise disaster risk throughout a society, to avoid (prevention) or to limit (mitigation and preparedness) the adverse impacts of hazards within the broad context of sustainable development. It is also a component of successful reconstruction. In fact, risk reduction can be implemented and is essential before, during and after disasters. See http://www.unisdr.org/eng/library/lib-terminology-eng%20home.htm. However, to limit the scope of the study, the term as used in this chapter pertains mostly to prevention, mitigation and preparedness measures in a developmental, pre-disaster context. The term risk financing describes measures to transfer or share risk, such as formal and informal disaster (self-)insurance.
3 Risk is defined by UNISDR as: 'The probability of harmful consequences, or expected losses (deaths, injuries, property, livelihoods, economic activity disrupted or environment damaged) resulting from interactions between natural or human-induced hazards and vulnerable conditions.' See http://www.unisdr.org/eng/library/lib-terminology-eng%20home.htm. Risk is expressed by Risk = Hazards × Vulnerability × Lack of coping capacity. Note that in some other existing definitions coping capacity is part of vulnerability.
4 Note that the term 'programme' is used as an umbrella term for programmes, projects, and other type of sector support/assistance. In the following, all terms are used synonymously.
5 Note that from now on, the umbrella term housing/planning organisations will be used for this type of organisation.
6 The research undertaken since 2003 analysed at different levels the interlinkages between disasters and the built environment (and related planning practices) and how it is tackled in practice. In fact, step by step the global, national, municipal and the household level were the focus of the enquiry. El Salvador in Central America, which is one of the most disaster-prone regions in the world (Lavell 1994), was the focus country for the case studies at the national, municipal and household levels. At all levels, the research methods included text review, group discussions, semi-structured interviews, walk-through analyses and observation. As regards the interviews, at the global level, 64 programme and project managers, operational or academic staff from 33 organisations were interviewed; at the national and municipal level around 70 project managers and operational

project staff from 40 organisations; and at the household level 62 households, comprising 331 persons, living in 15 disaster-prone 'slum' communities. Research trips were made, amongst others, to Geneva, Switzerland; Stockholm, Sweden; Washington DC, USA; Rio de Janeiro, Brazil; various locations in the United Kingdom; Manizales Colombia, San Salvador, El Salvador; and Manila in the Philippines.

7   The sources of all tables are Wamsler (2004, 2006a–g, 2007b,c), complemented with additional research outcomes gained during 2007, as well as information obtained from Worldwatch (2007) and UNDP (2004). Note that the different aspects listed separately in the tables are interconnected, as they are partly causes and/or effects of other aspects mentioned. In Wamsler (2006a) system analyses of different aspects, including feedback loops, were carried out for the household level in El Salvador.

8   See note 7.

9   Note that the terms 'urban planning', 'planning', 'settlement development planning' and 'settlement planning' in the text are used synonymously. They mainly refer to social housing, settlement upgrading, new settlement development and urban governance programming.

10   Definitions of the terms risk reduction and risk financing can be found in note 2.

11   This statement is based on Wamsler (2006g: 155). In addition, Wamsler (2004: 21) states that disaster risk management professionals repeatedly referred to the planning/construction sector as a 'bad experience' and/or a 'nightmare' to work with.

12   As described in Wamsler (2006e), partial integration was to some extent also achieved as regards national and municipal legislation, as well as the organisations' operational instruments, and institutional and organisational structures.

13   The socio-economic, environmental and institutional vulnerabilities were seldom considered.

14   This also relates to the fact that the participatory and community-based approaches used, which also utilise capacity analysis, generally relate only to the construction process and not to disaster risk management. Hence, people's coping strategies are not looked at.

15   'Slum' dwellers reported, for instance, on neighbours downhill felling trees or excavating the slopes below their houses, or neighbours uphill building latrines close to the declivity and allowing waste and storm water to flow onto their land.

16   For a more thorough description see Wamsler (2007a). Important reference literature is Tannerfeldt and Ljung (2006), Jenkins et al. (2007), and Mumford (1961/98).

17   Meurman was the first teacher of urban planning at Helsinki Technical University (1936) and the first professor of the discipline (1940).

18   Note that others such as Mumford (1961) offered a more pessimistic perspective of urbanism, referring to the development of cities racked by war, famine and disease.

19   See, for instance, www.unisdr.org/eng/about_isdr/isdr-mission-objectives-eng.htm.

20   'Mainstreaming' is a specific type of integration. Generally, 'mainstreaming' signifies the modification of a specific type of core work (e.g. within a specific type/sector of development assistance) in order to take a new aspect/topic into account and to act indirectly upon it. Thus, the term 'mainstreaming' does not mean to completely change an organisation's core functions and responsibilities, but instead to view them from a different perspective and carry out any necessary alterations, as appropriate. Other types of (disaster risk management) integration are described under Strategies I and II. These integration strategies are commonly

confused with mainstreaming measures, partly resulting in competition and the duplication of efforts of organisations that specialise in different humanitarian and development sectors.

21 Compared with most tools already in existence, it was developed in close collaboration with practitioners with a focus on sector-specific, project-level implementation. Based on growing experiences in the field, it can create over time a bottom-up development that can nurture the development of proper monitoring and evaluation tools for assessing progress in disaster risk management at both the national and international levels. Note that the Operational Framework has been revised and a second version published by Benfield Hazard Research Centre at the end of 2007.

22 From now on called the 'Integration Model'.

23 To date, housing/planning organisations have used capacity analysis; however, this is used only in respect of people's existing capacities for housing financing and construction and not for coping with risk and disaster occurrence.

24 Offering related mechanisms that work or come into effect after programme implementation is crucial, given the incremental development processes in 'slums'.

25 Wamsler (2006c) presents an analysis framework and methodology for viewing local disaster risk. Wamsler (2007b) includes a framework for analysing and supporting local coping strategies (assisting in the selection of adequate programme measures).

26 Note that the development of the Integration Model and related frameworks, concepts and guidelines to stimulate the integration of disaster risk management in sector-specific development programming is not sufficient in itself. In fact, two important key factors for 'translating' technical (policy) instruments such as the Integration Model into practice are (a) related scientific input and (b) political will/commitment. The former refers, for instance, to information on existing hazards, the development of past disaster impacts, as well as knowledge on how to adequately construct disaster resistant structures. As regards the latter, the political commitment of national and municipal authorities, civil society as well as international and national aid organisations for disaster risk management and its integration in settlement development planning/programming is a pre-requirement for implementing the Integration Model. However, the model itself, i.e. the proposed conceptual shift and resulting activities could assist in this regard (see also Wamsler 2007a).

References

Benson, C. and Twigg, J. (2007) *Tools for Mainstreaming Disaster Risk Reduction: Guidance Notes for Development Organisations*, Geneva: ProVention Consortium.

Hamdi, N. and Goethert, R. (1997) *Action Planning for Cities: A Guide to Community Practice*, Chichester: John Wiley & Sons.

Jenkins, P., Smith, H. and Wang, Y. P. (2007) *Planning and Housing in the Rapidly Urbanising World*, London: Routledge.

Koenigsberger, O. (1964) 'Action planning', *Architectural Association Journal*, May, London: Architectural Association.

Lavell, A. (1994) 'Prevention and mitigation of disasters in Central America: vulnerability to disasters at the local level', in Varley, A. (ed.) *Disasters, Development and Environment*, Chichester: John Wiley & Sons, pp. 49–64.

Meurman, O. (1947) *Asemakaava-oppi [City planning]*, Helsinki, Otava.

Mumford, L. (1961/89) *The City in History: Its Origins, Its Transformations, and Its Prospects*, New York: Harcourt, Brace & World.

Newman, O. (1972) *Defensible Space: Crime Prevention Through Urban Design*, New York: Macmillan.

Rossetto, T. (2006) 'Reducing disaster risk through construction design, building standards and land-use planning', *TRIALOG*, 91(4): 9–14, special issue on 'Building on Disasters'.

Tannerfeldt, G. and Ljung, P. (2006) *More Urban Less Poor: An Introduction to Urban Development and Management*, Sida publication, London: Earthscan.

UNDP (2004) *Reducing Disaster Risk: A Challenge for Development*, New York: UNDP.

UN-HABITAT (2003) *The Challenge of Slums*, Global report on human settlements, London: Earthscan.

UNISDR (2006) *Disaster statistics 1991–2005*. Online. Available HTTP: http://www.unisdr.org/disaster-statistics/introduction.htm (accessed 7 June 2006).

Wamsler, C. (2004) 'Managing urban risk: perceptions of housing and planning as a tool for reducing disaster risk', *Global Built Environment (GBER)*, 4(2): 11–28.

Wamsler, C. (2006a) 'Understanding disasters from a local perspective: insights into improving assistance for social housing and settlement development', *TRIALOG*, 91(4): 4–8, special issue on 'Building on Disasters'.

Wamsler, C. (2006b) 'Building on disasters' (editorial), *TRIALOG*, 91(4): 2, special issue on 'Building on Disasters'.

Wamsler, C. (2006c) 'Tackling urban vulnerability: an operational framework for aid organisations', *Humanitarian Exchange*, 35: 24–6.

Wamsler, C. (2006d) *Operational Framework for Integrating Risk Reduction for Aid Organisations Working in Human Settlement Development*, Benfield Hazard Research Centre (BHRC)-HDM Working Paper No. 14, London: BHCR.

Wamsler, C. (2006e) 'Integrating risk reduction, urban planning and housing: lessons from El Salvador', *Open House International (OHI)*, 31(1): 71–83.

Wamsler, C. (2006f) 'Managing urban disasters' (editorial), *Open House International (OHI)*, 31(1): 71–83.

Wamsler, C. (2006g) 'Mainstreaming risk reduction in urban planning and housing: a challenge for international aid organisations', *Disasters*, 30(2): 151–77.

Wamsler, C. (2007a) 'Managing urban disaster risk: analysis and adaptation frameworks for integrated settlement development programming for the urban poor', PhD dissertation, December 2007, Lund: Lund University.

Wamsler, C. (2007b) 'Bridging the gaps: stakeholder-based strategies for risk reduction and financing for the urban poor', *Environment and Urbanization*, 19(1): 115–42, special issue on reducing risks to cities from climate change and disasters.

Wamsler, C. (2007c) 'Coping strategies in urban slums', in: *State of the World 2007: Our Urban Future*, New York: Worldwatch Institute, p. 124.

World Bank (1993) *Housing: Enabling Markets to Work*, World Bank Policy Paper, Washington, DC: World Bank.

Worldwatch (2007) *State of the World: Our Urban Future*, Chapter 6 on 'Reducing natural disaster risk in cities', by Chafe, Z., Worldwatch Institute, New York: Norton.

Part IV
Conclusions

Chapter 17

Afterword

Integrating resilience into construction practice

Andrew Dainty and Lee Bosher

Introduction

In recent years the concept of 'resilience', the capacity of human and physical systems to respond to extreme events, has become increasingly prominent in disaster research. Indeed Tierney and Bruneau (2007) argue that the concept has largely supplanted the concept of 'resistance' with its focus on pre-disaster mitigation. This may reflect the realisation that the changing nature of natural and human-induced threats are such that built assets can never really be future-proofed to be totally resistant. The contemporary focus therefore has shifted to ensuring the capability of the built environment to both resist *and* recover rapidly following a disaster event. Despite the theoretical attractiveness of this proposition however, the structure of the construction industry and the nature of the interaction between those who plan, design, construct, operate and maintain the built environment provides a problematic context within which to integrate disaster risk management concepts. The socio-political landscape of the industry and professions arguably act as fundamental impediments to the achievability of this goal. Building-in resilience will therefore demand a paradigm shift in the way that built environment professionals integrate their activities and interact with the communities within which built assets reside. The preceding chapters have explored the challenges facing the built environment and have examined the strategies that must now be taken if built-in resilience is to be realised in the future and built assets safeguarded. This concluding chapter seeks to reflect on these contributions and examine the implications for future practice.

This chapter has a dual purpose. Initially, it seeks to summarise some of the cross-cutting themes which emerge from the individual contributions in order to synthesise the issues and challenges that the built environment faces. In addition, it seeks to join-up some of the preceding contributions by attempting to draw some broad conclusions with respect to the ways in which construction practitioners might adapt their *modus operandi* to better respond to the risks and threats previously described. This is a tough remit, particularly given the multifaceted and eclectic perspectives of the contributors to this book. Each author has drawn from their own considerable experience and research and many of these contributions are rooted in empirical insights and evidence from a number of different socio-political contexts. They view the concept of disaster risk management and resilience through a range of different theoretical lenses. However, whilst they provide a rich understanding of how risks might be mitigated, the challenge for the reader is to assimilate these different world-views and to relate them to their personal context. In this chapter we have drawn upon the 16 contributions and perspectives selectively in order to examine how built-in resilience might be achieved in the future. It is important to recognise that the ideas advocated here would play out very differently in different contexts. Hence, whilst we have attempted to pool the collective insights in order to construct a platform for future research and practice, we do not lay claim to having found any kind of panacea for building-in resilience to hazards in the built environment. Rather, we seek to provide an overview of the challenges faced and the new directions that researchers and practitioners must take to address these issues in the future.

The susceptibility of the built environment to natural and human-induced hazards

Everyone interacts with, and is affected by, the built environment; as Ofori (Chapter 3) asserts, the built environment accounts for '*most of every nation's savings*'. By the same token however, the impacts of disasters on the built environment can be so profound as to wipe out years of development and investment. In the past two decades alone, direct economic losses from disasters associated with natural hazards totalled US$ 629 billion (see World Bank 2004). The scale of the threats facing the built environment have clearly escalated in recent years as a result of demographic, economic and socio-political phenomena including increasing global population/urbanisation, climate change and terrorist threat. It is against this backdrop that this book was commissioned.

In the opening chapter, Bosher questions whether 'natural' disasters really exist. Clearly the way in which the built environment has expanded over the past 30 years, with little apparent regard to the evolving climatic conditions, has placed much development in a precarious position. It seems clear that an unrelenting desire to build has contributed to many disasters and/or has exacerbated their effects. In Chapter 16, Wamsler describes the interrelationship between disasters and the built environment (and related planning, design and construction practices). Her analysis reveals the reciprocal two-way and multifaceted relationship which exists between development and disasters, which in some respects determines their vulnerability. For example, Soetanto *et al.* (Chapter 7) suggest that the number of people living at risk of devastating floods worldwide is set to double from one billion in 2004 to two billion by 2050 (cf. United Nations

University 2004). Within the UK, if house-building rates were to increase to the level recommended in the Barker report (Barker 2004), almost 200,000 homes would be built each year on previously developed land over the next 10 years, requiring over 70,000 hectares of brownfield land, much of which will be located in the floodplain (DEFRA 2004). Thus, as Mileti (1999) points out, many emergencies are not unexpected, but stem from the predictable result of interactions between the physical environment, the built environment and the communities that experience them. Indeed, it is important to recognise that new threats can even arise out of the solutions to old problems (Kletz 1996).

Commentators continue to suggest that the impact of natural and anthropogenic climate change will further increase the threat from natural hazards in the future (Munich Re 2003). A report by the Association of British Insurers (ABI 2005) states that although some advances have been made concerning the protection of the built environment, development pressures, technological changes and climate change will continue to challenge the built environment in the future. In recent years, and particularly since the 9/11 attacks in the USA, the threat of terrorism has become even more acute. As Coaffee explains in Chapter 15, '*it is generally accepted that the nature of potential terrorist threats has changed, requiring an alteration in counter-responses from governments and security agencies at all spatial scales of defence, from the local area to the development of global coalitions to fight the "war of terrorism" (Coaffee 2004)*'. Thus, it would seem that new forms of human-induced threats further exacerbate the vulnerability of the built environment.

Although the scale of the resilience challenge confronting the built environment is not contested within the chapters of this book, there is a plethora of evidence within this text to suggest that a differential exists in the threat that natural hazards pose to developed and developing economies. As Green (Chapter 11) states, vulnerability to geophysical hazards are not uniform, but are concentrated in economically less-developed countries who also experience far greater mortality rates per capita. Green explains that many cities in Latin America, the Middle East, and Central and Southern Asia face the dual challenge of susceptibility to rapid-onset hazards and a loosely controlled construction industry. Thus, the prime locations for natural hazards are increasingly less-economically developed nations (Dilley *et al.* 2005). By the same token, the link between disasters and poverty seems irrefutable. In Chapter 4, Norton and Chantry highlight the social and economic costs of recovery from disasters in developing countries. They suggest that reducing the vulnerability of homes from the economic shocks that loss or damage creates is as important as preventing pandemics in this respect.

What emerges from the preceding chapters is a picture of a built environment under increasing threat from a multiplicity of different factors, some well established but escalating, with others emergent and unpredictable. Ofori's excellent summation of the vulnerability of the built environment to these threats (Chapter 3) in some respects reveals the scale of the challenge facing the built environment. For example, the immutable nature of built assets, the inability to accurately test them for resilience to hazards, the legislative and socio-economic requirements of development, requirements for ongoing maintenance, adaptation and redevelopment, and potential appropriation by the end-user all render built assets vulnerable to a wide range of hazards which will change over time. Lewis' fascinating exposition (Chapter 12) also reveals the vulnerability of the construction industry to corruption and the implications that this has for resilience of the built environment. The contributors to this book are unanimous in suggesting that such threats are likely to become more significant in future years and so it has become incumbent upon those responsible for planning, designing and constructing the built environment to take account of these threats as a core part of their professional activity. The decisions that are taken now will determine the burden that future generations inherit with regard to their resilience to a range of hazards.

From disaster management to disaster risk management

Most of contributors to this book articulate a powerful case for the built environment to undergo the paradigmatic shift from 'disaster management' to 'disaster risk management' (DRM). Although the concept of disaster risk management is traceable back to the beginning of the twentieth century (cf. Rollnick 2006), as Wamsler (Chapter 16) points out, it is gradually becoming institutionalised. This, she argues, is a product of social science research perspectives leading to the realisation that the impact of a natural hazard mostly depends on the capacity of people to absorb the impact and quickly recover from loss or damage. The resultant shift of focus has been towards understanding social and economic vulnerability. Wamsler further contends that the perception that disasters are the uncontrollable cause and the destruction of the built environment is the effect, is widespread amongst those who have a tendency to see the management of disasters in a purely physical way. Their responses are thus very limited and mainly focused on the post-disaster context. Accepting that disasters stem from the predictable result of interactions with the physical environment (see above), the disaster management perspective appears flawed and outdated.

It is useful in the context of this chapter to recap on the definition of disaster risk management (DRM). Several authors have drawn upon the definition proffered by the United Nations International Strategy for Disaster Reduction: '*a systematic process of using administrative decisions, organization, operational skills and capacities to implement policies, strategies and coping capacities of the society and communities to lessen the impacts of natural hazards and related environmental and technological disasters'* (UN/ISDR 2004). Thus, as Bosher explains in Chapter 1, the perspective has contributed to the shift towards 'bottom-up', community-based and sustainable long-term developmental initiatives in ensuring resilience. Such a perspective cross-cuts the contributions of this book which are almost all concerned with the appropriateness of resilience measures in relation to their compatibility with the context within which they are applied.

Resilience is not only applicable to natural and technological threats, as is discussed in detail by Coaffee in Chapter 15. He argues that the concept of resilience first emerged in research concerned with how ecological systems cope with stresses or disturbances caused by external factors, but has more recently been applied to human social systems, economic recovery, and disaster recovery in cities and urban planning. Coaffee's ensuing exposition is essential reading for understanding how the term has become an all-pervading metaphor for making sense of the counterterrorist challenge and is worth abridging here to elucidate the substantive meaning of the term. He argues that there are three key dimensions which differentiate it from traditional notions of disaster planning and recovery. First, the emphasis is on preparedness rather than post-disaster management. Second, there has been a widening of the emergency planning agenda to embrace security challenges in addition to natural hazards and technological accidents, despite the fact that the latter categories have a far greater collective impact. The third dimension concerns the role of institutional resilience to protect key infrastructural systems. This has necessarily broadened out the range of experts and professions whose input must now be garnered and integrated into the resilience effort. Thus, despite the theoretical attractiveness of disaster risk management as a concept, its effective delivery in practice is likely to be highly challenging.

Challenges and impediments to achieving built-in resilience

As alluded to above, the scale and expanding nature of the challenges facing the built environment, both 'natural' and human-induced, is a recurring theme within all of the contributions to this text. The question is therefore, how can disaster risk management be integrated into the processes of designing, locating, building, operating and maintaining the built environment in a way that enables built assets and their users to withstand the impacts of natural and human-induced hazards? It is perhaps surprising that so few of the chapters within the book have explored the role and importance of the 'construction industry' in designing and building in resilience. In order to establish the principles for building in resilience, it is firstly important to problematise the construction industry as an arena within which to embed disaster risk management principles.

Construction is frequently cited as the epitome of a project-based industry. Cherns and Bryant (1984) coined the term 'temporary multiple organisation' (TMO) to describe the unique team created on every project, each of which comprises a complex and temporary set of inter-organisational relationships, governed by project-defined interactions (see Bresnen *et al.* 2004). Thus, construction projects tend to be planned, designed and constructed by a combination of firms and individuals, many of whom will not have worked together before, and are not likely to work together again. The inevitable consequences of this temporal interaction is fragmentation and a separation of design and construction processes. The multiple interfaces which exist between the initiator of a construction project and the delivery of the finished product render construction supply chains some of the most complex and difficult to manage. Competing needs and objectives naturally lead to feelings of discord and tension, which in turn raise the possibility of conflict within the construction project team (Dainty *et al.* 2006). Built environment professionals are involved in an endless process of trading-off the objectives of the firm (typically bottom-line performance) with those of the project.

As well as the complexity of the delivery effort involved in the production of buildings and infrastructure, it is also important to recognise how the nature of its products complicate the production effort. For example, most new-build construction development tends to be large and requires expensive capital investment (Hillebrandt 1988). Built assets tend to be fixed in location with products assembled to a unique specification. The fact that each project is different, both in terms of the product and in terms of the people involved, makes it difficult to achieve the degree of repetition and routinisation achieved in other industries (Bresnen and Marshall 2001). Furthermore, the relatively low skill levels required for many construction operations and the low barriers to entry lead to an informal labour market and an industry that is difficult to regulate. This exacerbates the widespread corruption from which the industry suffers, which in turn increases the vulnerability of buildings and structures (see Chapter 12 for some fascinating examples in this regard).

Of the structural issues described above however, it is perhaps the length and complexity of the development process which compromises built environment resilience the most. The temporary multiple organisation which is formed for each project renders the effective coordination of their efforts around the resilience agenda problematic. Adding expertise in mitigating the effects of disasters to an already cluttered delivery effort further complicates the apportionment of responsibilities for the performance of the built artefact. Professional fragmentation is a hallmark of the industry, with architects, surveyors and engineers usually employed from outside construction firms as independent consultants (Morton 2002: 97). Propagating resilience through an integrated planning/design/construction effort appears a problematic notion given these structural constraints.

Towards a framework for building-in resilience

The preceding discussion has revealed the scale of the challenge in developing a more resilient built environment. It has also revealed the need for multiple, mutually reinforcing strategies to be developed concurrently if any real impact on resilience is to emerge. In Chapter 3, Ofori breaks down the components of the development process within which resilience must be built-in. These include building regulation and development control, procurement practices, design processes, construction and the operation of the built facility. The plethora of international examples described in relation to these areas suggests that frameworks already exist in both developing and developed country contexts. However, there remains an absence of any structured way to ensure that this guidance and promising practice is combined effectively.

To attempt to try and draw a single set of conclusions from this collection of contributions would not do justice to the chapter authors or the contextually embedded and nuanced insights that they provide. As discussed in more detail below, there is a clear need to move away from instrumentally rational solutions and to recognise that the way in which resilience should be built-in is entirely contingent on context. However, whilst the derivation of a single model for building in resilience is neither feasible nor desirable, broad principles can be drawn from the chapter contributions which can be used as a point of departure for the development of context-sensitive resilience frameworks in the future. In formulating the seven guiding principles outlined below we have drawn upon all of the contributions within this book, regardless of whether they are rooted in a developed or developing country context. The seven guiding principles are:

1   Adopt a holistic perspective;
2   Develop and appropriately apply resilient technologies;
3   Engage communities in resilience efforts;
4   Utilise appropriate existing guidance and frameworks;
5   Exploit opportunities to build-in resiliency measures post-disaster;
6   Integrate built environment and emergency management professionals into the DRM process;
7   Mainstream resilience into built environment curricula.

In reviewing these principles below we have signposted the reader to the chapters within which examples of the approaches advocated are to be found.

Adopt a holistic perspective

In Chapter 2, Alexander explains the factors impeding the application of DRM including the failure to apply knowledge, the failure to agree common standards, the need for professionalisation, problems with knowledge transfer and the apparent focus on technological over social issues in disaster risk management practice. Whilst not insurmountable, it is arguably the systemic and mutually reinforcing nature of these barriers which render the achievement of built-in resilience so problematic. Breaking down individual dimensions will arguably do little to overcome the ingrained problems in adopting a disaster risk management perspective, particularly in hazardous areas. Rather, a more holistic perspective is required within which the systemic impacts of such factors are considered.

A second issue concerns the need to view resilience and response as interrelated and mutually intertwined concepts. Response activities that do not take due regard of, or learn from, reconstruction, hazard mitigation and preparedness requirements are likely to miss key opportunities to attain not only physical resilience, but also social and economic resilience. For instance inappropriate responses to post-disaster situations can take a number of forms, such as:

1   Responses that do not reach out to the most affected people but are targeted towards favoured communities such as socio-political elites. It is in these situations that some sectors of society, such as the elderly, the poor, and marginalised communities (based for example on religious, political, or caste related grounds[1]) may be excluded from receiving not only post-disaster relief aid but also assistance towards recovery and reconstruction activities. It is therefore important to appreciate that when already existing socio-economic vulnerabilities are not being reduced, a key consequence is that social resilience is unlikely to be increased.

2   Post-disaster responses and reconstruction efforts that are overly influenced by a political and economic will to 'reconstruct quickly' (with the misguided belief that this will help society to 'bounce back') are not conducive to the attainment of physical resilience. As Menoni (2001: 105) notes, '*Market forces put pressures to reconstruct as quickly as possible transportation networks to long distances and commercial and office buildings, hampering efforts to implement lessons learnt from the disaster in the attempt to reduce ... vulnerability*'. Therefore, hazard mitigation and preparedness activities need to be intertwined with response and reconstruction activities; it is through this holistic approach that physical resilience is more likely to be attained.

Thus, by viewing resilience and response as interrelated and mutually intertwined concepts it should be possible to embed hazard mitigation and preparedness considerations into post-disaster response and reconstruction, to seize the opportunity to learn and implement the lessons learnt. Although this might sound like a straightforward suggestion, it presents many theoretical and practical challenges. The task of reconstruction after a major event requires coordinated efforts of all stakeholders for effective and efficient recovery of the affected community. Without developed frameworks for legislation and procurement, reconstruction and new development will be carried out on an ad-hoc basis with little regard for the needs of the society.

Develop and appropriately apply resilient technologies

Developing resilient technologies is, of course, also of paramount importance to mitigating threats to the built environment. This book is replete with excellent examples of how technologies, both traditional and contemporary, can be applied to reduce vulnerability. However, a recurring theme in the chapters of this book concerns the need to develop resilient technologies which are sensitive to the socio-economic environment within which they are to be used. Petal *et al.* (Chapter 10) critique the top-down, technologically-driven reconstruction projects that typify many post-disaster reconstruction efforts. These typically engage outside engineers and builders and use technologies which supplant both local knowledge and local labour. There are several implications of applying inappropriate technologies and processes in this context, particularly in terms of the likely disengagement of local stakeholders with the development process. As Petal *et al.* state *'People who have homes built for them – without consultation, without information and without choice – will naturally adopt a fatalistic view of the product'*.

A good example of appropriate technology is provided by Jigyasu (Chapter 5) who explains how the seismic-resistant construction systems found in Gujarat (the *bhungas*) have withstood earthquakes by virtue of their circular form and traditional construction. They are also sensitive to the locally available resources, the environment in which they are constructed and the spatial requirements of those who inhabit them. Pampanin (Chapter 6) also extols the benefits of learning from traditional techniques in the ways in which seismic-resistant structures are designed and Green (Chapter 11) notes the disaster resistance of many traditional construction techniques (e.g. seismic belts, lintel bands, thru stones, cross beams, and timber bracing).

Engage communities in resilience efforts

The nature of threats to the built environment is such that there will always be unintended consequences associated with attempts to improve resilience. Norton and Chantry (Chapter 4) show how improvements in socio-economic circumstances and building practice have actually contributed to the increased vulnerability of some communities. It has to be recognised therefore, that future threats cannot be predicted any more accurately than the future socio-economic circumstances of the countries within which they will occur and, as Alexander (Chapter 2) suggests, vulnerability cannot be separated from the social and cultural conditions under which it exists.

The challenge for researchers and practitioners is to design flexible and responsive solutions which can adapt to the changing physical threats that they face and the evolving institutional context within which they are embedded. As Norton and Chantry state, '... *if one does not address the developmental context in which risks develop and disasters take place, then there is a strong probability that the victims will be equally or even more vulnerable next time round*'. It is here where traditional knowledge is key in ensuring the long-term resilience of new built assets. This is exemplified in the example provided by Jigyasu (Chapter 5) of traditional bamboo housing in Majuli which can be dismantled and relocated to take account of local floods. This provides an excellent example of how traditional methods have evolved to respond to threats in a way which circumvents the hazards rather than attempting to resist them. In a similar vein McCarthy *et al.* (Chapter 8) caution that a multiplicity of local issues will always need to be engaged with when considering flood-protection measures. They advocate using a variety of methods to create a bespoke approach towards meeting local community expectations, supported through meaningful consultation methods. This enables the local socio-economic status of the population at risk to be considered. Engaging the user community is of course key in gaining acceptance for resilient measures, especially when these impact on the environment or utility of the built asset in any way. In this vein, McCarthy *et al.*'s empirical investigation into the community acceptance of flood-resistant structures shows how they can be convinced to embrace effective measures as long as support is garnered through a robust consultation process.

Community-based disaster risk reduction, as espoused by Petal *et al.* (Chapter 10), appears to offer a potential solution for overcoming the failures of legislative 'sticks' to generate more resilient solutions. As they state '*This approach shifts away from punishment as a primary motivator and instead points toward a community-based imperative that emphasises users and builders who are educated sufficiently to take the lead in voluntary compliance and in developing a critical mass providing leadership from the grass roots up.*' The framework that they present suggests that shared information, local ownership, positive relationships founded on dialogue, capacity building and robust evaluation can provide a framework for embedding this type of community involvement.

Utilise appropriate existing guidance and frameworks

As Alexander suggests in Chapter 2, the challenge of disaster risk reduction is largely a matter of how to *apply* existing knowledge (investment and adaptation plus reducing the 'implementation shortfall') rather than generating new knowledge *per se*. The research challenge, therefore, is grounded in the process of technology transfer and diffusion. Adopting selected elements of existing guidance for building in resilience is a sensible starting point for determining methods for embedding the principles of disaster risk management within existing professional activities. Many of these frameworks are sufficiently flexible and reconfigurable to enable the user to appropriate them for their own requirements and contexts in any case.

Some interesting frameworks are discussed and developed within this book which offer suitable starting points for achieving the principles set out above. For example in Chapter 14, Fox summarises the contents and implications of the Civil Contingencies Act (CCA 2004). The principles enshrined within are to enable continuity of service before, during and after critical incidents by ensuring the structural and collaborative conditions are in place to respond effectively. The Act encourages bilateral and cross-border cooperation to be developed outside the forum framework. From a built environment professional's point of view this is important because, as Fox argues, it is inconceivable that CCA responders would not call upon engineers in some shape or form when developing emergency plans. The six-step planning process seems to offer an excellent framework for embodying the recommendations of many of the contributors to this text. Wamsler's model (Chapter 16, Figure 16.4) provides a point of departure for those seeking to better integrate disaster risk management into construction planning practices. Based on seven complementary strategies, it could help to translate disaster risk reduction strategies into tangible working practices. This provides an excellent starting point for those wishing to integrate disaster risk management into the planning process.

Exploit opportunities to build-in resiliency measures post-disaster

Whilst the focus of this book is not post-disaster reconstruction, there appears to be an acceptance of the need for more resilient measures post disaster. Soetanto *et al.*'s work for example (Chapter 7) revealed that professionals are appreciative of the need to incorporate resilient measures into the repair of flood-damaged property. Thus, those with responsibility for post-disaster reconstruction and retrofitting post event are often amenable to taking on board resilient technologies given that they have witnessed the effects of the initial threat. This is supported by Le Masurier *et al.* (Chapter 13), who contend that addressing the contractual and legislative approaches to post-disaster reconstruction prior to any event will help to build-in resilience for a community by assisting post-event recovery and reconstruction efforts. Le Masurier *et al.* note that existing regulatory provisions may constrain reconstruction efforts by causing difficulties in apportioning multi-agency responsibilities, co-ordination and resource allocation.

Integrate built environment and emergency management professionals into the DRM process

There is undoubtedly an urgent need for the construction industry to adopt a disaster risk management (DRM) perspective. Ofori (Chapter 3) suggests that professional institutions and trade associations should enhance the awareness of their members of the need to assess the risks of disasters in order that they take the necessary precautions at all stages of the planning, design and construction processes. However, the difficulties of trying to do so in an environment with typically fragmented relationships between the various actors renders this a problematic notion (see Trim 2004; Lorch 2005; Bosher *et al.* 2006). Achieving built-in resilience demands that traditional demarcations in roles and responsibilities are reconstituted in order to propagate the free-flow of knowledge between the stakeholders of the built environment. Glass' contribution (Chapter 9) focuses on the role and contribution of a particular stakeholder to building-in resilience. The role of the architect is particularly interesting because of their ability to profoundly influence the design process through their interpretation of a brief and specification and configuration of materials. Glass views the role as comprising two concurrent but fundamentally different roles, that of an 'information manager' and that of a 'creative individual'. This leads to a requirement to blend subjectivity and objectivity throughout the design process which places them in an excellent position for coalescing other influential participants around them in support of the resilience effort.

The need for integration extends beyond the need to join-up built environment professionals' activities. Glass (Chapter 9) also argues that there is a disconnect between emergency managers and the construction industry. This view is congruent with that of Lorch (2005), who believes that some of the non-technological problems of emergency planning are a demonstration of the disciplinary boundaries within the scientific community and between the scientific community and the policy community. Policy makers, practitioners and the academic community must develop a collective approach for embedding hazard risk reduction and emergency management into the mainstream risk-management process. Lorch believes that higher education and training can play a major part in the integration of sustainable development and hazard, vulnerability and risk-reduction principles into the domain of built environment students (see below).

Mainstream resilience into built environment curricula

In the longer term there has to be a real and sustained commitment to mainstreaming the need for a resilient built environment into the education programmes of those who are charged with planning, designing, constructing and operating/maintaining it. Ofori (Chapter 3) suggests the redesign of the curricula of professional programmes to cover the relevant aspects of disaster management. This is more problematic than it may first appear given the need for professionals to respect the local context within which their disaster risk management takes place. This requires a change in attitudes towards construction development as well as knowledge of how to design and construct for hazard mitigation. Not all of the chapter authors share this view however. Glass (Chapter 9) suggests that rather than mainstream resilience into the architectural profession, there may well be a role for architectural practices to decide to specialise in resilient building design. She suggests that this is more plausible given that there is simply too much to learn to treat resilience as just another facet of regular building design. Petal *et al.* (Chapter 10) similarly suggests that championing old and new community-based construction approaches for disaster risk reduction will require acculturating a generation of technical specialists and practitioners. This will require higher education techniques and on-the-job training in order to equip those responsible for engaging local communities with the skills necessary for buy-in and involvement in resilient solutions.

In a developing countries context, Jigyasu (Chapter 5) presents a persuasive case for avoiding the categorisation of traditional and scientific knowledge into mutually exclusive domains. Rather, attempts should be made to reconcile the two; science can enable traditional knowledge systems to be easily understood by the professionals, and traditional knowledge enables scientific concepts to be translated into modes of communication that are locally understood. The theoretical attractiveness of this proposition is obvious, but its practical realisation requires both an open-mindedness on the part of built environment professionals and local communities to embrace traditional/contemporary methods, and the safeguarding of the traditional skills and knowledge but informed by scientific understanding of future hazards. As Jigyasu states '... *the crucial challenge is about overcoming highly defensive positions taken by the heritage professionals on the one hand and the engineers on the other, when neither of them are willing to make any compromise on their rigid professional standpoints*'. However, overcoming these challenges is necessary to safeguard and diffuse the knowledge of traditional techniques which have demonstrably led to hazard-resistant buildings in the past. Adapting and reapplying this knowledge will ensure that it evolves in a way which accords with the changing nature of threats, as it will be informed by the involvement of those with knowledge of the local context.

Conclusions

During the last few decades the growth in disasters has stimulated a growth in theoretical developments in relation to the way in which disasters are avoided and managed. A paradigmatic shift has led to focus on disaster preparedness, hazard mitigation and vulnerability reduction rather than disaster management and relief. The discourse of resilience now resonates throughout the disciplines involved with the mitigation of disasters. This is reflected within the contents of this book, where a persuasive case for embracing disaster risk management provides the common thread through an eclectic set of contributions to the resilience debate. Collectively, these chapters portray a resilient built environment as providing the essential foundation upon which the technical, organisational, social and economic frameworks so necessary for societal resilience can be founded.

This chapter has summarised some of the key themes emerging from the diverse collection of contributions contained within this book. By drawing upon the multiple insights provided by the authors, a range of principles have been identified which have the potential to address or circumvent the social, structural, economic and process-related barriers to achieving built-in resilience. In attempting to join-up and synthesise the various insights, advice and perspectives provided in this book, it has been necessary to problematise the notion of achieving built-in resilience. Two factors stand out in this regard. Firstly, the complexity and interrelatedness of the array of continually evolving threats facing the built environment and its users, and secondly, the institutional resistance to change which pervades the construction industry. Although the recursive nature of these factors renders the integration of disaster risk management into all construction activity an aspirational goal, this book contains a panoply of examples of how elements of such resistance can be broken down. Together these offer a point of departure when configuring methodologies for embedding disaster risk management principles in the future.

The authors should be commended for providing the research and practice communities with an excellent, thought-provoking text. Their contributions both reflect and support the shift towards more proactive risk mitigation and should help to cement the burgeoning profile of this perspective within the built environment and disaster risk management research communities.

Note

1 From more information on this topic refer to Bosher (2007).

References

ABI (2005) *Safe as Houses? Flood Risk and Sustainable Communities*, April 2005, London: Association of British Insurers.

Barker, K. (2004) *Review of Housing Supply: Delivering Stability. Securing our Future Housing Needs*, Office of the Deputy Prime Minister, London: HMSO.

Bosher, L.S. (2007) *Social and Institutional Elements of Disaster Vulnerability: The Case of South India*, Bethesda, USA: Academica Press.

Bosher, L.S., Dainty, A.R.J., Carrillo, P. M., Glass, J. and Price, A.D.F. (2006) 'The construction industry and emergency management: towards an integrated strategic framework', Paper presented at the Information and Research for Reconstruction (i-Rec) Third International Conference on 'Post-disaster Reconstruction: Meeting Stakeholder Interests', University of Florence, Florence, Italy, 17–19 May.

Bresnen, M. and Marshall, N. (2001) 'Understanding the diffusion and application of new management ideas in construction', *Engineering, Construction and Architectural Management*, 8(5/6): 335–45.

Bresnen, M.J., Goussevskaia, A. and Swan, J. (2004) 'Embedding new management knowledge in project-based organizations', *Organization Studies*, 25(9): 1535–55.

Cherns, A.B. and Bryant, D.T. (1984) 'Studying the client's role in construction', *Construction Management and Economics*, 2: 177–84.

Coaffee, J. (2004) 'Rings of steel, rings of concrete and rings of confidence: designing out terrorism in Central London pre and post 9/11', *International Journal of Urban and Regional Research*, 28(1): 201–11.

Dainty, A.R.J., Moore, D.R. and Murray, M.D. (2006) *Communication in Construction: Theory and Practice*, Oxon: Taylor and Francis.

DEFRA (2004) *Study into the Environmental Impacts of Increasing the Supply of Housing in the UK*, DEFRA Statistics, April, London: Department for Environment, Food and Rural Affairs.

Dilley, M., Chen, R.S., Deichmann, U., Lerner-Lam, A.L., Arnold, M., Agwe, J., Buys, P., Kjekstad, O., Lyon, B. and Yetman, G. (2005) *Natural Disaster Hotspots: A Global Risk Analysis*, Washington, DC: World Bank Hazard Management Unit.

Hillebrandt, P.M. (1988) *Analysis of the British Construction Industry*, London: Macmillan.

Kletz, T. (1996) 'Disaster prevention: current topics', *Disaster Prevention and Management*, 5(2): 36–41.

Lorch, R. (2005) 'What lessons must be learned from the tsunami?', *Building Research and Information*, 33(3): 209–11.

Menoni, S. (2001) 'Chains of damages and failures in a metropolitan environment: some observations on the Kobe earthquake in 1995', *Journal of Hazardous Materials*, 86(1–3): 101–19.

Mileti, D.M. (1999) *Disasters by Design: A Reassessment of Natural Hazards in the United States*, Washington, DC: Joseph Henry Press.

Morton, R. (2002) *Construction UK: Introduction to the Industry*, Oxford: Blackwell.

Munich Re (2003) *Topics: Annual Review: Natural Catastrophes*, Munich: Munich Re Group.

Rollnick, R. (2006) 'The aftermath of natural disasters and conflict', *Habitat Debate*, 12(4): 4–5.

Tierney, K. and Bruneau, M. (2007) 'Conceptualizing and measuring resistance: a key to disaster loss reduction', *TR News 250*, May–June, pp. 14–17.

Trim, P. (2004) 'An integrated approach to disaster management and planning', *Disaster Prevention and Management*, 13(3): 218–25.

UN/ISDR (United Nations International Strategy for Disaster Reduction) (2004) *Living with Risk: A Global Review of Disaster Reduction Initiatives*, Geneva: UNISDR.

United Nations University (2004) 'Two billion people vulnerable to floods by 2050', News release, 13 June. Online. Available: HTTP: http://www.unu.edu/news/ehs/floods.doc (accessed 17 September 2007).

World Bank (2004) 'Natural disasters: counting the cost', Press release 2 March. Online. Available: HTTP: http://go.worldbank.org/NQ6J5P2D10 (accessed 1 September 2007).

Index